Technology, Environment, and Human Values

Technology, Environment, and Human Values

Ian G. Barbour

Westport, Connecticut
London

In order to keep this title in print and available to the academic community, this edition was produced using digital reprint technology in a relatively short print run. This would not have been attainable using traditional methods. Although the cover has been changed from its original appearance, the text remains the same and all materials and methods used still conform to the highest book-making standards.

Library of Congress Cataloging in Publication Data

Barbour, Ian G
Technology, environment, and human values.

Includes indexes.
1. Environmental policy – United States. I. Title.
HC110.E5B37 363.7'00973 80-12330
ISBN 0-275-90448-2
ISBN 0-275-91483-6 (pbk.)

Library of Congress Catalog Card Number: 80-12330
ISBN: 0-275-91483-6

First published in 1980

Praeger Publishers, 88 Post Road West, Westport, CT 06881
An imprint of Greenwood Publishing Group, Inc.

Printed in the United States of America

The paper used in this book complies with the Permanent Paper Standard issued by the National Information Standards Organization (Z39.48-1984).

20 19 18 17 16 15 14 13

to DEANE

ACKNOWLEDGMENTS

This volume is an outgrowth of a series of interdisciplinary seminars in the program in Science, Ethics, and Public Policy at Carleton College. The program was initiated under a grant from the National Endowment for the Humanities and continued with the assistance of the General Service Foundation. I am indebted to faculty and student participants from many departments for discussions of most of the topics in these chapters. The first draft was written while on a fellowship from the National Endowment for the Humanities. An earlier version of Chapter 2 appeared in Warren Reich, ed., *Encyclopedia of Bioethics* (Macmillan, 1978).

I have benefited greatly from the suggestions offered by friends and colleagues who read the first draft of the manuscript: Rodger Bybee, John Compton, J. Patrick Dobel, Frederick Ferré, Edward Langerak, Arthur Parsons, Thomas Pender, David Peters, Roger Shinn and Norman Vig. I am grateful to Deane Barbour for helping with the readability of the text—and above all for sharing the years of my life.

CONTENTS

Chapter | Page

PART THREE: SCARCE RESOURCES

Technology, Environment, and Human Values

1
INTRODUCTION: THE HARD CHOICES

For 200 years U.S. industrial growth has been propelled by cheap fuel, abundant resources, and an environment that seemingly could absorb unlimited wastes. Early in the 1970s we became aware of environmental degradation, though we still were confident that technology could overcome resource constraints. But by the end of the decade the escalating environmental and human costs of technological solutions were evident.

Around the world there is an environmental crisis, an energy crisis, a food crisis and a population crisis. Growing populations compete for limited resources. Expanding industries consume raw materials and generate wastes at unprecedented rates. The United States has taken significant steps to slow environmental deterioration, but it has scarcely begun to acknowledge resource scarcities and issues of global justice.

Today we face difficult trade-offs among environmental preservation, economic growth, jobs, and human health. This book examines the conflicting values and political forces in these hard choices. It represents a second generation of environmental literature that recognizes the necessity of balancing environmental and human costs and benefits.

The 1960s started with high confidence in technology. The United States set out to put a man on the moon and the space program was a triumph of technical ingenuity, but by the late 1960s there was growing uneasiness about the social and environmental impacts of technology. In the early 1970s landmark legislation was adopted to control air and water pollution and to protect natural areas. But after the oil embargo and the economic recession of the mid-1970s many people considered environmental protection less urgent than additional energy and economic growth. By the end of the 1970s the economic costs of environmental preservation were frequently cited as reasons for postponing or relaxing regulations.

Imagine a discussion among four persons: an industrial engineer, an environmentalist, a steelworker, and a member of a rural commune. The engineer is defending a huge expansion in the use of coal to provide the energy needed for economic growth. To do this, he argues, we will have to accelerate the strip-mining of western coal and postpone air standards until more economical pollution abatement technologies are developed. The environmentalist is insisting on stricter legislation for land reclamation after strip-mining and the installation of sulfur-removal equipment on coal-burning plants to meet air quality standards. Even more strenuously, he is arguing for conservation to eliminate the need for additional generating capacity.

The steelworker speaks up in support of the engineer's view. He has heard that steel companies have closed some plants partly because of the cost of pollution control. He is willing to accept a degradation of air quality for the sake of employment. He considers the prevention of job layoffs and inflation to be of higher priority than the protection of western lands. The rural commune member says he agrees with many of the environmentalist's objectives but does not think they can be achieved by legislation. He tries to live a simpler life that stresses authentic personal relationships rather than the consumption of goods that are energy intensive and resource intensive. He advocates the use of solar energy and renewable resources. With the environmentalist, he thinks that people should live in harmony with nature, but he holds that this can be achieved only by alternative life-styles, not by political processes. The differences among these four persons reflect the conflicting priorities in U.S. life today and the interlocking character of the problems we face.

The developing countries confront a far more dramatic convergence of crises. The 1960s were heralded as the decade of development, and the Green Revolution held out the prospect of major increases in agricultural production. But by the 1970s it was clear that economic growth mainly had benefited a privileged minority and had done little to meet the basic needs of the vast majority. Population had grown almost as fast as food production, and several seasons of bad weather brought severe hunger, malnutrition, and famine. The new agricultural technology was dependent on fertilizer and water; developing countries were hard hit by the quadrupling of the price of the oil that was needed to make fertilizer and run tractors and irrigation pumps. Intensive farming, overgrazing, and the cutting of trees for firewood resulted in soil erosion in many parts of the world. Here again, the problems of energy, economic growth, and environment were seen to be interdependent.

The disparities between rich and poor countries are enormous and continue to grow. Affluent nations consume a grossly disproportionate share of the world's resources. A U.S. citizen, on the average, uses energy at 100 times the rate of a person in the poorest quarter of humankind.[1] The ratios differ for various resources, but it is estimated that overall the United States, with 5% of the world's population, accounts for 35 to 40% of the world's consumption of resources.[2] The doubling of living standards in rich countries would increase the consumption of world resources six times as much as a doubling of population in poor countries.[3] Today 500 million people bear severe effects of malnutrition, such as mental retardation and lowered resistance to disease; many people in the United States have the diseases of overeating, such as heart disease. An American dog gets more protein than a child in Bangladesh.[4]

In the past, industrial and economic growth in Europe and North America was fueled by abundant resources, obtained either domestically or from less developed countries. Water, air, and land were adequate to absorb industrial wastes. The American dream envisioned technological progress, the conquest of nature, and ever-growing consumption. But by the early 1970s environmentalists were warning that within the lifetime of our children the world's population would exceed the global carrying capacity—not only because of food shortages, but also because of resource depletion and the generation of pollution. Technological improvements in agricultural productivity, resource extraction, and pollution abatement might extend the earth's carrying capacity somewhat, but not indefinitely, and not without mounting environmental and human costs.

The hard choices that we face can be formulated in terms of conflicting values. If we subdivide human values into material and social values, we can list three groups of values that are considered in this volume: material, social, and environmental. First are the material values that correspond to basic human needs:

- *Survival.* Life is the precondition for the realization of all other human values, and it assumes preeminent importance when it is threatened. There are a few events that actually might endanger humankind as a species, such as the massive release of radioactivity in an all-out nuclear war or the irreversible destruction of crucial components of the biosphere. The survival of millions of individuals is indeed at stake if exploding populations continue to put additional burdens on the environment. There are other possible dangers on which scientific data are very limited, such as global climate changes from the burning of fossil fuels (the greenhouse effect), or increased skin cancer from solar radiation if the ozone layer is depleted. How should we take into account such uncertain but potentially catastrophic effects?
- *Health.* Agricultural and medical technologies have reduced hunger and disease and have added to life expectancy. But health has been jeopardized by other aspects of technology, such as air and water pollution, pesticides, and cancer-producing chemicals (carcinogens). The health and safety of workers, consumers, and citizens are issues in many policy decisions today. How can low-level, long-term risks be weighed against short-term economic benefits? What should we do when experts disagree about risks? How can public health be protected without excessive regulation? Have we in some cases purchased small gains in safety at exorbitant costs?
- *Material welfare.* Beyond survival and health, people seek a higher level of material well-being, including housing, transportation, and a variety of goods and services. Technology has been the main instrument for achieving these

goals. It has increased dramatically the speed of transportation and created new materials and products to enhance physical well-being and comfort. But many of these technologies have been energy intensive and resource intensive, and their side effects on nature and society have been more and more apparent.

• *Employment.* The opportunity to have a job is essential for self-respect and income. Unemployment is widespread today, especially in developing countries, and it will grow if for a given output capital-intensive and energy-intensive techniques replace labor-intensive ones. In industrial nations, the fear of unemployment is a major obstacle to pollution control and resource conservation measures. In technological societies, productivity has risen and back-breaking toil has been reduced, but meaningful work roles and opportunities to participate in work-related decisions have been rare. What policies might enhance both the quantity and the quality of jobs?

In addition to these material values, there is a group of social values that are central in the chapters that follow:

• *Distributive justice.* Inequalities in the distribution of food, energy, and resource consumption have been increasing. The impacts of pollution fall unequally on diverse income groups; the owners of a factory seldom live downwind of its smokestacks. Radioactive wastes from nuclear plants today may be a risk to people thousands of years in the future. In any policy decision, we must ask not only about the total costs, risks, and benefits, but also about how they are distributed—and who decides. Large-scale technology has accelerated the trend toward the concentration of economic and political power and the centralization of control. How might technology be used to diminish rather than to increase the gaps between rich and poor nations, and between rich and poor within a nation?

• *Participatory freedom.* In an interdependent world, freedom must be understood not individualistically as license to do as one wants, but socially as participation in the decisions that affect one's life. But how can the democratic control of technology be achieved? How can citizen participation and the consent of the governed be reconciled with the technical expertise needed in complex policy decisions? Does governmental regulation always entail a high price in bureaucracy and inefficiency? Should we seek decentralized technologies that are amenable to local control? Or do we confront threats to survival that justify coercive measures (such as compulsory sterilization of men or women after two children, or the withholding of food aid from starving nations that are unwilling to control population growth)? Will governments inevitably become more authoritarian in a world of resource scarcities and recurrent crises?

• *Interpersonal community*. The search for new life-styles is partly a reaction to the alienation, impersonality, and loss of community in a technological society. Often community cohesion has been sacrificed to other values. An urban freeway cuts through an ethnic neighborhood and pollutes its air—for the benefit of suburban commuters. Developing countries experience the breakdown of cultural traditions and social solidarity when Western technologies are imported. How might technology contribute more positively to human relationships? Do intermediate-scale technologies facilitate community interaction along with lower environmental impacts? How might a future society foster cooperation and diversity rather than competition and uniformity?

● *Personal fulfillment*. The dominant images of our society identify happiness with material possessions and higher levels of consumption. A broader view of human fulfillment would stress self-realization, the full development of distinctive human potentialities. It would include education, lifelong learning, and artistic creativity. It would encourage sources of satisfaction, such as personal relationships and the enjoyment of nature; which do not depend on high levels of energy and resources. These various facets of personal fulfillment together constitute an image of authentic human existence, a vision of "the good life."

Third, a group of *environmental values* is at stake in the choices we make:

● *Resource sustainability*. Sustainability can be judged only in an extended temporal and spatial frame. Our actions have consequences remote in space and time. Do we have obligations to future generations? How might a long-term global perspective be encouraged? Sustainability in a finite world clearly requires population equilibrium and greater reliance on renewable resources. Conservation and restraint in consumption by affluent nations are necessary for the sake of posterity as well as for the sake of developing nations today, yet current economic and political institutions are oriented almost exclusively toward short-run costs and benefits.

● *Ecosystem integrity*. Ecology has made us more aware of the interconnectedness of the web of life. Human actions have unintended and indirect repercussions; many ecosystems are fragile and vulnerable to large-scale human intervention. We have reduced greatly the ecological diversity that contributes to the stability and balance of natural systems. Endangered species represent an irreplaceable genetic heritage that might be of considerable scientific importance. What are the grounds of respect for nonhuman nature and loyalty to the wider community of life of which we are a part? Does the natural order have intrinsic value? Do we have obligations toward other creatures apart from our own self-interest?

● *Environmental preservation*. Control of the pollution of air, water, and land is important for human health, for the integrity of the ecosystem, and also for aesthetic enjoyment. But there are trade-offs among environmental quality, economic growth, social justice, and individual freedom. Intense pressures from industry have resulted in postponement of standards for auto emissions, stationary source emissions, and water effluents. We will examine alternative strategies for pollution control, and methods of assessing the environmental impact of proposed dams, highways, power plants, and so on. What decision-making processes are most promising for weighing these diverse values? How can both private and public interests be represented in land-use decisions? How can natural beauty be preserved amidst the relentless drive to develop and extract natural resources?

I am particularly concerned about ways of combining environmental preservation and distributive justice. This double goal has been called ecojustice.[5] The gap between humanity and nature and the gap between rich and poor are equally serious. Until recently, the environmental movement has given relatively low priority to justice. Since the days of Thoreau and John Muir, the retreat to nature often has implied turning one's back on the city and its problems. Enjoyment of the wilderness has been limited to the comparatively affluent. The membership of environmental organizations has been drawn mainly from middle-class, white suburbanites. Only in the last few years have urban pollution and world hunger been more prominent on the environmental agenda.

For the poor, both at home and abroad, food, health, and employment are more immediate concerns than environmental preservation. Yet environmental degradation has harmed agricultural productivity and human health in all parts of the world. Poverty, population, and pollution are as interlinked in developing nations as in industrial ones. Often the same economic institutions are responsible for the exploitation of people and the exploitation of nature, whether it be in Appalachia or in Africa. Uncontrolled technologies can be as destructive of human as of environmental values. I will maintain that we must at the same time seek human liberation and environmental stewardship.

The various levels at which these values are involved in decisions about technology and the environment are taken up in successive parts of the book. *Part One* ("Conflicting Values") starts by examining the cultural assumptions about nature (Chapter 2) and technology (Chapter 3) that have contributed to our present predicament. These fundamental attitudes are the product of ideas and beliefs, on the one hand, and of social and economic institutions on the other. So too, both ideas and institutions are sources of change for the future. Chapter 4 explores the philosophical and religious grounds for justice, freedom, and other social values. Chapter 5 takes up environmental values, the justification of wilderness preservation, and duties to nonhuman creatures and future generations. Some proposals for deriving ethics from ecology also are considered.

Part Two ("Environmental Policies") deals with value issues in environmental policy decisions. Chapters 6 and 7 explore the political system and the politics of environment (especially pollution and land use). Are there effective political mechanisms for combining the inputs of technical experts, elected representatives, agency administrators, and concerned citizens? I will argue that, despite the powerful political pressures exerted by industrial and bureaucratic interests, there are significant opportunities for the citizen to work within the political system to effect change. The topics of Chapter 8 are the uses and limitations of cost-benefit and risk-benefit analysis in administrative decisions (such as proposals for dams and public projects) and in setting regulatory standards (for pollutants, toxic substances, food additives, and the like). Broader interdisciplinary assessment methods, including environmental impact statements and technology assessments, are examined in Chapter 9. These procedures offer promising ways of considering a broad range of environmental and human values.

They also raise some interesting questions about the relationship between scientific judgments and value judgments in policy decisions.

Part Three ("Scarce Resources") applies the conclusions of previous chapters to specific resource decisions in the context of global needs. Energy options (Chapter 10) are analyzed both in terms of technical and economic feasibility and in terms of the diverse values outlined earlier. Chapter 11 looks at the ethical issues in food and population policy, and their relation to alternative development strategies. In Chapter 12, the limits-to-growth controversy and the possibilities of selective growth are discussed. The focus is on U.S. policies, but within a global context. The final chapter presents some new directions in appropriate technology and individual life-styles. People can express their value commitments not only by political action but also by patterns of daily life, such as choice of a job or the adoption of a simpler and less resource-consumptive life-style.

Because these issues are multidimensional, this book is by design interdisciplinary. Part One refers extensively to the humanities—history, philosophy, religion, and occasionally literature—in examining cultural attitudes and ethical principles. Part Two draws heavily from the social sciences—political science, economics, and sometimes sociology—in dealing with the political context of decision making and the social impact of alternative policies. The natural sciences are the basis of technological and environmental knowledge throughout the book, but they are particularly relevant to the case studies on energy, food, and resources in Part Three. Whatever the dominant mode of analysis, it is evident that every policy decision about technology and the environment involves ethics, politics, and science.

Let me close on a more personal note. I have felt keenly the conflict among these values in my own life. My undergraduate training was in science and engineering, followed by a Ph.D. in physics. After teaching and pursuing research in physics for several years, I returned to graduate school in philosophy and religion in order to explore the relation of science and technology to human values. During the 1970s I have been increasingly concerned with the environmental and social effects of technology. As director of a program in science, ethics, and public policy, I have worked with faculty colleagues in a series of interdisciplinary seminars; I am indebted to professors and students from many departments for discussions of most of the topics in this volume.

Throughout the book I will survey a range of current views and also indicate my own position. To simplify the diversity of views, I often will present two extremes and a middle ground. The first extreme is more commonly found among technologists and has been predominant in the U.S. past; the opposite extreme is prevalent among environmentalists and has gained adherents in the 1970s. Attitudes toward nature, for instance, range from "domination" to "unity with nature"; "stewardship" is an intermediate position. Between protechnology and antitechnology extremes is the view that technology is neutral in itself but becomes an instrument of social power. Again, there is a polarization between advocates of nuclear energy and advocates of solar energy; those in the middle seek a mixture of energy strategies. Between "progrowth" and "no-growth" positions there is an intermediate option, "selective growth." Between optimism and pessimism is a cautious hope.

The intermediate position often seems to be reasonable as a middle path between extremes, or as a synthesis of opposing views. It also offers a promising prospect for consensus or compromise that might emerge from mediation between conflicting parties. In general, however, my own conclusions lie between the middle and the environmentalist pole (though I assign higher priority than many environmentalists to employment, distributive justice, and the material welfare of the poor). As I see it, some weighting toward the environmentalist side is called for by the seriousness of the crises we face, and by the continuing influence of the assumptions, habits, and institutions that have promoted domination over nature and ever-increasing consumption. I have tried to describe other viewpoints fairly, but I also have taken a stand myself, indicating my own position within the range of views set forth. I have used the first-person pronoun to distinguish advocacy from description and analysis.

I believe the nation confronts unprecedented challenges in seeking justice, participation, and sustainability today.[6] Some changes will be forced on us by external pressures. The oil embargo, inflation, and the falling dollar show how vulnerable we are to events in other countries, and the bargaining power of raw material exporters will be greater in the future. Other changes toward greater frugality and reliance on renewable resources might be made voluntarily. Moral exhortations or appeals to altruism are not likely to be effective; new directions will come from new assumptions and new ways of seeing the world (which I later will describe as "paradigm shifts"). They will reflect alternative images of human fulfillment and visions of the future. They will foster stewardship of nature rather than conquest, the satisfaction of basic human needs rather than increased material consumption, community goals rather than individual gains.

Are there any grounds for hope that our society will curb its escalating levels of environmental destruction, energy use, and resource consumption? Can massive starvation in the developing world be averted? We must acknowledge the realities of economic and political power and the strength of past habits, assumptions, and institutions. Yet there are significant sources of change that are indicated throughout the book, and especially in its concluding pages. Education can change perceptions and values. Political coalitions can change institutional structures. Crises and disasters can challenge basic assumptions. Visions of alternative futures can motivate new directions in individual and social action. For our day, I will suggest, only the global vision of a just, participatory, and sustainable society is adequate. In looking to the future of our small planet, this spaceship Earth, we must find ways to bring together technological creativity, environmental stewardship, and a commitment to human values.

NOTES

1. Robert McNamara, "The Third World: Millions Face Risk of Death," *Vital Speeches* 41 (October 15, 1974): 17.

2. *The United States and the Developing World: Agenda for Action, 1973* (Washington, D.C.: Overseas Development Council, 1973), p. 3.

3. Calculations by Gunnar Adler-Karlsson cited in Anne-Marie Thunberg, "The Egoism of the Rich," *Ecumenical Review* 26 (1974): 459–68.

4. See Frances Moore Lappé and Joseph Collins, *Food First* (Boston: Houghton Mifflin, 1977).

5. For example, Dieter T. Hessel, "Eco-Justice in the Eighties," in idem, ed., *Energy Ethics: A Christian Response* (New York: Friendship Press, 1979).

6. See Paul Abrecht, ed., *Faith, Science and the Future* (Philadelphia: Fortress Press, 1979).

PART ONE: CONFLICTING VALUES

2
ATTITUDES TOWARD NATURE

We begin the exploration of conflicting values in contemporary policy decisions by examining the broad cultural assumptions about nature that have influenced the way we treat the environment. In both the past and the present, attitudes toward nature have been extremely diverse.[1] To simplify within a brief chapter the representation of this many-faceted diversity, some of these views have been grouped under three headings: Domination over Nature, Unity with Nature, and Stewardship of Nature (intermediate between the first two positions). Within each of these groups there are many variations, yet there are similar assertions concerning human relationships to nonhuman nature, and similar implications for environmental behavior.

An attitude is a set of beliefs and values organized around a specific object or situation.[2] Attitudes toward some objects or situations are relatively transitory and inconsequential. But attitudes toward nature involve fundamental beliefs and values that have far-reaching consequences. We will see that attitudes toward nature are correlated with distinctive attitudes toward resources, technology, and growth. We will be particularly concerned about attitudes as they are expressed in social institutions and in the structures of economic and political power. An analysis of the cultural attitudes and the institutions that have contributed to our present predicament also can help us in looking toward sources of change for the future.

I. DOMINATION OVER NATURE

Western exploitative attitudes toward nature have roots in ancient history. Other important influences can be traced to the rise of modern science, technology, and industrial capitalism. Finally, the American experience produced a distinctive version of the theme of domination over nature.

Biblical and Classical Roots

The first chapter of Genesis includes the commission to "be fruitful and multiply, and fill the earth and subdue it, and have dominion over the fish of the sea and over the birds of the air and over every living thing."[3] Humanity alone is said to be created "in the image of God," and is set apart from all other forms of life. Moreover, nature is desacralized in biblical religion. Ancient Israel believed that God had revealed Himself primarily in historical events rather than in the sphere of nature. More than any other religion of antiquity, the Hebrew tradition stressed God's transcendence of nature and the distinctiveness of the human.

A number of recent authors have claimed that *the biblical idea of dominion* was the main historical root of environmentally destructive attitudes in the West. In a widely quoted article, Lynn White points to the separation of humanity and nature and the rights of humanity over nature in biblical thought. Holding that ideas and attitudes are significant influences in history, White concludes that, because it has been so anthropocentric (man-centered) and arrogant toward nature, Christianity "bears a huge burden of guilt" for the environmental crisis.[4]

The Christian tradition as it developed in the Middle Ages did indeed stress divine transcendence at the expense of immanence, and it dwelt on the contrast between the human and the nonhuman (even though it had an organic rather than a mechanistic view of the natural order). But White's critics insist that it was only in later centuries that the dominion theme was misused to justify exploitative practices. The Bible itself speaks not of unlimited domination but of stewardship, which implies restraint and responsibility (see Sec. III below). Furthermore, the anthropocentric slant of Western thought is a product of Greek as well as biblical ideas. Other critics say that we must look to later institutions, such as industrial capitalism, as the main determinants of environmentally destructive behavior—even though those institutions were themselves partly the product of basic cultural beliefs.[5] In short, a one-sided emphasis on unqualified dominion contributed to Western irresponsibility toward nature, but other historical factors are equally important.

From the many strands of *Greek thought*, Plato, Aristotle, and the Stoics may be singled out for their influence on the West. Their writings portrayed a gulf between humans and all other beings, based on the unique human capacity for reason. Aristotle stated that other creatures are devoid of the contemplative activity in which humans are most akin to God; plant and animal life exists solely for the sake of human life. Cicero, drawing upon Stoic writings, insisted that we have no obligation to respect animals because they are not rational beings. To the neo-Platonists of the Hellenistic era, the eternal forms are embodied only imperfectly in the world of nature. To the Gnostics and Manicheans, nature is the realm of evil from which the human soul seeks to escape. Greek and Roman views were indeed extraordinarily diverse; some pantheistic authors who were more appreciative of the natural world will be mentioned later. But the classical sources that were taken up in the early church, the Middle Ages, and subsequent Western thought seem to have stressed the differences between humankind and nature.[6]

The Growth of Science and Technology

With the rise of *the scientific world view* in the seventeenth century, the domination theme assumed increasing prominence. To Francis Bacon, the conquest of nature is the goal of science, for "knowledge is power." "Let the human race recover the right over nature which belongs to it by divine bequest." Bacon's *New Atlantis* called for a state-funded research establishment and a scientific elite through which humanity's rightful supremacy would be systematically extended.[7] René Descartes similarly extolled practical knowledge that would make us "the lords and masters of nature." With the Stoics and the Scholastics, he thought that our unique rationality justified such sovereignty. Descartes elaborated a sharp dualism of matter and mind. Apart from the human mind, he said, the world consists of particles in motion, and mathematics is the key to understanding it. Animals are machines without minds or feelings. The human knower is set over against the whole world of matter. The gap between mind and matter in Descartes's anthropocentric outlook posed the central problems for modern philosophy.

The *mechanistic interpretation of nature* was developed further by Isaac Newton. In earlier centuries the world had been viewed as a hierarchy of organisms, each with its place and purpose in an overall plan. Newton and his followers said that nature is constituted by impersonal masses and forces, operating according to deterministic laws. The measurable "primary" qualities, such as mass and velocity, were said to be objective characteristics of the real world; all other "secondary" qualities, such as color and temperature, were relegated to subjective responses in the observer's mind. Here was the "objectification" of nature as a realm essentially alien to the human. While Newton himself respected the cosmic watch as the product of the Divine Watchmaker, it is not surprising that his more secular successors had no scruples about exploiting it. If nature is a machine, it has no inherent rights, and we need not hesitate to manipulate and use it.[8]

John Locke provided the *political philosophy* within which such domination could be justified and encouraged. Locke maintained that the political order is necessary primarily to protect the rights of the individual to "life, liberty and property." When a person exploits a natural resource, he "mixes his labor with it" and therefore has a right to it, since there is enough left for other people. Unfettered private ownership and use of resources, he said, would lead to economic and technological growth. By protecting the property rights of the rising middle class, the state would be encouraging industrial development and prosperity. Locke's writings thus endorsed individualism, the accumulation of wealth, technological development, and the subjugation of nature, all supported by the structures of government.[9]

In the emerging *industrial technology* of the eighteenth century, domination was increasingly achieved in practice as well as in theory. To the leaders of the Industrial Revolution, the environment was primarily a source of raw materials. In the new capitalism, the private ownership of resources fostered the treatment of the natural world as a source of commercial profit. Along with rising standards of living came increasing burdens on the environment. Since antiquity there had been deforestation, overgrazing, and soil erosion; but the technologies that developed in the last two centuries produced pollution and consumed natural resources at unprecedented rates. Here were combined the influences of biblical religion, dualistic philosophy, mechanistic science, and, above all, capitalist economics and industrial technology.

The American Experience

In addition to beliefs and institutions brought from Europe to America, there were distinctive features of the American experience that reinforced the attitude of domination. For the New England Puritans the new surroundings were strange and threatening, and it is not surprising that their writings often refer to nature as *an enemy to be subjugated*. As the pioneers moved progressively westward, much that they encountered was hostile, a threat to survival, an obstacle to be overcome. Forests were cleared and wilderness destroyed to make way for civilization. The advancing frontier was interpreted in the light of the nation's "manifest destiny" to "conquer a continent." The New World offered apparently endless stretches of good land and seemingly unlimited natural resources. If timber was used up in one area, it could always be found somewhere else. Nature was treated as a source of raw materials in inexhaustible quantities. Air, water, and land appeared ample to absorb the waste products of a burgeoning civilization.[10]

In the early days of the nation, there was considerable support for Thomas Jefferson's ideal of a nation of farmers. The agrarian, pastoral vision that stressed the virtues of a rural society contrasted with Alexander Hamilton's goal of *an urban, mercantile nation* in which commerce and manufacturing would thrive. The tensions between these two ideals continued, but it was clearly the industrial vision that prevailed. By the time of the Civil War the steam locomotive, the Iron Horse, had become a symbol of both technology and the conquest of nature in the growth of the nation. As industry thrived, urbanization accelerated and the United States changed from a rural and small-town nation to a predominantly urban one. The new technologies and the private ownership of land and resources led to the concentration of economic power, the amassing of personal fortunes, and the rise of giant corporations for which the environment was a source of wealth.

To these geographic and economic factors may be added some *characteristic American values* that encouraged exploitative attitudes. Faith in technology, confidence in the expert, and trust in American know-how to solve any problem have been pervasive. An optimistic belief in inevitable progress has until recently been shared by all levels of society. The United States also has been obsessed by the goal of growth, the assumption that bigger is better, the dream of ever-growing consumption. Impressive increases in productivity and gross national product certainly have occurred, but with social and environmental costs that were only slowly recognized. It also can be argued that a male-dominated society has admired in public life the aggressive, competitive, rational qualities that it calls "masculine," rather than the nurturing, conserving, intuitive qualities that it associates with women and family life. Images of male domination have closely paralleled images of nature as an object of domination. The exclusion of women from positions of economic and political power and the separation of the realms of work and home have accentuated this polarity.[11]

The degradation of the environment is in large part the product of decisions by corporations or individuals in the interest of profit or narrowly defined efficiency. But such actions are supported by the *faith in technological progress* that is held by many Americans. In the next chapter we encounter examples of the confidence that human ingenuity can provide "technical fixes" to problems of pollution and resource scarcity. Some "future planners" maintain that humankind can get along quite well without nature in a totally man-made environment—of which the astronaut's space capsule might be a forerunner. Here anthropocentrism has taken the final step: the declaration of human independence from nature (except as a source of materials).

Recent attacks on the environmental movement have reflected such *technological optimism*. For example, in *The Doomsday Syndrome*, John Maddox gives a protracted critique of "exaggeration and alarmism" in environmentalist writings. He holds that the hazards from pesticides, the threat of famine, the impending death of the oceans, and other ecological catastrophes have been grossly overstated. In a historical perspective, he says, we can see that pollution is nothing new, and wildlife species have become extinct throughout the past. Maddox urges us to have confidence in the ability of science and technology to solve future problems and extend our control over nature. We have to be "hard-headed," he writes, if we are to manage the natural environment most efficiently for humanity's benefit.[12] The world is an abundant storehouse of natural resources that technical ingenuity can use for human welfare. Similar statements can be found among other critics of environmentalism.[13]

Note, finally, that *Marxism* shares the confidence in technology and industrial growth and the idea of "mastery of nature" that have been dominant in the West. To be sure, the young Marx wrote about the alienation of humanity from nature, and he occasionally did mention the environment as an object of aesthetic satisfaction. Many ideas in Engel's *Dialectics of Nature* are compatible with an ecological outlook. In Marx's later writings, however, these themes were overshadowed by his central concern, the alienation between people. He viewed nature, mastered by technology, only as an instrument for the fulfillment of human needs. Marx was keenly aware of the importance of economic institutions and technology in the exercise of power over people. He held that changes in "ownership of the means of production" would end the exploitation of human beings, but he said little about the exploitation of nature.[14] Of the three factors in production—capital, labor, and natural resources—capitalism emphasized the first and Marx the second, but neither contemplated the prospect of resource scarcities. In twentieth-century Russia, the injunction to "transform nature" even has been coupled with a denial of any environmental limits. The environmental record of the Soviet Union is no better than that of capitalist countries.[15]

II. UNITY WITH NATURE

The opposing idea of humanity's unity with nature can be found in three rather different versions: Eastern and Native American religions, the Romantic Movement in literature and philosophy, and evolutionary biology and ecology. All of these views stress humanity's harmony with and participation in the world of nonhuman nature.

Eastern and Native American Views

In contrast to ancient Israel, the other cultures of *the Near East* sought the harmonious integration of human life within the life of nature. Their rituals and festivals celebrated the annual cycle of the seasons and the fertility of nature (rather than celebrating historical events, as the biblical religions always have done). In nations around the Mediterranean, the earth-mother assumed differing forms, including Athene, Artemis, Demeter, and Isis. Myths of dying and rising gods (such as Adonis and Osiris) were associated with the rebirth of life in the spring; other gods and goddesses represented the power of a variety of natural forces.[16]

Greek popular religion often made reference to the sacred in nature, such as the sacred grove or mountain and the spirit of the river or rock. Other versions of human interconnectedness with nature can be found among the Epicureans. Lucretius, for example, held that the world was not designed for human use; in his nature poetry he reflected on the beauty and interdependence of the world as a natural process.[17] But neither the nature religions of Asia Minor nor the naturalistic or pantheistic philosophies of Greece were as influential on later Western thought as the biblical and Greek themes mentioned earlier.

A brief summary cannot begin to do justice to the rich diversity of Eastern thought concerning nature, but it can at least show some contrasts with traditional Western thought. *Taoism* in China portrays the world as an organic, interdependent system. There is an underlying unity behind "the world of ten thousand things." Nothing exists in isolation; the parts of the whole are interpenetrating and interfused. Every particular being is a manifestation of the Tao, the nameless unity that exists before differentiation into multiplicity. The human is part of a wider cosmic order. To achieve a harmonious relationship to the natural world, we must respect it and adjust to its demands. The path to the recovery of harmony and wholeness is surrender, tranquility, nonattachment, the ability to "let things be." The love of nature in traditional China is evident in its poetry and painting (especially of trees, mountains, and landscapes). But even agriculture and land management were represented as cooperation with nature rather than as conquest.[18]

Zen Buddhism arose first in China from the confluence of Taoism and Mahayana Buddhism, and then developed further in Japan. Here, too, human kinship with nature was stressed. According to Zen, the merging of self and other is known in immediate experience. Intuition and personal awareness, not analytic rationality or conceptual abstraction, reveal the unity of subject and object. In the Zen tradition, nature is to be contemplated and appreciated rather than mastered. Humankind should act on nature with restraint, bringing out the latent beauty and power of the natural world. These aesthetic elements in the Zen outlook entered many Japanese cultural expressions such as flower arrangement, the tea ceremony, and the miniature garden. In the short poems known as haiku, events in nature typically provide moments of insight into the beauty, harmony, and dynamic flow of reality. The path of spiritual awareness, it is held, can liberate a person from the obsessive drives of the ego and from preoccupation with material possessions. The Buddhist ethic thus encourages humility, simplicity and frugality.[19]

We must recognize, however, that there were *diverse strands* in each of these cultures. The China that produced Taoist nature-mysticism also accepted principles of hierarchical order, praised emperors for massive feats of engineering, and built walled cities following geometrical patterns. Moreover, there are gaps between the ideals and the practices of every culture. Behavior, we have said, is the product of social and economic forces and institutional structures as well as beliefs. Environmental destruction was by no means absent from classical China. For instance, wood was in great demand for building, fuel, and charcoal used in metallurgy; deforestation resulted in widespread soil erosion.[20]

Nevertheless, it appears that in previous centuries Eastern countries have on balance treated nature with somewhat greater respect than Western ones, and that changes in attitudes during the twentieth century have been mainly the result of Western influences. *Chinese communism* shares the Marxist assumption that nature is an object of conquest. Mao Tse-tung called for "a war against nature." In many official pronouncements nature was explicitly presented as an enemy against which the people must struggle and achieve victory. When environmental measures have been taken, the motive was primarily the improvement of sanitation and health, or the recovery of industrial wastes for reuse. There have been substantial efforts, such as reforestation and water management projects, in response to pressing human needs. Considering the extent of human deprivation in China, it is understandable that a revolutionary government has seen nature mainly as a resource to be exploited for the good of society. But this does represent a major shift from the characteristic cultural attitudes of the Chinese past.[21]

In the case of *modern Japan*, industrialization had few indigenous roots and arose in large measure from the impact of the West. The attitudes engendered have remained in considerable tension with traditional views of nature. During the last decade, rapid economic growth and population concentration in very limited land areas produced some of the worst air and water pollution in the world. The paternalistic relation of management to employees and the strong political ties between government and industry have hindered citizen action and government regulation. Yet it is possible that the recent measures for environmental preservation in Japan will receive stronger support not only from recognition of self-interest but also from the legacy of earlier traditions of respect for nature. Japan is spending 2% of its gross national product on pollution control, the highest percentage in any nation today.[22]

Native American culture provides a final example of unity with nature in a religious framework. Here again there is a strong sense of the interrelationship of all living things within a cosmic pattern. Human beings are related to the moon, sun, and mountains as part of a balanced order. The spiritual forces that constitute the reality of the world are present in all things. Human and animal beings are kin, and all forms of life must be respected. Native Americans sought to understand themselves in relation to the landscape and the sacredness of particular places and ancestral origins. When an animal was killed, it was not done thoughtlessly; there were prayers of apology to the spirit of the animal. The Native American tries to live in harmony and balance with the environment, and many tribal stories suggest humility and dependence on other creatures.[23]

It would be misleading, however, to interpret the Native American as an incipient ecologist. In this world view, the ultimate source of power is *the spiritual world* that controls natural phenomena. The vision quests of the Plains peoples, the tribal ceremonies of the Pueblo peoples, and the rituals associated with corn planting or hunting were all ways of influencing nature for human benefit. The spirit beings, if properly respected, would provide people with all that they need. (For instance, the belief that buffalo had been given in inexhaustible supply for human use lies behind the seemingly ruthless and wasteful practice of driving herds of buffalo over cliffs.) In short, as in the case of the Taoist and the Zen Buddhist, the Native American has a strong sense of the interdependence of humankind with other living things, which can be understood only as part of a distinctive set of beliefs about the cosmic order.[24]

The Romantic Movement

The idea of unity with nature, so common in Eastern cultures, found few exponents in the early centuries of Christendom. In the medieval period one might point to *St. Francis* with his deep love of the natural world and his sense of union with it. He saw nature as a living whole and all creatures as objects of God's love, and hence as significant in their own right. He spoke of our sister the earth and greeted the birds as brothers, extending the family relationship and the circle of God's love to include all created beings sharing a common dignity and equality under God. Humanity, he said, is part of a wider community, and each creature has its own integrity that must be respected. St. Francis has continued to appeal to the popular imagination, but he was hardly typical of medieval thought.

Apart from some of the Christian mystics, the unity motif found few advocates in Europe in the early modern period either. Only after the Industrial Revolution—and largely as a reaction to it—did this theme come into prominence in *the Romantic literature* of the late eighteenth and early nineteenth centuries. For Blake, Wordsworth, and Goethe, nature is not an impersonal machine but an organic process with which humanity is united. God is not the remote watchmaker but a vital force immanent in the natural world. Not rational analysis but feeling and imagination are the highest human capacities. Intuition grasps the unity of organic wholes, the interrelatedness of life. In natural settings a person can find a healing power, a sacramental presence, an experience of peace and joy. Other Romantic writers extolled wild, sublime, untouched landscapes, forests, and rivers. They idealized the "noble savage" uncorrupted by civilization, and they exalted the "natural" and the "primitive."[25]

The *Transcendentalists* in New England referred in similar terms to the presence of the sacred in the realm of nature. Henry Thoreau held that nature is a source of inspiration, vitality, and spiritual renewal; it can teach us humility and simplicity. "In Wildness is the preservation of the World," he wrote. Thoreau criticized the frantic pursuit of progress and affluence, the growth of technological industrialism, and the pressures of an impersonal urban life. His year and a half of living alone at Walden Pond made him more aware of the interrelationships among creatures and the natural stability upset by humans; in solitude he found serenity and peace. Unspoiled nature was for him both a symbol of qualities that he valued (freedom, courage, vitality) and an environment that would bring out these qualities in us. But he did not advocate giving up civilization. He sought a simplification of life and an alternation and balance between life with nature and civilization.[26]

Starting in the 1870s, the writings of John Muir gave wide circulation to a philosophy of *wilderness preservation*. With the disappearance of the frontier, wilderness areas could be saved only by deliberate national policy. Muir, like Thoreau, found a divine harmony in nature, a freedom not possible in the artificial constraints of civilization, a source of serenity in a decadent urban society. He was a founder of the Sierra Club (1892), and through his writing campaigned tirelessly for legislation to protect wild areas of the U.S. West from human intervention.[27] In a later chapter we will examine more recent arguments for the preservation of wilderness.

Some of the themes of the Romantic Movement have recurred in the *youth counterculture* since the late 1960s. As in Romanticism, feeling, imagination, and immediacy of experience are valued more than intellectual reasoning. The counterculture has been disenchanted with technology, holding that preoccupation with efficiency, productivity, and rational control has alienated us from nature and our fellow humans. Some young people dropped out of a competitive society to form communes in rural settings within which they hoped to recover harmony with nature and with each other. In such communes, nature-mysticism, the use of organic foods, and frugal living have been common. Other people have sought alternative life-styles that would encourage personal, social, and environmental harmony without a radical break from the prevailing social order. The quests for personal fulfillment, interpersonal relatedness, and ecological awareness have been combined in a variety of new patterns of individual and group life that are discussed in Chapter 3.

The Ecology Movement

The unity of humanity and nature has received a very different kind of support from the sphere of science. An early forerunner of some of the ideas of ecology was George Perkins Marsh's *Man and Nature* (1864), the first systematic study of the destructive influence of people on their environment. Marsh was an accomplished linguist familiar with ancient history and literature. As ambassador to Italy he traveled in Mediterranean countries and saw barren deserts where great cities and civilizations had once flourished. He traced the effects of deforestation and overgrazing on soil erosion, as well as the destruction of land by salination from excessive irrigation. He described how the domestication of some species of plants and animals had resulted in the attrition or extinction of other species. He provided careful documentation of the fragility of the environment and the disruptive and often irreversible effects of civilization in disturbing the natural equilibrium.

In Charles Darwin's *The Origin of Species* (1859), and especially in *The Descent of Man* (1871), the human species is treated as a part of nature in continuity with other forms of life. No sharp discontinuities separate human from animal life, either in evolutionary history or in present morphology and behavior. The theory of evolution seemed to undermine humanity's unique status; close parallels to most human capacities could be found among lower forms. Moreover, Darwin's studies brought out the interconnectedness of the web of life and the complex balance of interactions in the biological world. Subsequent research in population dynamics has underscored the importance of the relation of organisms to their environment, including habitat, food sources, and predator-prey relationships. Recognition of the interdependence, diversity, and vulnerability of biological species prepared the way for ecology.

Within the twentieth-century science of ecology, four concepts have significant implications for our relation to the environment:

1. The Ecosystem Concept

The interdependence of the forms of life in biotic communities and the complex interactions among organisms have been traced in detail by ecologists. There are food chains linking diverse species, interlocking cycles of elements and compounds, and delicate balances that are easily upset. The interconnected web of life must be considered as a system; a change at one point can have far-reaching repercussions at other points.

2. Finite Limits

There are limits to the growth of any population; the environment has a finite carrying capacity. In nonhuman populations, there are mechanisms such as territorialism, interspecies competition, and food depletion that limit population growth. In the human case, neither population size, nor resource consumption, nor pollution generation can grow indefinitely. One of the most significant implications of ecology is its challenge to prevailing assumptions about unlimited growth.

3. Ecological Stability

Equilibrium is not a static concept, but a dynamic balance of inflow and outflow. Stability is not incompatible with change and adjustment to constantly altering conditions. Balance often is achieved by homeostatic feedback mechanisms that exert some control and regulation (that is, a change in one direction brings into play a counterinfluence in the opposite direction). Diversity in an ecosystem contributes to its stability and adaptability; systems with a very small number of plant or animal species usually are more vulnerable to disease, predators, and changing conditions.

4. Long Time-Spans

The ecologist is concerned about repercussions and indirect consequences that may be distant in time as well as space. Continuity and sustainability are sought, not short-term benefits at the expense of long-term costs. Ecologists are used to studying population changes over many generations; they urge us to think about the consequences of our actions on future generations as well as on other species.[28]

Recognition of the vulnerability of the biotic community leads ecologists to urge *restraint in human intervention*. Acknowledging human dependence on the biosphere, they advocate humility and cooperation with nature in place of attempted mastery and control. Since some consequences of our actions may be irreversible, or may be reversed only at great cost over an extended time period, we should err on the side of caution when our knowledge is so limited. There also has been extensive work in animal behavior, anthropology, psychology, and sociology that illuminates the genetic and cultural basis of our relation to the rest of nature. Of particular interest are studies of overcrowding, environmental perception, and the effects of natural environments on human cultures.[29]

Some authors have found more specific *ethical guidance* in ecology. Writing in the 1930s, Aldo Leopold was one of the first scientists to suggest that our ethical concern should be extended to the land and nonhuman life. The good, he proposed, is "that which preserves the integrity, stability and beauty of the biotic community," whose destiny is in the long run inseparable from ours. The goal, he believed, is to be humble and responsible citizens of the earth rather than its conquerors.[30] Rachel Carson's early writings combined scientific knowledge with a sense of spiritual unity with nature reminiscent of Romanticism. In her influential book, *Silent Spring* (1962), the call for an ecological conscience was tied to scientific studies of the effects of pesticides on bird populations. By the early 1970s a large number of popular writings were delineating with increasing urgency the effects of a variety of pesticides, phosphates, nitrates, lead, mercury, asbestos, food additives, radioactive wastes, and air pollutants on both man and the environment. The word "ecology" had entered the public vocabulary.[31]

In recent years, many scientists have focused on *growth* as the crucial ecological problem. For Paul Ehrlich and Garrett Hardin, the control of population overshadows all other issues. The Club of Rome study, *Limits to Growth*, and the British *Blueprint for Survival*, assert that there are limits to industrial growth that are set by resource reserves and pollution levels.[32] These studies advocate policies of "no growth" or "steady state" for industrial production as well as population. In opposing growth and in underscoring the finite carrying capacity of the environment, they adopt an ecological viewpoint. In ecology, then, the reasons for emphasizing the interdependence of the human and the nonhuman are scientific rather than religious. Yet here, too, there is a humility and respect for the environment that strongly contrasts with prevalent Western views.

III. STEWARDSHIP OF NATURE

The third outlook on nature, stewardship or responsible use, represents a middle ground between domination and unity. As one would expect in dealing with a continuum, some versions are closer to the former and others to the latter. Examples in this section are taken from biblical thought, the conservation movement, and the writings of some recent theologians and philosophers.

Stewardship in Biblical Thought

We have seen that the first chapter of Genesis speaks of dominion over nature. But according to the second chapter (the Adam and Eve story, which is held by most scholars to be considerably older), human beings are put in the garden "to till it and keep it." Throughout the Bible, humankind does not have absolute and unlimited dominion, but is responsible to God. "The earth is the Lord's" because He created it. The land belongs ultimately to God; we are only trustees, caretakers, or stewards, responsible for the welfare of the land that is entrusted to us, and accountable for our treatment of it. In the last analysis, the biblical outlook is neither anthropocentric nor biocentric, but theocentric.

In the biblical view, the created world is *valued in itself*, not simply as an instrument of human purposes. In several of the Psalms (including numbers 19, 89, and 104) God is said to delight in the earth and the manifold variety of life quite apart from humanity. Even in the first chapter of Genesis, each form of life is pronounced good before humankind is on the scene. The Sabbath is a day of rest for the earth and other living things, as well as for humanity. Every seventh year the fields are to lie fallow; the land deserves respect and will cry out if misused. Human life and nature stand together jointly as God's creation.[33]

Many biblical passages express *appreciation and wonder* in response to nature. Job is overwhelmed by the majesty of natural phenomena. Jesus spoke of the lilies of the field and God's care for the sparrow. Value pervades all life, not just human life. Furthermore, nature is part of the drama of redemption and will share in the ultimate harmony, as portrayed in the symbolic vision of the coming kingdom, when "the wolf shall dwell with the lamb, and the leopard shall lie down with the kid."[34] Paul imagines that "the whole creation has been groaning in travail together until now," but it will all take part in the final fulfillment.[35] While the focus of attention was on human history, the world of nature was not neglected within the Bible as it was in much of later Christian thought.[36]

St. Benedict might be taken as an early model of the stewardship perspective. Compared with St. Francis's deep feeling and sense of union with the natural world, St. Benedict's response was more practical, using nature with care and respect. The Benedictine monasteries combined work and contemplation. They developed sound agricultural practices, such as crop rotation and care for the soil, and they drained swamps and husbanded timber all over Europe. Benedictines were creative in practical technologies related to nature.

One also might mention the continuing literary tradition that has extolled *the pastoral ideal* of cultivated nature, intermediate between the expanding city and the inhospitable wilderness. In antiquity, the pastoral poems of Theocritus, Virgil, and Horace celebrated the beauties of the country landscape and the virtues of the simple rural life in contrast to the growing urbanization of Greek and Roman life. The pastoral motif continued through the history of literature, and was prominent in eighteenth-century Europe. Leo Marx has traced in a number of nineteenth-century American novels the role of the "middle landscape," which combines the values of civilization and nature.[37]

The Conservationist Outlook

From more recent times, Gifford Pinchot could serve as an example of a *conservationist* rather than preservationist viewpoint at the level of policy choice. Opposing John Muir's campaign for the preservation of untouched wilderness, Pinchot advocated "wise use" and "scientific management" of federal lands, and eventually secured Theodore Roosevelt's support. In 1905, the U.S. Forest Service was established, with Pinchot at its head. Its policy was intermediate between unlimited exploitation and absolute protection, namely, rational planning for the maximum sustained yield that would conserve resources for future use. If the Forest Service can be criticized for opening up federal lands too rapidly to mining, grazing, and lumber companies, it also can be appreciated for the dedicated service of the new professional foresters.[38]

Other *conservation agencies*, such as the Reclamation Service (1903) and the Soil Conservation Service (1935), interpreted "wise use" and "scientific management" even more narrowly in terms of long-term economic yield and technical efficiency. The National Park Service Act (1916) did specifically defend aesthetic, recreational, and wildlife values. But not until the National Wilderness Preservation Act of 1964 was there legislation defending preservationist ideals against the increasingly utilitarian outlook of federal conservation agencies and their state counterparts, which have been even more subject to pressures from local financial and industrial interests.

The biologist René Dubos gives a contemporary rendition of a philosophy of *respectful use* and *creative intervention*. He advocates managing and transforming the earth, but with awareness of the consequences and limitations of human activity. It is not true, says Dubos, that "nature knows best"; intelligent human action can bring about changes that benefit other species as well as humankind. Humanity has unique capacities and potentialities, yet it is dependent on the biotic community; there is no radical disjunction of man and nature. Dubos writes:

> We certainly must reject the attitude which asserts that man is the only value of importance and that the rest of nature can be sacrificed to his welfare and whims. But we cannot escape, I believe, an anthropocentric attitude which puts man at the summit of creation while still a part of it.[39]

As another example, consider Barry Commoner's interest in both human and environmental values. He analyzes the relative roles of population, affluence, and technology in environmental pollution. He does not accept the thesis of Ehrlich and others that population is the main problem; instead, he lays the blame on heavily polluting technologies (especially synthetic plastics, detergents, fertilizers, and pesticides). He argues that continued economic growth is possible with an ecologically sound agriculture and technology. But Commoner's rejection of a no-growth, antitechnology policy is based also on his concern for social justice and global inequalities. He holds that improvement of living standards of the poor at home and abroad requires the growth of the right kind of technology.[40] Commoner looks at the political and economic structures that perpetuate inequalities and at issues of distributive justice that are ignored by many environmentalists.

Advocates of unity with nature have in the past tended to neglect the issue of *social justice.* From Thoreau to Muir to countercultural communes, the retreat to nature often has involved turning one's back on the city and "the dirty institutions of men." (To be sure, Thoreau advocated the abolition of slavery and Muir campaigned for wilderness legislation, but their areas of political activity were limited.) The political viewpoint of environmentalists has reflected their predominantly middle-class background and interests. But this situation seems to be changing, and there are many within the environmental movement today who are deeply concerned about inequalities among nations in a world of limited resources. There also is greater interest in the environmental problems of the poor and the structures of political and economic power that support environmental degradation (see Chapter 5).

Advocates of this stewardship position often will join advocates of unity with nature in *political action* to protect the environment. Together they recognize that by treating air and water as free commodities, the market economy has not charged industry for the true costs. Any "public goods" or unrestricted "commons" will be overused. These indirect costs ("externalities") can be controlled only by legislative action (such as effluent standards, taxes, or subsidies). But persons in this stewardship group are more likely to emphasize the human consequences of environmental legislation and to assign greater weight to employment and social justice in policy decisions.

Contemporary Theology and Philosophy

A number of theologians recently have presented variants of the stewardship motif. In *Brother Earth,* Paul Santmire maintains that both "compulsive manipulation of nature" and "romantic retreat to nature" ignore the demands of social justice. The biblical God, he insists, is ruler of both nature and history. God establishes the created realm, takes delight in it, and values it for its own sake. Humanity, in turn, has the role of caretaker, extending God's care, preserving the life of other creatures and defending their rights. God has purposes in history, too, and the ultimate goal is a kingdom of human as well as cosmic redemption. Santmire arrives at an ethic of responsibility and dedication to both distributive justice and care of the earth.[41]

The majority of recent *writing by theologians* probably could be placed in the stewardship camp, though leaning toward domination. Thomas Derr, for instance, says that human beings are emancipated from nature into history.

"Technology sets us free from the tyranny of nature." Human welfare requires "the subjugation of nature." Derr thinks the environmental movement has diverted attention from more serious issues, such as war, famine, poverty, and racial oppression.[42] But there are a few Christian writers who lean in the other direction, and would belong on the scale between stewardship and unity. Frederick Elder sides with those for whom humanity is fundamentally a part of nature. He cites Albert Schweitzer's principle of "reverence for life," and Loren Eisley's sense of involvement in the web of life, and his awe and wonder in response to natural phenomena. Elder calls for a modern asceticism, restraint in consumption, and respect for the integrity of the natural world.[43] A comparable range of perspectives can be found in contemporary Judaism.[44]

A similar diversity can be found among *contemporary philosophers*. John Passmore has written an excellent study of attitudes toward nature in Western thought, and in a concluding chapter has developed his own views.[45] He rejects nature-mysticism and pantheism as inimical to the scientific spirit. He argues that nonhuman nature has no rights, and he defends a utilitarian ethic of enlightened anthropocentrism. Passmore proposes a secular stewardship of nature for the sake of human welfare. Any responsibility for nature is derivative from responsibility toward humanity. We can transform and humanize nature for our own purposes, but we should do so with an informed recognition of the long-run consequences of our actions. Similar views have been advanced by other philosophers.[46]

Process philosophy, on the other hand, supports attitudes intermediate between stewardship and unity. Alfred North Whitehead and his followers rejected the separation of humankind and nature that had dominated much of nineteenth- and twentieth-century philosophy, especially in Europe. Kant had contrasted the human realm of freedom and history with the mechanical realm of nature. Existentialism had drawn an even stronger distinction between the subjective sphere of personal selfhood and the objective world of impersonal nature. Whitehead tried to bring these two realms back together within a coherent intellectual system, a single set of categories applicable to all beings. As he saw it, nature is a creative process, a community of interacting organisms, not a deterministic machine. Every being is constituted by its relationships and dependent on its environment, but each is at the same time in its own way a center of experience. Mind and matter are not two opposing substances, as Descartes held, but two aspects of events in systems having many levels of organization.[47]

For *process thinkers*, all beings are intrinsically valuable and worthy of respect as centers of at least rudimentary experience. Moreover, humans are to be understood in the same categories as other beings. There are no metaphysical discontinuities, though the importance of any given category varies widely among differing levels of being. God transcends nature but is also immanent in the creative process; He does not intervene coercively from outside, but participates throughout cosmic history. Process thought thus avoids the sharp separation of the human and the nonhuman, and of God and nature, which has in the past encouraged environmentally destructive attitudes.[48] But it also provides a rationale for giving higher priority to human needs than to the needs of other forms of life (see Chapter 5).

Conclusions

The classification of attitudes into three groups in this chapter is an oversimplification, introduced only to trace some broad patterns among the bewildering variety of historical and contemporary outlooks. There are diverse ideas within each group, and there is a continuum of intermediate positions. On a scale running from domination through stewardship to unity, I would place myself between stewardship and unity. This conclusion is based on the beliefs that I hold and the value priorities I will be defending. It also is influenced by strategic considerations. In our society, the stewardship position tends to be compromised in the direction of the domination view, which has been so widespread in the past and continues to be supported by powerful institutional forces. Some weighting toward ideas of unity is required to challenge prevailing attitudes and interests.

The term *nurture* perhaps may suggest a care for the earth that goes beyond stewardship or conservation in attributing an independent value to the nonhuman. At the same time nurture implies a more active human role than most forms of unity, and a greater dedication to human as well as environmental values.[49] Let me indicate such an intermediate position on six issues running through this chapter; each is elaborated in later chapters.

1. Religious Beliefs

The domination theme has been historically associated with a strong emphasis on divine transcendence and the contrast between God and nature. The religious versions of unity, on the other hand, have stressed immanence (variously expressed in terms of nature gods, a pantheistic principle, nature-mysticism, an experience of cosmic unity, or a Romanticism that finds spiritual qualities in nature). The stewardship view combines transcendence and immanence. It holds that nature is to be neither despoiled nor worshipped, but respected and appreciated. I submit that in our culture the recovery of such neglected ideas in the biblical heritage is more promising than the attempt to transplant ideas from Eastern religions with which most people are unfamiliar. But I believe special attention should be given to immanence in a theology of nature today. The celebration of nature should be given more prominence in the liturgy of the church, and care for the earth should have a larger place in its ethical teachings.[50]

2. The Human and the Nonhuman

In the domination view, there is a radical disjunction, an absolute gulf, between human and other forms of life. Nonhuman nature is valued only for its contribution to humanity. Exponents of unity, by contrast, often submerge humanity in nature, and minimize any differences between the human and the nonhuman. Neither extreme seems to me tenable. There is overwhelming scientific evidence for the evolutionary continuity of human life with lower forms, and for behavioral similarities; but there also are distinctively human dimensions of language, culture, personal life, self-consciousness, and historical existence.[51] Stewardship leads to respect for nature as a unified, created order of which we are a part and for which we are responsible. I would go further in stressing our dependence on the wider community of life and our obligations toward nonhuman creatures (see Chapter 5).

3. The Lessons of Ecology

The findings of ecology concerning ecosystem interdependence, diversity, and stability are crucial in any policy decisions concerning the environment. Ecologists have made us aware of finite limits, challenging our previous assumptions about unlimited growth. They have encouraged the adoption of the long time-scale essential to the discussion of resource sustainability. Here I find myself close to the scientific exponents of unity, though I will suggest that many environmentalists have neglected the possibilities for intelligent and ecologically informed intervention to extend natural limits in the interest of human welfare. Recognition of the far-reaching repercussions of human actions can engender humility and caution in place of the attitudes of mastery, subjugation, and technological arrogance typical of the past.

4. The Role of Technology

Whereas people in the first group have great confidence in industrial technology, many of those dedicated to unity are ready to curtail or even abandon it because of its environmental and human costs. A middle position seeks political mechanisms to control technology in order to reduce its environmental impacts, redirect it to basic human needs, and distribute its benefits more equitably. I would go further in advocating appropriate technologies scaled to local control and local needs, in both developed and developing countries. These topics are taken up in subsequent chapters.

5. The Problem of Growth

If people who favor domination are progrowth, and those who favor unity urge no growth, those who favor stewardship will ask "Whose growth?" and "What kind of growth?" Global justice requires that economic growth in developed countries be channeled toward services that are not resource intensive. Ecological wisdom requires research on technologies for recycling and waste reduction. Restraint in consumption in affluent nations will require changes in personal values and life-styles, and also in national priorities and policies (see

Chapters 12 and 13). But the progrowth mentality is so deeply ingrained that those who favor selective growth and resource sustainability will have to devote most of their efforts to limiting the inequitable demands that the industrial West places on the world's resources and environment.

6. Political and Economic Institutions

Environmentalists have recognized that the concentration of economic and political power in large-scale industries has been a central factor in environmental degradation. But many of them have been less concerned about the exploitation of people that, along with the exploitation of nature, is a product of economic institutions designed to maximize profit and efficiency. Those in the stewardship camp usually look to legislative regulations and government action to protect both people and nature. I support such efforts to work through existing political channels for both environmental preservation and human welfare, as will be evident in Part Two. But I also believe that more far-reaching institutional changes will be necessary to achieve a just, participatory, and sustainable society, as will be indicated in Part Three.

NOTES

1. Collections of readings on attitudes toward nature include Ian G. Barbour, ed., *Western Man and Environmental Ethics* (Reading, Mass.: Addison-Wesley, 1973); Robert Roelofs, Joseph Crowley, and Donald Hardesty, eds., *Environment and Society* (Englewood Cliffs, N.J.: Prentice-Hall, 1974); Robert Detweiler, Jon Sutherland, and Michael Werthman, eds., *Environmental Decay in Its Historical Context* (Glenview, Ill.: Scott, Foresman, 1973); and Roderick Nash, ed., *Environment and Americans* (New York: Holt, Rinehart and Winston, 1972).

2. Milton Rokeach, *The Nature of Human Values* (New York: Free Press, 1973), p. 18. See also his *Beliefs, Attitudes and Values* (San Francisco: Jossey-Bass, 1968). Values are defined and discussed in Chapter 4 below.

3. Gen. 1:28. Quotations throughout this volume are from the Revised Standard Version of the Bible (copyright 1946 and 1952, National Council of Churches).

4. Lynn White, Jr., "The Historical Roots of Our Ecologic Crisis," *Science* 155 (1967): 1203-07, reprinted (with a reply to his critics by White) in Barbour, op. cit. See also David and Eileen Spring, eds., *Ecology and Religion in History* (New York: Harper & Row, 1974); and John Black, *The Dominion of Man* (Edinburgh: Edinburgh University Press, 1970).

5. Lewis W. Moncrief, "The Cultural Basis of Our Environmental Crisis," *Science* 170 (1970): 508-12.

6. Clarence Glacken, *Traces on the Rhodian Shore* (Berkeley: University of California Press, 1967); and idem, "Man's Place in Nature in Recent Western Thought," in *This Little Planet*, ed. Michael Hamilton (New York: Scribner's, 1970). See also John Passmore, *Man's Responsibility for Nature* (New York: Scribner's, 1974).

7. See William Leiss, *The Domination of Nature* (New York: George Braziller, 1972), chap. 3.

8. The view of nature associated with the rise of modern science is traced in Edwin A. Burtt, *The Metaphysical Foundations of Modern Science*, rev. ed. (New York: Humanities

Press, 1951); Alfred N. Whitehead, *Science and the Modern World* (New York: Macmillan, 1925); R. G. Collingwood, *The Idea of Nature* (Oxford: Clarendon Press, 1945); and E. J. Dijksterhuis, *The Mechanization of the World Picture* (Oxford: Clarendon Press, 1961).

9. See Victor Ferkiss, *The Future of Technological Civilization* (New York: George Braziller, 1974), chaps. 3 and 4.

10. Arthur Ekirch, *Man and Nature in America* (New York: Columbia University Press, 1963); Hans Huth, *Nature and the American: Three Centuries of Changing Attitudes* (Berkeley; University of California Press, 1957); Perry Miller, *Errand into the Wilderness* (Cambridge, Mass.: Belknap Press, 1956); Peter Carroll, *Puritanism and the Wilderness* (New York: Columbia University Press, 1969); Henry N. Smith, *Virgin Land: The American West as Symbol and Myth* (Cambridge, Mass.: Harvard University Press, 1950); and Chester Cooper, ed., *Growth in America* (Westport, Conn.: Greenwood Press, 1976).

11. Rosemary Ruether, *New Woman, New Earth* (New York: Seabury Press, 1975); Susan Griffin, *Woman and Nature* (New York: Harper & Row, 1978); and Carolyn Merchant, *The Death of Nature: Women, Ecology and the Scientific Revolution* (New York: Harper & Row, 1979).

12. John Maddox, *The Doomsday Syndrome* (New York: McGraw-Hill, 1972), p. 181.

13. For example, M. J. Grayson and T. L. Shepard, *The Disaster Lobby* (Chicago: Follett, 1973). See other references in Chapter 3.

14. Alfred Schmidt, *The Concept of Nature in Marx* (New York: Humanities Press, 1972); and Howard Parsons, *Marx and Engels on Ecology* (Westwood, Conn.: Greenwood Press, 1977).

15. Marshall Goldman, "The Convergence of Environmental Disruption," *Science* 170 (1970): 37-42; and Donald Kelley, Kenneth Stunkel, and Richard Wescott, *The Economic Superpowers and the Environment: The U.S., the Soviet Union and Japan* (San Francisco: W. H. Freeman, 1976).

16. Mircea Eliade, *The Sacred and the Profane* (New York: Harcourt Brace, 1959), chap. 3; and idem, *Patterns in Comparative Religion* (New York: Sheed & Ward, 1958), chaps. 8 to 10.

17. See Glacken, op. cit., pp. 62ff.

18. Huston Smith, "Tao Now: An Ecological Testament" in *Earth Might Be Fair*, ed. Ian G. Barbour (Englewood Cliffs, N.J.: Prentice-Hall, 1972); and Chung-yuan Chang, "The Meaning of Tao," in *Traditional China*, ed. James Liu and Wei-ming Tu (Englewood Cliffs, N.J.: Prentice-Hall, 1970).

19. D. T. Suzuki, *Zen and Japanese Culture* (New York: Pantheon, 1959); idem, *Studies in Zen* (New York: Philosophical Library, 1955), chap. 8; H. Byron Earhart, "The Ideal of Nature in Japanese Religion and its Possible Significance for Environmental Concerns," *Contemporary Religions in Japan* 11 (1970): 1-26; Masao Watanabe, "The Conception of Nature in Japanese Culture," *Science* 183 (1974): 279-82; and Hajime Nakamura, "Environment and Man: Eastern Thought," in *Encyclopedia of Bio-Ethics*, ed. Warren Reich (New York: Macmillan, 1978).

20. Yi-Fu Tuan, "Our Treatment of the Environment in Ideal and Actuality," *American Scientist* 58 (1970): 246-49.

21. Rhoads Murphey, "Man and Nature in China," *Modern Asian Studies* 1 (1967): 313-33; and Leo Orleans and Richard Suttmeier, "The Mao Ethic and Environmental Quality," *Science* 170 (1970): 1173-76.

22. Kelley et al., op. cit.; and Norie Huddle and Michael Reich, *Island of Dreams: Environmental Crisis in Japan* (New York: Autumn Press, 1975).

23. John Neihardt, *Black Elk Speaks* (Lincoln: University of Nebraska Press, 1961); T. C. McLuhan, *Touch the Earth* (New York: Outerbridge and Dienstfrey, 1971); Murray Wax, "The Notions of Nature, Man and Time of a Hunting People," *Southern Folklore Quarterly* 26 (1962): 175-86; Walter Capps, ed., *Seeing with a Native Eye* (New York:

Harper & Row, 1976); Dennis Tedlock, ed., *Teachings from the American Earth* (New York: Liveright, 1975); and Fred Fertig, "Child of Nature: The American Indian as an Ecologist," *Sierra Club Bulletin* 55 (August 1970): 4–7.

24. Clara Sue Kidwell, "American Indian Attitudes toward Nature," in *Contemporary Native American Address*, ed. John R. Maestas (Provo, Utah: Brigham Young University, 1976).

25. Majorie Nicolson, *Mountain Gloom and Mountain Glory* (Ithaca, N.Y.: Cornell University Press, 1959); and Huth, op. cit., chap. 3.

26. Henry David Thoreau, *Walden* (1854; reprinted, New York: W. W. Norton, 1951); and idem, *Cape Cod* (1864; reprinted, New York: Thomas Y. Crowell, 1961).

27. Roderick Nash, *Wilderness and the American Mind*, rev. ed. (New Haven: Yale University Press, 1973); idem, *The American Environment: Readings in the History of Conservation* (Reading, Mass.: Addison-Wesley, 1968); and Peter J. Schmitt, *Back to Nature: The Arcadian Myth in Urban America* (New York: Oxford University Press, 1969).

28. Among many examples are Paul Shepard and Daniel McKinley, eds., *The Subversive Science* (Boston: Houghton Mifflin, 1969); G. Tyler Miller, *Living in the Environment* (Belmont, Calif.: Wadsworth, 1975); William Murdoch and Joseph Connell, "All About Ecology," in *Western Man and Environmental Ethics*, ed. Barbour, op. cit.; and Charles H. Southwick, *Ecology and the Quality of the Environment*, 2d ed. (New York: D. Van Nostrand, 1976).

29. Yi-Fu Tuan, *Topophilia: A Study of Environmental Perception, Attitudes and Values* (Englewood Cliffs, N.J.: Prentice-Hall, 1974); William H. Ittelson, *An Introduction to Environmental Psychology* (New York: Holt, Rinehart and Winston, 1974); and William Burch, Neil Cheek, and Lee Taylor, eds., *Social Behavior, Natural Resources and the Environment* (New York: Harper & Row, 1972).

30. Aldo Leopold, "The Land Ethic," in *A Sand County Almanac* (New York: Oxford University Press, 1949). Leopold is discussed further in Chapter 5.

31. Useful histories of the environmental movement are Carroll Pursell, ed., *From Conservation to Ecology: The Development of Environmental Concern* (New York: T. Y. Crowell, 1973); and Donald Fleming, "Roots of the New Conservation Movement," *Perspectives in American History* 6 (1972): 7–91.

32. References for the debates on population, resources, and industrial growth are given in Chapters 11 and 12.

33. Bernard Anderson, "Human Dominion over Nature," in *Biblical Studies in Contemporary Thought*, ed. Miriam Ward (Cambridge, Mass.: Greeno Hadden, 1975); W. Lee Humphreys, "Pitfalls and Promises of Biblical Texts as a Basis for a Theology of Nature," in *A New Ethic for a New Earth*, ed. Glenn Stone (New York: Friendship Press, 1971); and H. Wheeler Robinson, *Inspiration and Revelation in the Old Testament* (Oxford: Clarendon Press, 1946), chaps. 1 and 2.

34. Isa. 11:6.

35. Rom. 8:22.

36. Eric Rust, *Nature and Man in Biblical Thought* (London: Lutterworth Press, 1953); idem, *Nature—Garden or Desert?* (Waco, Tex.: Word Books, 1971); and C. F. D. Moule, *Man and Nature in the New Testament* (Philadelphia: Fortress Press, 1967).

37. Leo Marx, *The Machine in the Garden: Technology and the Pastoral Ideal in America* (New York: Oxford University Press, 1964).

38. Samuel Hays, *Conservation and the Gospel of Efficiency* (Cambridge, Mass.: Harvard University Press, 1959).

39. René Dubos, "A Theology of Earth," in *Western Man and Environmental Ethics*, ed. Barbour, op. cit., p. 53; and idem, *A God Within* (New York: Scribner's, 1972).

40. Barry Commoner, *The Closing Circle* (New York: Knopf, 1971).

41. H. Paul Santmire, *Brother Earth* (New York: Thomas Nelson, 1970); and Paul Lutz and H. Paul Santmire, *Ecological Renewal* (Philadelphia: Fortress Press, 1972). Compare Joseph Sittler, *Essays on Nature and Grace* (Philadelphia: Fortress Press, 1972).

42. Thomas Derr, *Ecology and Human Need* (Philadelphia: Westminster Press, 1975). A more anthropocentric attack on the environmental movement in the name of social justice is Richard Neuhaus, *In Defense of People* (New York: Macmillan, 1971). For a reply to Lynn White from the viewpoint of evangelical theology, see Francis A. Schaeffer, *Pollution and the Death of Man* (Wheaton, Ill.: Tyndale House, 1970).

43. Frederick Elder, *Crisis in Eden* (Nashville, Tenn.: Abingdon Press, 1970).

44. E.g., Erick Freudenstein, "Ecology and Jewish Tradition," *Judaism* 19 (1970): 406-14; Jonathan Helfund, "Ecology and the Jewish Tradition: A Postscript," *Judaism* 20 (1971): 330-35; and Robert Gordis, "A Basis for Morals: Ethics in a Technological Age," *Judaism* 25 (1976): 20-43.

45. Passmore, op. cit., chap. 7.

46. E.g., William Blackstone, "Ethics and Ecology," in *Philosophy and Environmental Crisis*, ed. idem (Athens: University of Georgia Press, 1974).

47. Whitehead, op. cit. His major work, *Process and Reality* (New York: Macmillan, 1929), is difficult but rewarding. For a brief introduction, see Norman Pittenger, *Alfred North Whitehead* (Richmond, Va.: John Knox Press, 1969).

48. John Cobb and David Griffin, *Process Theology* (Philadelphia: Westminster Press, 1976); and John Cobb, *Is It Too Late? A Theology of Ecology* (Beverly Hills, Calif.: Bruce, 1972). See other references in Chapter 5.

49. Wendell Berry, *The Unsettling of America: Culture and Agriculture* (San Francisco: Sierra Club, 1977).

50. These topics are explored in Ian G. Barbour, *Issues in Science and Religion* (Englewood Cliffs, N.J.: Prentice-Hall, 1966), chap. 13; idem, *Earth Might Be Fair*, op. cit., chap. 9; and idem, *Science and Secularity* (New York: Harper & Row, 1970), chap. 2.

51. See Barbour, *Issues in Science and Religion*, op. cit., pp. 91-93 and 357-64.

3
ATTITUDES TOWARD TECHNOLOGY

The hard choices we face today involve attitudes toward technology as well as attitudes toward nature. New technologies have contributed to health, material welfare, and higher standards of living, but they also have resulted in mounting environmental and human costs. What has been the impact of technology on nature and on people? What are the human and environmental values at stake in future technological decisions?

In the past it usually was assumed that science and its offspring, technology, were inherently good and contributed to progress. But in many cases it has been the very success of a technology in achieving its limited objectives that has created unanticipated problems. Large-scale technologies also have accelerated the concentration of economic and political power and have proved difficult to control. In addition, the technological goal of narrowly defined efficiency, which tends to pervade all aspects of life in industrial societies, has come under increasing attack. In this chapter the general characteristics of modern technology are examined to establish a framework for the particular technological decisions discussed in later chapters.

Technology may be defined as "the organization of knowledge for the achievement of practical purposes."[1] It is a set of skills, techniques, and activities for the shaping of materials and the production of objects for practical ends.[2] Traditional technologies arose not from science but from trial-and-error and accumulated experience in crafts and trades. Since about 1850, however, technology has been increasingly based on science. Today there is a continuous spectrum from pure science (basic research aimed at understanding), through applied science (research aimed at practical applications), to technology (operational design aimed at construction and production). We will be using the term technology to refer to a broad set of human activities, and not simply to the machines and tools that constitute the hardware of technology.[3]

Attitudes toward technology will be grouped according to a threefold scheme similar to that used in the preceding chapter. This is an oversimplification since there is considerable diversity within each group, but it will make it easier to see the broad outlines of the debate and the issues at stake. The first section presents some optimistic views of technology and its effects. The next section examines the critics of technology, including those in the youth counterculture. The third section develops the position that technology is neither inherently good nor inherently evil; the consequences depend entirely on its social context and its relation to economic and political power. Whereas proponents of the first view think that technology does not need to be controlled, and adherents of the second often hold that it cannot be controlled, the third group concentrates on possible mechanisms of social and political control.

I. TECHNOLOGY AS LIBERATOR

Throughout its history, modern technology often has been enthusiastically welcomed as our liberator from famine, disease, and poverty. It has been celebrated as the source of material progress and human fulfillment. Optimistic views of technology were common in nineteenth-century America, and also have been forcefully presented by a number of contemporary writers.

The Promise of Technology

The *Industrial Revolution* of eighteenth-century England was the first major development of industrial technology. It was based on the availability of coal as a new source of energy (replacing water and wind power) and on iron as the strategic material. The practical harnessing of coal and iron depended on Watt's invention of the steam engine (1769), which made possible the new factory system. Apologists for technology grant that working conditions in the early Industrial Revolution were deplorable and provoked justified reactions (which we will note later); but they maintain that legislation corrected the worst abuses, and that higher living standards eventually resulted, not just for the owners of capital but for the whole populace. The chief benefit of the new industrial technology was its remarkable productivity.

In the previous chapter, some of the factors favorable to the rapid *growth of technology in the United States* were mentioned: vast resources, a virgin land, the frontier spirit, and an economic system that encouraged the accumulation of investment capital. Technology was to be the instrument for "subduing a continent," "taming the wilderness," and "controlling the forces of nature." The sense of newness, freedom from traditional patterns, and dedication to the creation of a nation helped to generate a spirit of innovation and invention. Faith in unlimited progress and growth took specific form in the dream of creating a productive civilization and material abundance founded on the machine.

As early as 1830, de Tocqueville wrote that confidence in the benefits of technological advance was a distinctive characteristic of the United States. Technology also was seen as a means of achieving democratic objectives, including new forms of freedom to act and a greater equality through improvement in the lot of the poor. Because technology was assumed to be beneficent and governed by an orderly law of inevitable progress, there seemed to be no need for deliberate direction and control.[4]

One can find many examples of *enthusiasm for technology in the United States.*[5] The locomotive, in particular, captured the popular imagination. By 1855, 700 steamboats were plying the Mississippi and its tributaries and were admired as an example of American creative genius. By the end of the century there was great interest in the work of individual inventors. Thomas Edison, George Westinghouse, Alexander Graham Bell, and the Wright brothers were cultural heroes. By World War I, the United States was the world's leading industrial nation, and technology was clearly the main source of national power. The official slogan of the Century of Progress exposition in Chicago in 1933 was an expression of hope, though we can now read it as a warning: "Science Finds–Industry Applies–Man Conforms."

World War II brought a great expansion of *industrial technology*, which was to prove decisive in the outcome of the war. Since then, the growth of chemical industries producing plastics and synthetics has yielded a whole array of new man-made materials and products. Even more far-reaching has been the development of electronic technologies, very different from the mechanical technologies dominant since the Industrial Revolution. Electronic devices provided new ways of processing and using information rather than materials, opening the door to an age of computers, communications, and automation. For many people the image of the first astronaut stepping onto the surface of the moon was a symbol of the triumph of technology. After that feat, it seemed that no task would be too difficult for mankind's inventive genius.

The Benefits of Technology

Four kinds of benefits can be distinguished if one looks at the recent history of technology and considers its future potential:

1. Higher Living Standards

New drugs, better medical attention, and improved nutrition have more than doubled the average life span in the United States within the past century. Machines have released us from much of the back-breaking labor that in previous ages absorbed most of people's time and energy. The ancient dream of a life free from famine, disease, and poverty is beginning to be fulfilled through technology. Material progress represents liberation from the tyranny of physical nature. The standard of living of low-income families in the United States has doubled in a generation, even though relative incomes have changed little. Many people in developing nations now look on technology as their principal source of hope.

2. Opportunity for Choice

Individual choice has a wider scope today than ever before because technology has produced new options not previously available and a greater range of products and services.[6] Social and geographical mobility allow a greater choice of jobs and locations. In an urban industrial society, a person's options are not as limited by parental or community expectations as they were in a small-town agrarian society. The dynamism of technology can liberate people from static and confining traditions to assume responsibility for their own lives. Birth control techniques, for example, allow a couple to choose the size and timing of their family. Power over nature gives greater opportunity for the exercise of human freedom.

3. More Leisure

Increases in productivity have led to shorter working hours. Automation and cybernation hold the promise of eliminating much of the monotonous work typical of earlier industrialism. Through most of history, leisure and culture have been the privilege of the few, while the mass of humanity was preoccupied with survival. In an affluent society there is time for continuing education, the arts, social service, sports and participation in community life. Technology can contribute to the enrichment of human life and the flowering of creativity. Labor-saving devices free us to do what machines cannot do. One estimate is that in a fully automated society only 2% of the population will need to work to produce abundant food and goods for the entire nation.[7]

4. Improved Communications

With new forms of transportation, one can in a few hours travel to distant cities that once took months to reach. With electronic technologies (radio, television, computer networks, and so on), the speed, range, and scope of communication have vastly increased. Visual image and auditory message combine to give an immediate total impact not found in the linear sequence of the printed word. In these new media there is a potential for instant worldwide communication and for greater interaction, understanding, and mutual appreciation in the "global village."[8] It has been suggested that by dialing coded numbers on telephones hooked into computer networks, citizens could participate in an instant referendum on political issues. Another proposal envisages home TV sets connected to information centers from which individuals could request films, library holdings, or data displays.[9] According to its defenders, technology, in short, brings psychological and social benefits as well as material progress.

Optimistic Views of Technology

Among recent versions of technological optimism are the writings of a number of engineers. Charles Susskind's *Understanding Technology* expresses confidence that technology itself can solve the problems it has created. "Technology, the source of the problem, will once again prove to contain within itself the germs of a solution compatible with the betterment of man's lot and dignity."[10] Susskind reviews a number of criticisms but dismisses them: "Don't blame technology (and by implication, the technologist) for society's shortcomings." A physicist, Alvin Weinberg, maintains that it is easier to find *technical fixes* for social problems than to try to change social attitudes. He proposes, for instance, that instead of trying to persuade people to use less water when it is scarce, we could provide enough water if we used nuclear energy to desalinate ocean water or to pump from deep wells. Weinberg concludes: "To technological optimists like me, the earth's resources are almost unlimited." Affluence will eliminate class conflict by producing material abundance for all.[11]

A similar hope is common among *futurologists*. There has been a long tradition of literary utopias and science fiction in which technology was portrayed as the agent of the golden age.[12] But recent visions of the future claim a scientific basis. Buckminster Fuller foresees a one-world town, an organically united civilization of abundance as more efficient technologies are developed. His own designs run the gamut from small geodesic domes to floating "tetra cities" of 1 million inhabitants. But politics will be obsolete in his future society run by experts whose "comprehensive anticipatory design" will meet all human needs. Technology is to be the instrument by which we can deliberately plan our future and shape our destiny in a totally controlled environment.[13] Other authors envisage the prospects of genetic engineering for redesigning humanity itself, or for making hybrid combinations of humans and machines (cybernetic organisms or "cyborgs").[14]

The sociologist Daniel Bell welcomes the postindustrial society, which he thinks is already beginning to emerge. In this new society, power will be based on knowledge rather than property. The dominant class will be scientists, engineers, and technical experts; the dominant institutions will be intellectual (universities, industrial labs, and research institutes). The economy will be devoted mainly to services rather than material goods. According to Bell, decisions in the emerging society will be made on rational-technical grounds, marking "the end of ideology." There will be a general consensus on social values; experts will coordinate social planning, using rational techniques such as decision theory and systems analysis. This will be a future-oriented society, the age of professional managers, the technocrats—though the general goals of society still will be set through political processes.[15] Zbigniew Brzezinski describes in similar terms the "Technetronic Era" which he believes the United States is entering. It will be the age of computers, communications, and leadership by an intellectual elite that takes a problem-solving approach to social issues.[16]

While most *theologians* have had more to say about the dangers of technology than about its promise, a few have been enthusiastic about the latter. Their views reflect not only concern for the material welfare of the underprivileged, but also the conviction that technology is the outcome of humanity's God-given powers, an instrument of creative reason. In the 1960s, Harvey Cox celebrated the "secular city" and the promise of planetary urbanization. Cox warns of the illusions of utopian expectations and the perils of a technical elite, but he affirms our freedom to master and shape the world through technology, which liberates us from the confines of tradition to assume responsibility for the future. It brings greater freedom of choice and delight in invention; humanity and God are cocreators of an unfinished universe.[17] Other theologians have praised the creative human powers evident in computers and space exploration, and welcomed the new possibilities of controlling nature.[18]

I find the views of the technological optimists inadequate at a number of points. First, they dismiss too rapidly the *environmental costs* of technology. The optimists are confident that technical fixes can be found for pollution problems. Of course, pollution abatement technologies can control many of the effluents of industry, but often there are unintended or delayed impacts. The effects of carcinogens may not show up for 25 years or more. We have discovered only recently the increased mortality among shipyard workers exposed to asbestos during World War II. Toxic wastes may begin to surface a couple of decades after they have been buried. Recent technology has had a history of unexpected risks to people and the environment. Moreover, technical fixes do not deal with the political and social dimensions of problems. In theory, a technological response to symptoms of crisis might buy time for the analysis and treatment of underlying causes. But in practice they create the comfortable illusion that no fundamental changes are necessary, as we shall see in the case of the energy crisis.

Second, the optimists have not realistically faced either *resource constraints* or issues of *global justice*. Most technologies today (with the exception of computers and electronic communications) are resource intensive and energy intensive. Let us grant that conservation, waste reduction, greater efficiency, new design standards, and better extraction and recycling techniques could reduce significantly the use of materials and fuels. But the finite limits of global resources call into question our grossly disproportionate consumption of energy and minerals, and cast doubt on the optimists' vision of continuing industrial growth and higher levels of affluence. The expansion of material desires can lead only to keener competition for scarce resources. It is the richest societies that are most powerful and grow most rapidly, violating the norms of both distributive justice and long-term sustainability (see Chapter 12).

Third, there are particular problems with the *large-scale technologies* typical of industrial nations today. They are capital intensive rather than labor intensive, and add to unemployment in many parts of the world. Large-scale systems tend to be vulnerable to error, accident, or sabotage. Because of one person's mistake in labeling bags at a chemical plant in Michigan, PBB (a fire retardant) was added to a batch of livestock feed. One-and-a-half million chickens, 29,000 cattle, and many other animals had to be killed to contain the contamination, and the harm to human health is still uncertain.[19] In 1977, a combination of a dozen natural events, human errors, equipment failures, design inadequacies, and unexpected system interactions plunged New York City into darkness—despite the protective measures taken after the 1965 blackout. The near-catastrophe at the Three Mile Island nuclear plant near Harrisburg, Pennsylvania in 1979 was the product of human errors, faulty equipment, poor design, unforeseen reactor behavior, and unreliable safety procedures. Nuclear energy is a prime example of a vulnerable and capital-intensive technology that accelerates the concentration of political and economic power and the centralization of control (see Chapter 10). One does not have to accept Murphy's Law ("If anything can go wrong, it will") to recognize that systems in which human or mechanical failures can be disastrous are risky even in a stable society—quite apart from additional risks under conditions of social unrest.

Fourth, I believe that greater *dependence on experts* would not be desirable. The technocrats claim that their judgments are value-free; the technical elite supposedly is nonpolitical. But those with power seldom use it rationally and objectively when their own interests are at stake. When social planners think they are deciding for the good of all—whether in the French or Russian Revolution, or in the proposed technocracy of the future—the assumed innocence of moral intentions is likely to be corrupted in practice. Social controls over the controllers are always essential. With respect to freedom, technology does present the consumer with new options, but the opportunities for choice are very restricted if the basic decisions are made by corporate management, government bureaucrats, and technical experts. I will suggest that the most important form of freedom is participation in the decisions affecting one's life; large-scale technology provides few opportunities for such participation.

II. TECHNOLOGY AS THREAT

An opposing theme also can be traced through the last two centuries: technology as a threat to human fulfillment. There were protests against technology during and after the Industrial Revolution. More recent critics have maintained that technology jeopardizes human and environmental values. They have pointed to the dehumanizing impact of a technological society and the dangers when a technological mentality pervades all areas of human life. An antitechnological outlook also is common in the counterculture today.

The Human Cost of Technology

The *Industrial Revolution* exacted a high toll in human misery and exploitation. With the advent of textile mills in central England, thousands of people crowded into grimy cities and noisy factories. To provide coal for the new industries, other thousands, including women and children, worked long hours underground. To be sure, some of the worst abuses were corrected eventually by protective laws. But many of the social consequences of industrialization were inescapable results of the technologies themselves and of the factory system that they produced. Urbanization was accelerated by the labor demands of factories. Mass production required a specialized work force, centrally managed. The rational organization of labor around the demands of the machine encouraged hierarchical control.

Reactions to industrial technology in England took many forms. In the Luddite campaign (1811–16), workers smashed the new machines that they blamed for unemployment and low wages. The movement was suppressed and the working classes were persuaded that mechanization would benefit them in the long run. The Romantic Movement in literature had a strong antitechnology component, in addition to the pronature component mentioned in the previous chapter. The Romantic poets defended natural beauty against the encroaching ugliness of industrialization; they also defended human imaginative and intuitive capacities against the dominance of technical rationality. They thought social and personal life, as well as the natural world, were in jeopardy from the new industrialism. More systematic attacks on technology appeared, such as the writings of Thomas Carlyle, or Samuel Butler's *Erewhon* (1872), which is a vision of a future society in which humanity has been liberated from slavery to the machine.[20]

Voices raised against technology in the United States were more muted and evoked little response. Thoreau spoke of the machine as a threat to the human spirit, but he did not want to turn his back on a growing technological civilization. He described how the railroad shattered the peace of Walden Pond, yet he was fascinated and elated by it. There is a similar ambivalence in many nineteenth-century reactions to "the machine in the garden"; both the promise and the dangers of technology were recognized.[21] Rural areas of the United States continued to be suspicious of industrialism and the city, but they knew that opposition was futile. World War I showed the destructive potential of military technology, but it did not evoke in the United States the uneasiness about technology itself that was evident in postwar Europe.

Events since World War II, however, have contributed to more widespread *anxiety about technology*. The destructiveness of the atom bomb dwarfed any previous human act, and the threat of nuclear annihilation has continued to hang over nations with missiles poised for launching. The allocation of the majority of federal research funds to military purposes and the use of scientific resources for an inhumane and futile war in Vietnam helped turn some of the younger generation against technology. The unintended and indirect consequences of technology became increasingly evident, both in the environment and in human life. The very rapidity of technological change contrasts with the slower changes in human values and attitudes. People can tolerate rapid change in some areas of their lives if there is stability in other areas; but technological change alters nearly every area of modern life.[22]

Four characteristics of modern technology seem to its critics particularly inimical to human fulfillment:

1. Uniformity

Mass production yields standardized products, and mass media tend to produce a uniform national culture. Individuality is lost and local or regional differences are obliterated in the homogeneity of industrialization. Nonconformity hinders efficiency, so passive and docile workers are rewarded. Even the interactions among people are mechanized and objectified. Human identity is defined by roles in organizations. Conformity to a mass society jeopardizes spontaneity and freedom. There is little evidence that an electronic, computerized, automated society really will produce more diversity than earlier industrialism did.

2. Efficiency

Technology leads to rational and efficient organization. Efficiency requires fragmentation, specialization, speed, the maximization of output. The criterion is efficiency in achieving a single goal or a narrow range of objectives; side effects and human costs are ignored when efficiency becomes an end in itself. Quantitative criteria tend to crowd out qualitative ones. The worker becomes the servant of the machine, adjusting to its schedule and tempo, adapting to its requirements. There are few meaningful work roles in industrial societies today. Advertising creates demand for new products, whether or not they fill a real need. The "hidden persuaders" stimulate artificial demands that allow a larger volume of production and consumption.

3. Impersonality

Relationships in a technological society are specialized and functional. Genuine community and interpersonal interaction are threatened when people feel like cogs in a well-oiled machine. In a bureaucracy, the goals of the organization are paramount and responsibility is diffused so that no one feels personally responsible. An outlook is engendered in which an Adolph Eichmann sending Jews to extermination camps can claim to be merely "arranging train schedules." Moreover, technology has created subtle ways of manipulating people, new techniques of human engineering, electronic surveillance, and behavior modification by drugs or psychological conditioning. When the technological mentality is dominant, people are viewed and treated like objects.

4. Uncontrollability

Separate technologies form an interlocking system, a total, mutually reinforcing network that seems to lead a life of its own. We speak of "runaway technology" as if it were a vehicle out of control, with a momentum that cannot be stopped. Technology is not just a set of adaptable tools for human use, but an all-encompassing form of life, a pervasive structure with its own logic and dynamic. Its consequences are unintended and unforeseeable. Like the sorcerer's apprentice who found the magic formula to make his broom carry water for him but did not know how to make it stop, we have set in motion forces that we cannot control. The individual feels powerless and alienated facing a monolithic system too complex to understand or influence.[23]

Recent Critics of Technology

These themes have been developed with variations by a number of contemporary authors. Lewis Mumford has written nearly two dozen books dealing with various aspects of *the interaction of technology and human culture.* He has traced the profound changes in society that have resulted from new sources of energy and materials. Writing in 1934, Mumford looked with considerable hope on twentieth-century technologies based on electricity and synthetic materials. Here, he thought, were new possibilities of adapting machines to people, rather than adapting people to the machine as had occurred in the mines and factories of the Industrial Revolution. But his later books are more pessimistic. The "megatechnics" of industrial and military power have proved more damaging to human purposes than he had anticipated. We are, he says, like passengers in a driverless auto "hurtling full speed toward doom." But Mumford does not adopt a total technological determinism. Humanity influences the machine even while the machine influences humanity. Mumford holds out at least the hope that if an organismic world view replaces our mechanistic outlook, there yet may emerge a "biotechnics" compatible with individual freedom and diversified community life.[24]

To Jacques Ellul, however, technology is *an autonomous and uncontrollable force* that dehumanizes all that it touches. The enemy is "technique"—a broad term Ellul uses to refer to the technological mentality and structure that he sees pervading not only industrial processes, but also all social, political, and economic life. Efficiency and organization, he says, are sought in all activities. The machine enslaves people when they adapt to its demands. Rational order is everywhere imposed at the expense of spontaneity and freedom. Ellul ends with a technological determinism, since technique is self-perpetuating, all-pervasive, and inescapable. Any opposition is simply absorbed. Public opinion and the state become the servants of technique rather than its masters. Technique is global, monolithic, and unvarying among diverse regions and nations. Ellul offers us no way out, since all our institutions and our personal lives are totally in its grip.[25]

Herbert Marcuse indicts technological rationality as *an instrument by which people are manipulated.* An industrial society has created false needs in order to generate a demand for its products. The pervasive power of the media has produced a "false consciousness." Choices among restricted ranges of alternatives (among TV programs, auto styles, or political candidates, for instance) gives the illusion of freedom, but the main choices already have been made by someone else. The repressive character of an affluent society is subtle because we willingly sacrifice our freedom in exchange for security and material comfort. People no longer can identify their "true needs" (food, clothing, housing, and the optimal development of the individual). Protest is rendered innocuous by a society organized to dominate people and nature in the interests of efficiency.[26]

Langdon Winner recently has given a sophisticated version of the argument that technology is *an autonomous system* that shapes all human activities to its own requirements. It makes little difference who is nominally in control—elected politicians, technical experts, capitalist executives, or socialist managers—if decisions are determined by the demands of the technical system. Human ends are then adapted to suit the techniques available, rather than the reverse. Winner says that large-scale systems are self-perpetuating, extending their control over resources and markets and molding human life to fit their own smooth functioning. Technology, in short, is not a neutral means to human ends, but an all-encompassing system that imposes its patterns on every aspect of life and thought. The machines that we thought were our slaves are really our masters.[27] We will turn shortly to some criticisms of such technological determinism.

Existentialist philosophers and theologians also have maintained that modern technology is *antithetical to human values.* Gabriel Marcel thinks that the technological outlook pervades our lives and excludes a sense of the sacred. The technician treats everything as a problem that can be solved by manipulative techniques without personal involvement. But this misses the mystery of human existence, which is known only through involvement as a total person. The technician treats other people as objects to be understood and controlled. Martin Buber has contrasted the I-It relation of objective detachment with the I-Thou relation of mutuality, responsiveness, and personal involvement. If the calculating attitude of control and mastery dominates a person's life, it excludes the openness and receptivity that is a prerequisite of relationship to God or other persons. Only in humility and surrender are reverence and respect possible.[28]

Other theologians have seen in recent technology a Promethean pride and *a quest for unlimited power.* To them, the search for self-sufficiency and omnipotence is a denial of creaturehood. Unqualified devotion to technology as a total way of life is a form of idolatry. Technology is finally thought of as the source of salvation, the agent of secularized redemption. In an affluent society a legitimate concern for material progress readily becomes a frantic pursuit of comfort, a total dedication to self-gratification. Such an obsession with things distorts our basic values as well as our relationships with other persons; we end as slaves of our own desires. Exclusive dependence on technological rationality also leads to a truncation of experience, a loss of imaginative and emotional life, and an impoverishment of personal existence.[29]

Technology as threat has been a continuing theme in literature. The existentialists from Dostoevski to Sartre and Camus have protested the conformity of mass society and passionately have defended human freedom and individuality from domination by the machine. The mechanization of an industrial civilization in which people feel like objects has been vividly dramatized in the Theater of the Absurd (such as Samuel Beckett and Harold Pinter). The earlier tradition of literary anti-utopias, picturing a world in which technological goals control the whole of human life, has continued in such novels as Huxley's *Brave New World*, Orwell's *1984*, and Vonnegut's *Player Piano*. Hostility to technology also is a common feature of contemporary poetry. For philosophers, novelists, and poets alike, the protest is made primarily in the name of individual freedom.[30]

The Countercultural Critique

The counterculture that emerged in the late 1960s and early 1970s was never a unified movement, and many of its expressions were short-lived.[31] But some of its characteristic attitudes, including disillusionment with technology, have continued among a significant portion of the younger generation. Amidst this diversity there have been several recurrent themes pertinent to our discussion.

1. Alternative Life-Styles

The counterculture disavows many of the goals of a technological society, such as efficiency, order, and rationality. It opts out of the competitive pursuit of success, affluence, material possessions, and obsession with work. Alternative patterns of life are sought in which there is more room for spontaneity, individuality, and freedom, which are excluded by the bureaucratic and hierarchical structures of society. The message is that one should take charge of one's own life rather than succumbing to overt or subtle pressures for conformity.

2. Harmony with Nature

The counterculture is critical of the technological goal of conquest of nature. It has turned to Eastern religions partly because they have stressed unity with nature and respect for life in all its forms. Many of the communes formed in the last decade were in rural settings in which a group could grow organic foods and try to acknowledge in practice the interdependence of the web of life. Frequently there is a keen concern for both unity with the natural world and a simpler life with a lower level of consumption. Communal living groups with similar ideals have been formed in many urban areas.

3. Interpersonal Relatedness

Many young people, reacting to our impersonal society, have sought the acceptance of a congenial and supportive group. In a true community there is belonging, affection, and solidarity, in contrast to the anonymity of urban mass culture. Surely we must admire these ideals of authentic human relationships—openness, honesty, freedom, and tenderness—even if they are only partially realized in practice.

4. Personal Experience

The process of self-discovery and the search for identity always have been part of early adulthood, but they take new forms today. There is a hunger for intensity of experience and for commitment, peace, and joy. In a world of mechanical routines, the capacity to experience more deeply and vividly is sought. A few years ago, psychedelic drugs seemed to promise heightened awareness and an expansion of consciousness. More recently, the spiritual search of many in the younger generation has led to the meditative practices of Eastern religions. There is interest in yoga and Zen and the teachings of a variety of gurus. The ultimate truth is sought within, by intuition, not by rational thought.[32]

5. Emotion versus Reason

Our culture in general, and the technological mentality in particular, tends to repress emotion. We exalt the intellect and seldom appreciate our feelings, our senses, or our bodies. The counterculture, by contrast, encourages nonverbal communication, awareness of the senses, and the celebration of vitality and feeling. It values the immediate, the concrete, and the subjective, and it distrusts abstract ideas. In its extreme form this becomes an anti-intellectualism, a rejection of reason, and an attack on disciplined thinking of any kind.

The counterculture has had an articulate spokesman in Theodore Roszak, who has vehemently attacked *the dehumanizing impact of industrial technology.* He holds that the evils of technology—uniformity, efficiency, and impersonality—can be traced back to the "objective consciousness" that modern science has engendered. He points out that the scientist reports impersonal data from which all personal elements have been removed. Ever since Descartes, the knower has been pictured as separate from the known, the detached observer at a distance from the object of knowledge. Roszak holds that it is this objectivity that has led to alienation from nature and from other people. If nature is a machine, dead and alien and external to us, it becomes an object to be conquered and used. The objective consciousness finally leads us to treat other persons like things. Roszak turns to the poets of the Romantic movement—Blake and Wordsworth, for instance, with their defense of imagination, intuition, and individuality. He claims that we have a lot to learn from alchemy, magic, and dreams. Roszak concludes that it is the visionary, not the technologist or the political activist, who can deliver us from the wasteland of industrial civilization.[33]

I find myself in agreement with many of *the positive ideals of the counterculture*: harmony with nature, interpersonal relatedness, and the importance of personal experience and human emotion. I can agree with much of its critique of technology. Life is indeed impoverished if the technological attitudes of mastery and power dominate one's outlook. Calculation and control do exclude mutuality and receptivity in interpersonal relationships and prevent the humility and reverence that religious awareness requires. The pursuit of rationality and efficiency can lead to impatience with individual differences and personal emotions. But I would submit that the threat to these areas of human existence comes not from technology itself but from a one-sided view of human fulfillment. The enemy is not technology, but an unqualified reliance on technology and a preoccupation with material progress and technical goals alone.

The technologist neglects human emotions, but the counterculture seems to have reacted by neglecting reason—perpetuating in a different form the very dichotomy that it criticizes. If our goal *is the recovery of the wholeness of experience*, reason must be combined with emotion, thought with feeling, critical inquiry with creative imagination. Emotion without reason can lead to individual caprice or group fanaticism. Self-criticism and reflective evaluation are correctives for the destructive potentialities of emotion. Perhaps the counterculture deliberately emphasizes dimensions of human life that often are suppressed today, and perhaps every critic has to exaggerate in order to be heard. But out goal, surely, should be a more balanced recovery of the whole person.[34]

The technocratic mentality identifies reason narrowly with technical reason, the processes of mathematical calculation and logical deduction that can be carried out by a computer. The critics of technology rightly reject exclusive reliance on technical reason, but often end in irrationalism. What is needed by both the technocrat and the critics is *a wider understanding of reason.* There are various kinds of critical inquiry and disciplined thinking in the sciences and in the humanities, each closely related to a distinctive type of human experience. The complex problems of contemporary society require interdisciplinary collaboration, which is greatly hindered by the continued polarization of "two cultures" that cannot communicate.

Finally, the counterculture's concern for a new consciousness has led to a neglect of *economic and political institutions* and the demands of social justice. Political passivity would allow ecologically and humanly destructive practices to continue unrestrained. The counterculture has been predominantly a middle-class white movement among youth from fairly affluent homes with a security to which they can return. It has understood freedom individualistically as independence and the absence of regulation, rather than as participation in social decisions. While many of its members have a genuine concern for the underprivileged, its goals are not those of the ghetto resident or the people of Asia and Africa who still lack the basic necessities of life. For most of the world today, technology of the right kind is the only hope of overcoming famine, poverty, and disease. My own conclusion is that changes in individual consciousness, alternative life-styles, and communal experiments that could be examples of a new kind of society are of great importance (they will be discussed further in Chapter 13); but we also must look at the social and political institutions through which technology might be redirected in the interest of human welfare and social justice as well as individual freedom.

III. TECHNOLOGY AS INSTRUMENT OF POWER

Optimistic views of technology as liberator and pessimistic views of technology as threat have been outlined. A third view sees technology as a tool or instrument, neither inherently creative nor inherently destructive, but essentially neutral until it is used in particular ways.

We are concerned here with technology as an instrument of social power. Its direction of development and social consequences depend primarily on the economic and political institutions that control it. Technology, according to these interpreters, has developed in the industrial West as part of a system that has had a cumulative and pervasive impact on life and thought. However, it is not an uncontrollable and autonomous force following an inevitable course. Social mechanisms for the control of technology are the key to its redirection in the service of human and environmental values.

Technology and Political Power

Let us first note some criticisms very similar to those of the preceding section. The writers below are critical of current technology. Yet by distinguishing technology from the economic order in which it is embedded, they offer some hope that it might be used for more humane ends, either by political measures for more effective control of technology within existing economic and political institutions, or by changes in the economic and political systems themselves. The following characteristics of technology in its industrial setting have been attacked.

1. The Concentration of Power

C. S. Lewis has written: "What we call Man's power over Nature turns out to be a power exercised by some men over other men with Nature as its instrument."[35] Technology tends to increase the power of those who already are powerful. It has increased the gap between rich and poor, both within nations and between nations. It has perpetuated patterns of domination and dependency between industrial and developing countries. The huge scale of many industries requires heavy investments and the centralization of planning and control. Technology reinforces existing power structures and becomes the tool of special interest groups unless measures are taken to prevent it.

2. Institutional Goals

The people who make decisions about technology today are not mainly a technical elite, or technocrats trying to run society rationally, or disinterested experts whose activity was to mark "the end of ideology." The decisions are made by managers dedicated to the interests of institutions, especially industrial corporations and government bureaucracies. The direction of research is determined by the goals of institutions: corporate profits, institutional growth, bureaucratic power, and so forth. In short, expertise serves the interests of organizations, and only secondarily the welfare of people or the environment.

3. The Alliance of Government and Industry

The interlocking structure of technologically based federal agencies and corporations, sometimes called the "technocomplex," is wider than the "military-industrial complex." Many corporations are virtually dependent on government contracts. The staff members of regulatory agencies, in turn, are mainly recruited from the industries they are supposed to regulate. We will see later that particular congressional committees, government agencies, and industries have formed a three-way alliance to promote such technologies as nuclear energy. Networks of industries with common interests form lobbies of immense political power. For example, auto manufacturers, insurance companies, the highway construction industry, oil companies, and labor unions joined forces to block legislation for the development of public mass-transit systems.

John McDermott says that the optimists who look for technical fixes to social problems are naive because they do not see that technology itself has become *the tool of special interests.* Decision-making power is concentrated in the hands of the managers in industry and government agencies. The liberal intellectuals' myth of the altruistic bureaucrat and the rational planner conceals the self-interest of the new managerial class and the authoritarian character of central management.[36] John Kenneth Galbraith maintains that industrial managers are administrators rather than innovative entrepreneurs, and that the new technostructure tends to stifle creativity. But these dangers are not products of technology itself but of the structures of corporate industry allied with government bureaucracy.[37]

Adherents of this position reject the technological determinism of the pessimists as well as the technocratic illusions of the optimists. Robert Heilbroner grants that technology is a potent influence on the socioeconomic order; it affects the composition of the labor force and the hierarchical organization of work. However, he insists that social policies can influence *the direction of technological development.* In the past, the expansion of technology in the context of laissez-faire capitalism had appeared automatic; even a socialism dedicated to maximizing production failed to control technology adequately. "Technological determinism," he writes, "is thus peculiarly a problem of a certain historic epoch—specifically that of high capitalism and low socialism—in which the forces of technical change have been unleashed, but when the agencies for the control or guidance of technology are still rudimentary."[38] Heilbroner concludes that we will need "a degree of public control over technology far greater than anything that now exists."

The Control of Technology

How can such control be achieved? In *The Revolution of Hope*, Erich Fromm vividly describes the dehumanizing impact of contemporary industrial society. Material affluence, he says, has not yielded the happiness it seemed to promise, and there is widespread dissatisfaction with a mechanized mass civilization dedicated to efficiency and production. But Fromm's book is subtitled "Toward a Humanized Technology"; he is convinced that technology can be directed toward optimal human development rather than maximum production. The goal of planning should be the unfolding of mankind's distinctive capacities; the process of planning should involve extensive participation by citizens rather than decisions by experts. Fromm maintains that a radical redirection of technology toward human fulfillment is possible within present political and economic structures without violent revolution.[39]

Victor Ferkiss also is hopeful about the control of technology. He thinks that both the optimists and the pessimists have neglected the diversity among different technologies and *the potential role of political structures* in reformulating policies. In the past, technology has been an instrument of profit, and decisions have been motivated by short-run private interests. Freedom understood individualistically became license for the economically powerful; individual rights were given precedence over the common good, despite our increasing interdependence. Choices that could only be made and enforced collectively—such as laws concerning air and water pollution—were resisted as infringements on free enterprise. But Ferkiss thinks that economic criteria can be subordinated to such social criteria as ecological balance and human need. He believes it is possible to combine centralized, systemwide planning in basic decisions with decentralized implementation, cultural diversity, and citizen participation.[40]

There is a considerable range of views among *contemporary Marxists*. Most share Marx's conviction that technology is necessary for the solution of social problems, but that under capitalism it has been an instrument of exploitation, repression, and dehumanization. The alienation of the working class in the Industrial Revolution arose because production was separated from consumption and workers had no control over working conditions. In modern capitalism, according to Marxists, corporations control the government, and political processes serve the interests of the ruling class. The technical elite likewise is subservient to the profits of the owners. The Marxist grants that absolute standards of living have risen for everyone under capitalist technology, as the optimists point out. But relative inequalities have increased, and advertising has created compulsive wants, so that class distinctions and poverty amidst luxury remain.

Marxists assign justice a higher priority than freedom. Clearly they blame capitalism rather than technology for these evils of modern industrialism. They believe that alienation and inequality will disappear, and technology will be wholly benign, when the working class owns the means of production. The workers, not the technologists, are the agents of liberation. Marxists are thus as critical as the pessimists concerning the past consequences of technology, but as enthusiastic as the optimists concerning its potentialities—within a proletarian economic order.

How, then, do Western Marxists view the human effects of *technology in the Soviet Union*? Reactions vary, but many would agree with Bernard Gendron that in the Soviet Union workers are as alienated, factories as hierarchically organized, experts as bureaucratic, and pollution and militarism as rampant as in the United States. But Gendron insists that the Soviet Union is not really a socialist country in Marx's sense. The means of production are controlled by a small group within the Communist party, not by the workers. Gendron maintains that in a truly democratic socialism, technology would be humane and work would not be alienating. He holds that in *mainland China* significant advances have been made in worker participation, local self-sufficiency, and small-scale technology. A minimum level of food, education, and health services is available to virtually all citizens because technology has been directed to basic human needs (rather than to rapid industrialization, as in the Soviet case). But Gendron acknowledges that a small group of party leaders in China has continued to have immense power without direct accountability, and that restrictions on individual freedom have been extensive.[41]

We have seen that a few *theologians* are technological optimists, while others (particularly those influenced by existentialism) have adopted pessimistic positions. A larger number, however, see technology as an ambivalent instrument of social power. As an example consider Norman Faramelli, an engineer turned theologian, who writes in a framework of Christian ideas: stewardship of creation, concern for the dispossessed, and awareness of the corrupting influence of power. He is critical of technology as an instrument of corporate profit and industrial growth, but he believes it can be reoriented toward human liberation and ecological balance. Technology assessment and the legislative processes of democratic politics, he holds, can be effective in controlling technology. But Faramelli also advocates restructuring the economic order to achieve greater equality in the distribution of the fruits of technology.[42] Similar calls for the responsible use of technology in the service of basic human needs have been issued by task forces of the National Council of Churches and the World Council of Churches.[43]

This third position seems to me more consistent with *the biblical outlook* than either of the alternatives. The machine is no more capable of saving man than nature is. Preoccupation with either technology or the environment becomes a form of idolatry, a denial of the sovereignty of God, and a threat to distinctively human existence. But technology directed to genuine human needs is a legitimate expression of mankind's creative capacities and an essential contribution to human welfare. In a world of disease and hunger, technology rightly used can be a far-reaching expression of concern for persons. The biblical understanding of human nature is realistic about the abuses of power and the institutionalization of self-interest. But it also is idealistic in its demands for social justice in the distribution of the fruits of technology.

There is some evidence that *the U.S. public* is no longer solidly in the camp of technological optimism. Confidence in the automatic benefits of technology has somewhat eroded. A recent survey found a significant minority, especially among younger respondents, who would belong with the pessimists; they agreed with a number of statements expressing hostility to technology. The majority of respondents, however, combined belief in the positive potential of technology with misgivings about its present impact on society and a conviction that it can be controlled. Most of this majority were able to distinguish between scientific research, which they said should not be regulated, and technological implementation, which they said should be. While they presumably would not accept the radical critiques of corporate industrialism advanced by some of the authors above, they revealed little confidence in the judgment of business and government leaders, and they wanted both scientific experts and the public to have a larger voice in technological decisions. The authors of this survey conclude that "the public has a distrust of the institutions associated with decision-making in technical policy areas."[44] It appears that many people have taken at least a few steps away from the first position and toward the third.

Conclusions

There is a rough parallel between the three sections of this chapter and the threefold classification in the previous chapter. "Technology as Liberator" is usually associated with "Domination over Nature." "Technology as Threat" often goes with "Unity with Nature"—as in the antitechnology, pronature stance of some environmentalists and the counterculture. The third group focuses attention on the structures of institutional power—over persons in the one case, over nature in the other. As before, I come out somewhere between the second and third groups. My own conclusions are summarized below.

1. Material Welfare, Freedom, and Justice

The technological optimists tend to emphasize material welfare. They welcome technology for its productivity and material benefits. They evaluate technology in a utilitarian framework (see Chapter 4), seeking efficiency in maximizing the total social good, the balance of benefits over costs. The pessimists, on the other hand, give priority to individual freedom (plus, in some versions, interpersonal community and personal fulfillment). They are concerned about individual rights and the dignity of persons. Those in the third group give prominence to social justice (and also to freedom understood as participation in social decisions). I believe that all of these values should be taken into account in technological decisions today, but that social justice is of crucial importance in a world of increasing scarcities and growing gaps between rich and poor. Later chapters will examine specific political mechanisms for the democratic control of technology in the interest of justice.

2. Defense of the Personal

Those in the second group have defended human values in a materialistic and impersonal society. I am sympathetic to the positive ideals of the counter-culture, even though I believe its attack on rationality and science is misdirected. In Chapter 13 I will urge that each of us should adopt individual life-styles more consistent with human and environmental values. There is much to be learned from those who have explored, in practice as well as in theory, a variety of alternative life-styles. The place to begin is indeed one's own life. Moreover, strong protest and vivid examples are needed to challenge the historical dominance of technological optimism and the disproportionate resource consumption of an affluent society. I agree with the critics of technology in defending individuality and choice in the face of standardization and bureaucracy, and join them in affirming the significance of interpersonal relationships and a vision of personal fulfillment that goes beyond material affluence.[45] But I also will defend the importance of working for change through political processes, which the pessimists usually neglect.

3. The Role of Politics

Differing models of social change and of man-machine interactions are implicit in the three basic attitudes toward technology.[46] The authors discussed in the first section of this chapter assume a "rational planning" model. Humans control the machine. Technology is predominantly beneficial, and the reduction of any undesirable side effects is itself a technical problem for the experts. Rational social planning can bring in the age of abundance in which expert administration will replace politics. Writers mentioned in the second section, by contrast, typically adopt a "technological determinism" model. The machine controls human beings. Technology is dehumanizing and uncontrollable. Runaway technology is an autonomous and all-embracing system that molds all of life, including the political sphere, to its requirements. The individual is helpless within the system. The views expressed in the third section presuppose a "social conflict" model. Here there is a two-way interaction between man and machine. Technology influences human life but is itself part of a cultural system; it is an instrument of social power serving the purposes of those who control it. It does systematically impose distinctive forms on all areas of life, but these can be modified through political processes. Whereas the first two groups are essentially nonpolitical, the third, with which I agree, holds that conflicts in the control of technology must be resolved in the political arena.

4. The Redirection of Technology

I believe that we should neither accept the past directions of technological development, nor reject technology in toto, but redirect it toward the realization of environmental and human values. In the past, technological decisions have been governed by narrowly economic criteria, to the neglect of environmental and human costs. In Part Two we will examine specific legislation for the protection of the environment, the worker, the consumer, and the public. We also will look at technology assessment, a procedure designed to use a broad range of criteria to evaluate the diverse consequences of an emerging technology—*before* it has been deployed and has developed the vested interests and institutional momentum that make it seem uncontrollable. I will argue in later chapters that new policy priorities concerning energy, resource allocation, and the redirection of technology toward basic human needs can be achieved within democratic political institutions. The key question will be: What decision-making processes and what technological policies can contribute to environmental and human values?

5. The Scale of Technology

Appropriate technology can be thought of as an attempt to achieve some of the material benefits of technology outlined in the first section, without the destructive environmental and human costs discussed in the second section, most of which are results of large-scale centralized technologies. Intermediate-scale technology encourages decentralization and local participation, in keeping with the goals of democratic control presented in the third section. The decentralization of production also allows greater use of local materials and often a reduction of impact on the environment. Appropriate technology does not imply a return to primitive and prescientific methods; rather, it seeks to use the best science available toward goals different from those that have governed industrial production in the West.

Technology in the United States was developed when capital and resources were abundant and labor was scarce, and we continue to assume these conditions. Automation, for example, is capital intensive and laborsaving. Yet in *the developing nations* capital is scarce and labor is abundant. Therefore the technologies needed must be relatively inexpensive and labor intensive. They must be intermediate scale so that jobs can be created in rural areas and small towns, to slow down mass migration to the cities. They must fulfill basic human needs, especially for food, housing, and health. There are alternative patterns of modernization, some of which may be less environmentally and socially destructive than the path that we have followed. But it is increasingly evident that many of these goals also are desirable in the United States. Small groups are springing up across the country to explore intermediate technologies (see Chapter 13).

The *redirection of technology* will be no easy task. Contemporary technology is so tightly tied to industry, government, and the structures of economic power that changes in direction will be difficult to achieve. As the critics of technology recognize, the person who tries to work for change within the existing order may be absorbed by the establishment. But the welfare of humankind requires a creative technology that is ecologically sound, socially just, and personally fulfilling. The challenge of our generation is to use technology in the service of these environmental and human values.

NOTES

1. Emanuel Mesthene, *Technological Change: Its Impact on Man and Society* (New York: New American Library, 1970), p. 25.

2. James K. Feibleman, "Pure Science, Applied Science, and Technology: An Attempt at Definitions," in *Philosophy and Technology*, ed. Carl Mitcham and Robert Mackey (New York: Free Press, 1972).

3. Among recent volumes dealing with attititudes toward technology are Herbert J. Muller, *The Children of Frankenstein: A Primer on Modern Technology and Human Values* (Bloomington: Indiana University Press, 1970); and Richard Dorf, *Technology, Society and Man* (San Francisco: Boyd & Fraser, 1974). Useful collections include Albert Teich, ed., *Technology and Man's Future*, 2d ed. (New York: St. Martin's Press, 1977); Melvin Kranzberg and William Davenport, eds., *Technology and Culture* (New York: New American Library, 1975); and Noel de Nevers, ed., *Technology and Society* (Reading, Mass.: Addison-Wesley, 1972). An excellent "Bibliography of the Philosophy of Technology," compiled by Carl Mitcham and Robert Mackey, appears in *Technology and Culture* 14, part 2 (1973). See also Philip Bearano, *Technology as a Social and Political Phenomenon* (New York: John Wiley, 1976).

4. See, for example, Leo Marx, *The Machine in the Garden* (New York: Oxford University Press, 1964), chap. 4; and Melvin Kranzberg and Carroll Pursell, eds., *Technology and Western Civilization* (New York: Oxford University Press, 1967), vol. 2. See also Hugo Meir, "The Ideology of Technology," in *Technology and Social Change in America*, ed. Edwin Layton (New York: Harper & Row, 1973).

5. Thomas P. Hughes, *Changing Attitudes toward American Technology* (New York: Harper & Row, 1975); John F. Kasson, *Civilizing the Machine: Technology and Republican Values in America, 1776–1900* (New York: Grossman, 1976); and Carroll Pursell, ed., *Readings in Technology and American Life* (New York: Oxford University Press, 1969).

6. Mesthene, op. cit., chap. 2.

7. Robert Theobald, *An Alternative Future for America II* (Chicago: Swallow Press, 1970); and idem, "Cybernetics and Problems of Social Reorganization," in *The Social Impact of Cybernetics*, ed. Charles R. Dechert (New York: Simon and Schuster, 1966).

8. Marshall McLuhan, *Understanding Media* (New York: McGraw-Hill, 1965). See also William Kuhns, *Environmental Man* (New York: Harper & Row, 1969); and John McHale, *The Future of the Future* (New York: George Braziller, 1969).

9. Irene Taviss, ed., *The Computer Impact* (Englewood Cliffs, N.J.: Prentice-Hall, 1970), part V; and C. C. Gotlieb and A. Borodin, *Social Issues in Computing* (New York: Academic Press, 1973).

10. Charles Susskind, *Understanding Technology* (Baltimore: Johns Hopkins University Press, 1973), p. 132; Ellis Armstrong, "The Future of a Challenging Profession," in

Bearano, op. cit.; Samuel Florman, "In Praise of Technology," *Harper's* November 1975; and idem, *The Existential Pleasures of Engineering* (New York: St. Martin's Press, 1976).

11. Alvin Weinberg, "Can Technology Replace Social Engineering?" *Bulletin of the Atomic Scientists* 22 (1966): 4-8; reprinted in Teich, ed., op. cit.; idem, "Prudence and Technology," *Bioscience* 21 (April 1, 1971); and idem, "In Defense of Science," *Science* 167 (1970): 141.

12. Frank Manuel, ed., *Utopia and Utopian Thought* (Boston: Houghton Mifflin, 1966); Robert Silverberg, ed., *Men and Machines* (New York: Universal, 1967); George Katep, *Utopia and Its Enemies* (Glencoe, Ill.: Free Press, 1963); and Muller, op. cit., pp. 369-83.

13. R. Buckminster Fuller, *Utopia or Oblivion* (New York: Bantam, 1969); idem, *Operating Manual for Spaceship Earth* (Carbondale: Southern Illinois University Press, 1969); and William Kuhns, *The Post-Industrial Prophets* (New York: Weybright and Talley, 1971), chap. 10.

14. Richard Landers, *Man's Place in the Dybosphere* (Englewood Cliffs, N.J.: Prentice-Hall, 1966); and Arthur Clarke, *Profiles of the Future* (New York: Bantam, 1972).

15. Daniel Bell, *The Coming of Post-Industrial Society* (New York: Basic Books, 1973); and idem, *The End of Ideology* (New York: Free Press, 1960).

16. Zbigniew Brzezinski, *Between Two Ages: America's Role in the Technetronic Era* (New York: Viking Press, 1971). See also Herman Kahn and Anthony Weiner, *The Year 2000* (New York: Macmillan, 1967); and Herman Kahn et al., *The Next 200 Years* (New York: William Morrow, 1976).

17. Harvey Cox, *On Not Leaving it to the Snake* (New York: Macmillan, 1967); idem, *The Secular City* (New York: Macmillan, 1965); and idem, "The Christian in a World of Technology," in *Science and Religion*, ed. Ian G. Barbour (New York: Harper & Row, 1968).

18. Myron Bloy, *The Crisis of Human Values* (New York: Seabury Press, 1962); and Kenneth Vaux, *Subduing the Cosmos: Cybernetics and Man's Future* (Richmond, Va.: John Knox Press, 1970).

19. Wendell Berry, *The Unsettling of America* (San Francisco: Sierra Club, 1977), p. 222.

20. See Henry L. Sussman, *Victorians and the Machine: The Literary Response to Technology* (Cambridge, Mass.: Harvard University Press, 1968).

21. Leo Marx, op. cit., chap. 4.

22. Alvin Toffler, *Future Shock* (New York: Bantam, 1971); and idem, ed., *The Futurists* (New York: Random House, 1972).

23. See, for example, Eugene Schwartz, *Overskill: The Decline of Technology in Modern Civilization* (New York: Ballantine, 1971); and Jack Douglas, ed., *The Technological Threat* (Englewood Cliffs, N.J.: Prentice-Hall, 1971).

24. Lewis Mumford, *Technics and Civilization* (New York: Harcourt Brace, 1934); and idem, *The Myth of the Machine, Vol. I: Technics and Human Development; Vol. II: The Pentagon of Power* (New York: Harcourt Brace Jovanovich, 1967 and 1970).

25. Jacques Ellul, *The Technological Society*, trans. J. Wilkinson (New York: Knopf, 1964).

26. Herbert Marcuse, *One-Dimensional Man* (Boston: Beacon Press, 1964). See also Jürgen Habermas, *Toward a Rational Society* (Boston: Beacon Press, 1970), chap. 6.

27. Langdon Winner, *Autonomous Technology* (Cambridge, Mass.: MIT Press, 1977).

28. Gabriel Marcel, "The Sacred in the Technological Age," *Theology Today* 19 (1962): 27-38; idem, "The Limitations of Industrial Civilization," in *The Decline of Wisdom* (New York: Philosophical Library, 1955); Hans Jonas, *Philosophical Essays: From Ancient Creed to Technological Man* (Englewood Cliffs, N.J.: Prentice-Hall, 1974), chaps. 1 and 3; and Martin Buber, *I and Thou*, trans. W. Kaufman (New York: Scribner's, 1970).

29. Paul Tillich, "The Person in a Technical Society," in *Social Ethics*, ed. Gibson Winter (New York: Harper & Row, 1968); Longdon Gilkey, *Religion and the Scientific*

Future,(New York: Harper & Row, 1970), chap. 3; A. G. Van Melsen, *Science and Technology* (Pittsburgh: Duquesne University Press, 1961), chaps. 12 and 13; and Scott Paradise, "Christian Mission and the Technician Mentality in America," in *Christians in a Technological Era*, ed. Hugh White (New York: Seabury Press, 1964).

30. See Mark Hillegas, *The Future as Nightmare* (New York: Oxford University Press, 1967); Wylie Sypher, *Literature and Technology: The Alien Vision* (New York: Random House, 1968); William H. Davenport, "Antitechnology Attitudes in Modern Literature," *Technology and Society* 8 (1973): 7–14; William Davenport, *The One Culture* (New York: Pergamon Press, 1970), chap. 2; and Irving Howe, "The Fiction of Anti-Utopia," *New Republic* April 13, 1962, p. 13. A bibliography on technology, literature, and the arts compiled by William H. Davenport appears in *American Journal of Physics* 43 (1975): 4–8.

31. Keith Melville, *Communes in the Counter Culture* (New York: William Morrow, 1972); and Rosabeth Kantor, *Communes: Social Organization of the Collective Life* (New York: Harper & Row, 1973).

32. Charles Reich, *The Greening of America* (New York: Bantam, 1971); and William Braden, *The Age of Aquarius: Technology and the Cultural Revolution* (New York: Pocket Books, 1971). For a personal defense of intuitive knowledge, see Robert Pirsig, *Zen and the Art of Motorcycle Maintenance* (New York: Bantam, 1975).

33. Theodore Roszak, *The Making of a Counter Culture* (New York: Doubleday, 1969); and idem, *Where the Wasteland Ends* (New York: Doubleday, 1972).

34. Ian G. Barbour, "Science, Religion and the Counterculture," *Zygon* 10 (1975): 380–97.

35. C. S. Lewis, *The Abolition of Man* (New York: Macmillan, 1965), p. 69.

36. John McDermott, "Technology: The Opiate of the Intellectuals," *New York Review of Books*, July 31, 1969, p. 25, reprinted in Teich, ed., op. cit.; see also Tom Bottomore, "Machines without a Cause," *New York Review of Books*, November 4, 1971, p. 12.

37. John Kenneth Galbraith, *The New Industrial State*, 2d rev. ed. (Boston: Houghton Mifflin, 1972).

38. Robert Heilbroner, "Do Machines Make History?" *Technology and Culture* 8 (1967): 345.

39. Erich Fromm, *The Revolution of Hope: Toward a Humanized Technology* (New York: Bantam, 1968).

40. Victor Ferkiss, *Technological Man* (New York: George Braziller, 1969); and idem, *The Future of Technological Civilization* (New York: George Braziller, 1974).

41. Bernard Gendron, *Technology and the Human Condition* (New York: St. Martin's Press, 1977). See also Willis H. Truitt, "Science, History and Human Values," in *Science, Technology and Freedom*, ed. Willis H. Truitt and T. W. Graham Solomons (Boston: Houghton Mifflin, 1974); J. D. Bernal, *The Social Function of Science* (Cambridge, Mass.: MIT Press, 1967); Richard England and Barry Blackstone, "Ecology and Social Conflict," in *Toward a Steady State Economy*, ed. Herman Daly (San Francisco: W. H. Freeman, 1973); and Charles Anderson, *The Sociology of Survival* (Homewood, Ill.: Dorsey, 1976).

42. Norman Faramelli, *Technethics* (New York: Friendship Press, 1971).

43. J. Edward Carothers, Margaret Mead, Daniel McCracken, and Roger Shinn, eds., *To Love or to Perish: The Technological Crisis and the Churches* (New York: Friendship Press, 1972). The report of a World Council of Churches' conference appears in *Anticipation*, no. 19 (November 1974); also Paul Abrecht, ed., *Faith, Science and the Future* (Philadelphia: Fortress Press, 1979).

44. Todd R. La Porte and Daniel Metlay, "Technology Observed: Attitudes of a Wary Public," *Science* 188 (1975): 121.

45. See also Ian G. Barbour, *Science and Secularity: The Ethics of Technology* (New York: Harper & Row, 1970), chap. 3.

46. See Don Shriver, "Man and his Machines: Four Angles of Vision," *Technology and Culture* 13 (1972): 549ff.

4
HUMAN VALUES

We have been looking at broad historical and cultural attitudes toward nature and technology. We now turn to some of the human values that are at stake in decisions about technology and the environment. In an era of resource scarcities, issues of distribution assume increasing importance. Who benefits and who pays the costs of policy decisions? By what criteria should we judge the consequences of our choices for individual and social life?

In Chapter 1, four material values were listed: survival, health, material welfare, and employment. Four social values that are the main topic of the present chapter also were listed: justice, freedom, community, and personal fulfillment. Environmental values, which often conflict with these human values, are taken up in Chapter 5.

We begin with several examples of studies of contemporary values by social scientists. After this, some philosophical concepts pertinent to public policy decisions are discussed: utilitarianism, rights and duties, and justice and freedom. The ethical principles developed here will be applied to specific policy choices throughout subsequent chapters. Finally, the perspectives of Western religions on these human values are explored, and some implications for environmental and technological decisions suggested.

I. THE NATURE OF VALUES

What are values and how can they be studied? What are the dominant U.S. values? How are values related to human needs? What can the social sciences tell us about the values that influence policy choices?

A Definition of Values

Following Kurt Baier and Nicholas Rescher, let us define *a value* as a general characteristic of an object or state of affairs that a person views with favor, believes is beneficial, and is disposed to act to promote.[1]

1. To hold a value is *to have a favorable attitude towards its realization.* In this respect values resemble preferences or desires. They often are associated with strong feelings and emotions.

2. To hold a value is *to believe that its realization would be beneficial.* In this respect values differ from preferences or desires, which make no claims beyond momentary individual feelings and offer no general rationale for the ordering of alternatives. Subscription to a value implies a claim of benefits or of moral obligations that can be invoked to justify, defend, or recommend it to others. Such beliefs are open to rational deliberation and public criticism.

3. To hold a value is *to be disposed to act to promote its realization.* Of course, circumstances may prevent such action (for instance, if one is in prison). A person may fail to act to promote a value because he or she is more strongly committed to another value that in a given situation conflicts with it. But one would question whether someone really holds a value if there are not circumstances in which that person would act to promote its realization by some investment of effort, time, and resources. In short, commitment to a value involves attitudes, beliefs, and dispositions to act.

So defined, the values that people hold can be studied *empirically* by social scientists. Data on attitudes and beliefs can be obtained through verbal testimony, questionnaires, and interviews; data on patterns of behavior can be obtained from research on allocation of time and effort, and on actual choices among alternatives.[2] The social scientist can note the behavior that a society praises and rewards, and the actions encouraged in various forms of writing and cultural expression. Sociologists have made numerous studies of the values held by scientists, college students, industrial workers, and other groups.

Values can be treated by social scientists as either *independent* or *dependent* variables. As independent variables, the values held by a group are viewed as initial attitudes, beliefs, and dispositions that influence later behaviors and institutions. Alternatively, one may ask how changes in social institutions have influenced a group's changing values, considered as dependent variables. Values thus can be viewed both as causes and as results of other social phenomena. We have seen, for instance, that Western cultural values influenced the development of technology, and also that technology influenced dominant cultural values.

Because values are beliefs about benefits and obligations, they can be the object of *rational reflection* as well as empirical research. Ascription of value goes beyond statement of preference by indicating the characteristics by virtue of which an object or end-state is preferred. The possession of particular qualities is put forward as a reason for preference. Defense of a choice in terms of values invokes general principles believed to hold universally, transcending individual preference. Values provide specifiable criteria for judgment and choice. They thus are of interest to the philosopher as well as to the social scientist. We can ask not only what people do in fact value, but also what they ought to value, or what is valuable.

Values are not held in isolation but as components of *a value system*, a hierarchy, or ordered set. Basic values are those that are seldom subordinated to derivative values in cases of value conflict, and that serve as the criteria for justifying derivative values. Distinctive constellations of values in a society are "ways of life" transmitted through the myths and images of particular historical communities. Ideals of self-realization and visions of the good life are influenced by fundamental philosophical and religious assumptions about nature, society, and God. Value systems both influence and are influenced by the prevailing economic, political, and social institutions.

The Social Sciences and the Study of Values

Social scientists have given definitions similar to that above and have described their research on the values that people hold. The anthropologist Clyde Kluckhohn, for example, has analyzed in detail the value systems of diverse cultures.[3] The sociologist Robin Williams defines values as "standards of desirability which provide criteria for judgment and selection in action"; he notes the presence of both affective and cognitive components. Williams's studies of American society find the following as its central values: achievement and success, activity and work, humanitarianism, efficiency and practicality, progress, material comfort, equality, and freedom.[4]

Studies of *the ranking of values* by various groups in U.S. society have been carried out by Milton Rokeach. Respondents were asked to rank a set of 20 values in order of importance as guiding principles in their life. In a national sample, men and women agreed on a high average ranking for world peace, family security, and freedom, while pleasure and an exciting life were at the bottom. But on other items men and women diverged; for instance, men put a comfortable life far ahead of salvation, whereas women ranked these in the opposite order. Rokeach then correlated value profiles with income, education, race, occupation, religious and political identification, and various behavioral characteristics. Not surprisingly, artists gave high ratings to beauty, professors to wisdom, and priests to salvation. Equality ranked number two for black respondents and number eleven for whites. Freedom was number three regardless of political affiliation, whereas equality assumed widely varying rank. Equality also was the value that correlated most strongly with views on political issues such as civil rights, medical care, and the Vietnam War.[5] Among the values central to the present volume we can expect justice (which is closely related to equality) to be more controversial than freedom.

Other authors have examined the constellation of values and beliefs characteristic of technological societies today. Dennis Pirages and Paul Ehrlich define a *dominant social paradigm* as "the collection of norms, beliefs, values and habits that form the world-view most commonly held within a culture and transmitted from generation to generation by social institutions."[6] Social paradigms influence the interpretation of social reality; they guide behavior and expectations. These authors hold that contemporary industrial societies give high priority to material affluence, economic growth, hard work, technological power, and the subjugation of nature. Another author lists these components of "the industrial era paradigm": competitiveness and individualism, unlimited material progress, growth and consumption, efficiency and productivity, faith in science and technology, and mastery of nature.[7] There has been some evidence of a recent shift toward "postmaterial values," especially among youth, which will be examined in Chapter 12. In Chapter 13, I will outline the "paradigm shift" that seems to me necessary for long-term sustainability in an era of resource constraints.

Psychologists as well as sociologists have carried out empirical studies of human values. Abraham Maslow has suggested a basis for the ordering of values in *a hierarchy of human needs*. Maslow finds five levels of need:

1. Survival (physiological needs): food, shelter, health.
2. Security (safety needs): protection from danger and threat.
3. Belonging (social needs): friendship, acceptance, love.
4. Self-esteem (ego needs): self-respect, recognition, status.
5. Self-actualization (fulfillment needs): creativity, realization of individual potentialities.

Maslow maintains that these levels form a hierarchy; lower levels must be satisfied before the individual can give attention to higher levels. People are preoccupied with survival and security if these are threatened. A starving person has little interest in artistic creativity or political liberty. The lower levels (which would correspond to what we have called material values) are more fundamental; at least a minimal fulfillment of them is a precondition for interest in higher levels. Once a need is satisfied, its role in motivation dwindles; the lowest level of ungratified need is the primary influence on behavior.[8]

Maslow suggests that *the good* can be defined in terms of the choices made by people at the highest level, the "self-actualizing" individuals. He selected for research a group of persons who are creative, spontaneous, and psychologically healthy; they are well-integrated and happy, with minimal internal and external conflicts. The free choices of such self-actualizing individuals at their peak moments tell us what is good for persons. Maslow proposes that in this manner a naturalistic value system can be derived from scientific research.

Maslow's *hierarchy of needs* is very helpful in any consideration of values. In his writing there is great sensitivity to the multiple dimensions of human experience. There is indeed considerable evidence that under conditions of extreme scarcity people do give priority to survival needs. I would submit, however, that higher-level needs are always present; they do not suddenly "emerge" when lower needs are satisfied. Hungry people miss those they love as much as the well-fed do, even if they can spend less time with them. Many creative artists have been very poor or have lived in impoverished societies. I will suggest in later chapters that policies should be designed to meet basic material needs in ways that at the same time promote community life and the fulfillment of human potentialities. The higher levels should not be looked on as the prerogative of the affluent.

I also am dubious about the possibility of justifying basic value commitments or ideals of human fulfillment by *scientific research* alone. There are such diverse ways of seeking self-actualization that additional criteria seem to be necessary in setting life goals at higher levels. The criteria by which Maslow selects his exemplary self-actualizers (spontaneity, psychological health, and so on) are respected in our culture, but they are not universal and cannot be justified by the methods of science. The social sciences can make an immense contribution to our understanding of the relationships between values, social institutions, and individual needs. Any realistic discussion of value change must take into account the structures of society and human personality. But we cannot expect the social sciences to tell us what values we should pursue.

II. VALUES AND PHILOSOPHY

Philosophy does not ask what values people in fact hold—but what values we ought to hold. The philosopher is concerned about conceptual clarity and the consistency and universality of ethical principles. We will start by looking at the strengths and weaknesses of utilitarianism as an ethical principle for policy choice. The concepts of justice and freedom then will be analyzed in some detail, along with implications for resource and environmental policy.

Utilitarianism and its Critics

Utilitarianism has been not only an important school of thought among philosophers, but also a major influence among social scientists. Cost-benefit analysis and other formal methods used in environmental and technological decisions are carried out within the framework of utilitarian assumptions. We can present here only the broad outlines of utilitarian philosophy and some of the issues that it raises for policy choice today.

The central principle of utilitarianism is *the greatest good for the greatest number*. That action should be chosen which produces the greatest net balance of good over evil consequences. For Jeremy Bentham, the good was identified with pleasure; one should select the alternative that maximizes the balance of pleasure over pain. John Stuart Mill maintained that happiness is a more inclusive and long-lasting good than pleasure.[9] The utilitarian economists in turn sought to maximize total social welfare, aggregated either from individual welfare or from subjective preferences and perceived satisfactions. There are significant differences among these versions, but some observations can be made about their common assumptions.

Many versions of utilitarianism are *anthropocentric*. "The greatest good for the greatest number" often has been taken to refer exclusively to human beings. Any harm to other creatures is to be considered only insofar as it affects humanity. We will examine in the next chapter a broader rendition of the principle that includes the good of all sentient beings. Moreover, the principle usually is taken to apply only to presently existing persons, though in itself it does not distinguish present from future generations. But there are difficulties when future persons are included, since at least in principle the largest total good might be achieved by having an enormous population at a low level of well-being. The question of how much weight to attach to future costs and benefits also is pursued in the next chapter.

Utilitarianism faces serious difficulties in attempting to *quantify* "the greatest good." If the good is identified with happiness, can it be measured on a single numerical scale? Utilitarian economists spoke of maximizing satisfactions or preferences. But do people really look on diverse kinds of satisfactions as equivalent and substitutable? Can preferences among persons be compared and then aggregated in order to determine whether the total for society has been maximized?[10] Many economists have concluded that the only practical way to measure people's preferences is by their willingness to pay. But the distribution of purchasing power is very uneven, and it is often misleading to assume that everything that is prized can be priced. We will see that the concern for quantification has tended to restrict attention to material values and measurable costs, benefits, and risks.

Another criticism is that in utilitarianism only the total good, and not its *distribution* among people, is relevant to moral choice. Suppose the extermination of a small minority would make the majority so happy that the total happiness is increased. Suppose a sheriff can prevent a riot that might endanger scores of people by acceding to demands to execute an innocent person. Suppose total national income can be increased if we accept greater poverty for one segment of society. The utilitarian can object to these actions only if it can be demonstrated that there are indirect repercussions that will harm the total welfare, for utilitarianism finds nothing inherently wrong with injustice or inequality as such. In many cases the long-term social costs of setting a precedent by unjust actions might be so serious that they would outweigh any short-term benefits. But such considerations would not always prevent the sacrifice of individuals for the social good. Many contemporary philosophers hold that utilitarian principles must be supplemented by a principle of justice.[11] If the total good were the only criterion, we could justify a small social gain even if it entailed a gross injustice. But if justice were the only norm, we would have to correct a small injustice even if it resulted in great suffering or social harm. It appears, then, that we need to consider both justice and the total good.

In Part Two I will suggest that *cost-benefit* and *risk-benefit* analyses, when supplemented by a principle of justice, often are useful techniques if one is comparing a small number of options and there is a narrow range of very specific objectives. But in most policy decisions today there are a large number of options and a broad range of impacts, many of which are difficult or impossible to quantify. The trade-offs are multidimensional and cannot be measured in a single unit or aggregated as a numerical total. They involve environmental and social as well as material values. I will maintain that environmental impact assessment and technology assessment allow a broader range of value considerations and escape some of the limitations of utilitarian calculations. I also will argue that policy choices usually entail value judgments among incommensurables, and therefore the basic decisions must be made through political processes, not by technical experts using formal analytic techniques.

There is a final objection of a more general kind. Utilitarianism judges entirely by consequences. But there are some acts, such as murder, which we do not condone even if they have good consequences. An alternative approach to ethics stresses *duty* and *obligation*, the choice of acts that are right in themselves, apart from the calculation of consequences. Such obligations are called *deontic* (from the Greek *deon*, "that which is binding").[12]

Historically there have been many variants of the idea that particular acts can be judged *right or wrong* according to universal principles or laws, without attempting to calculate their consequences. The Stoics said that people have a duty to act in accordance with the natural law, the rational and moral order expressed in the structure of the world. Traditional Judaism and Christianity stressed obedience to the divine law revealed in scripture. Immanuel Kant held that the right is determined by the unconditional obligation of rational moral law, apart from any considerations of motive or consequence; an action is right if the principle it expresses could be universally applied. For Kant, the demand for freedom and justice is based on the equality of persons as autonomous and rational moral agents; individual persons should never be treated as means to social ends.

Whereas utilitarianism emphasizes the social good, deontic ethics typically defends *individual rights*. Natural rights, it claims, must not be violated even in the interest of beneficial social consequences. Rights are in general correlated with *duties*; my right to life implies your duty not to violate my life. The language of rights appears to be absolutist, and often does lead to inflexible positions. If rights are "inalienable" and "inviolable," and duties are "categorical," there seems to be no room for compromise. However, it is possible to employ a deontic approach with considerable flexibility. One may start with universal rules but interpret them to allow for exceptions in special cases. A duty specifies an obligation, but it may be outweighed by other duties. When two rights conflict, one of them may be assigned priority. So rights and duties should not be regarded as absolute.[13]

The extreme deontic position asserts that individual acts should be judged right or wrong in themselves, entirely apart from their consequences. But in an ecologically interdependent world, *direct and indirect consequences* may be far-reaching and should not be neglected. Utilitarians, at the other extreme, judge only by consequences, and they usually restrict attention to quantifiable effects. I would argue that policy decisions should be examined both in terms of individual rights and duties, and in terms of consequences for the social good. I also take the social good to include a wider range of values than are recognized by most utilitarians.

The defense of *individual rights* is indeed important in a technological age in which governments wield vast powers and frequently defend their actions by pointing to benefits for society. Only a basic respect for persons can lead us to protect a minority from exploitation for the benefit of the majority. Both the protection of the individual and the good of society must be considered in the complex decisions we face today, and there is no simple formula for combining them. In some cases, such as population control, individual rights and the future welfare of society are very difficult to reconcile.

In several of the political controversies we will be considering, there are groups that have adopted *absolutist positions.* One can, of course, reach an unambiguous conclusion starting from utilitarian premises. For example, one might conclude that the harm from even very low levels of carcinogens in pesticides or food additives outweighs any possible benefit, so that an absolute ban is justified. However, uncompromising positions are more often defended by deontic principles taken as absolutes; a particular right or duty is emphasized and conflicting rights or duties are ignored. "Human rights" or "the right to life" seem to allow for no compromise, only for acceptance or rejection. We will see that environmental values are sometimes similarly treated as absolutes not subject to trade-off or compromise.

Uncompromising stands may be adopted on grounds of strategy as well as principle. They often are effective in confrontational politics and the mobilization of opposition. They avoid the dangers of the "slippery slope"; when you give an inch you may end by giving a mile. Once some uses of wilderness are admitted, or low levels of a carcinogen are allowed, it is harder to draw a clear line. Nonquantifiable values, such as those associated with the preservation of an endangered species, are easier to defend in absolute terms than in a utilitarian framework. Opponents of a project may seek confrontation in the political arena rather than risk being co-opted by participating in a planning process that appears more rational, but in which their concerns are treated as secondary considerations. Absolute positions can be effective in protest, but they seldom are effective in developing positive alternatives that could realize a diversity of values. Here again I will advocate a broad evaluation of consequences that goes beyond utilitarianism and cost-benefit analysis, together with a defense of rights and duties that avoids absolutism.

The Concept of Justice

In dealing with policy decisions we are primarily concerned with *distributive justice*, rather than with procedural justice as we would be in a legal context (due process, fair trial, equality before the law, and so on). In particular, we will be asking about the fairness of the distribution of costs and benefits in decisions about technology and the environment and in the allocation of finite resources.

Most concepts of justice start with an assumption of *the fundamental equality of persons*. For some people this may be based on a religious conviction of the equal worth of every individual in God's sight ("all men are created equal"). For others it may derive from a doctrine of equal intrinsic human rights ("natural rights"), or the requirements of a harmonious social order. Some philosophers have argued from the common nature of persons as rational beings or the universality of basic human capacities. Others take respect for human beings and belief in their equal dignity to be unanalyzable ultimate attitudes. Distributive justice, then, starts with the idea that people should be treated equally because they are fundamentally equal.[14]

But it is immediately apparent that in many respects people are *unequal*. Some of these differences, listed below, have been held to justify unequal treatment.

1. Rewards for Merit or Virtue

Special rewards would be fair only if everyone had an equal initial chance of achieving merit or virtue. But native endowments and social conditions vary widely; people do not start with equal chances. There are appropriate forms of reward (such as prizes or honors) for very specific competitive achievements that deserve recognition. But the distribution of economic resources to meet basic human needs that all people share cannot be considered a reward for a specific achievement.

2. Recompense for Past Contributions

It might be said that the person who contributes more to society deserves to receive more from society; this would be a "just dessert." This principle assumes that individual contributions can be isolated and compared. But the wealth created by any person is dependent on the work of many others and on the social structures provided by society. In any case, how can one really compare the relative social contributions of an artist, an engineer, a farmer, or a parent in a home? And what about the child or the mentally retarded adult who has made no material contribution to society?

3. Incentives for Future Productivity

Differences in income often have been defended as incentives for work and greater total productivity—from which everyone supposedly benefits. Conversely, it is claimed that welfare payments aimed at greater equality undercut incentives to work. The trade-offs between equality and efficiency in the economic system are complex, but it appears dubious that inequalities of the magnitude that exist in U.S. society are necessary to encourage productivity.[15] There also are non-monetary incentives that free-market societies have neglected. The radical inequalities between nations never could be justified by the need for work incentives.

4. Provision for Special Needs

Unequal treatment for the person with special needs or handicaps is a common practice. The allocation of public funds for health care for the ill, or for special education for the deaf, is in itself unequal, but it aims at achieving a greater equality by restoring health or the chance to learn to those deprived of them. Special attention to those with greater needs or disabilities is an effort to provide them with an equal opportunity for a good life.

5. Selection for Special Positions

Equality is impossible when there is only a limited number of positions available, but one can at least seek *equal access* for those with appropriate qualifications. Not everyone can be president, but no one should be arbitrarily excluded. Not everyone can be treated on a kidney dialysis machine, but anyone with suitable qualifications should have the same chance (for example, by a random drawing among applicants who need treatment). Not everyone can be enrolled in a university, but access is equal if criteria relevant to the educational process alone are used in selection. In all these cases there is equality of opportunity even if not equality of results. Discrimination is unjust precisely because it involves criteria unrelated to qualifications for the position itself.

Unequal treatment is justified, in short, only if it helps to correct some other form of inequality or if it is essential for the good of all. Inequalities of authority are necessary for the maintenance of the social order, but there can be equal access to the positions and offices that carry such authority. But are there limits to the degree of inequality that we will tolerate for the sake of other social benefits? Are inequalities in some goods and services more significant than in others? Questions of inequality assume added urgency if technology tends to increase the gap between rich and poor, and if global scarcities limit the resources available for distribution.

I would maintain that equality is a more compelling value in the distribution of resources to meet *basic human needs* (such as food, health and shelter) than in the distribution of other goods and services. Food to meet minimum protein and calorie requirements is necessary for life itself; justice in the production and distribution of food to meet these requirements should have the highest priority. Access to health care also is crucial since it so strongly affects life prospects. But some margin beyond bare survival is a prerequisite for a minimally decent human life. Estimates of the minimal material levels for human dignity and self-respect are of course historically and culturally relative; there is no sharp line between physical needs and psychological desires influenced by changing expectations. In the United States, a poverty line has been established for entitlement to food stamps and health care benefits; unemployment insurance and social security also were instituted in the name of justice rather than charity. A project sponsored by the United Nations has tried to establish quantitative measures for basic needs and standards on which there is an emerging world consensus.[16] Policies aimed at justice in meeting basic human needs will be examined in Part Three.

The most influential recent treatment of the relation between justice and equality is John Rawls' *A Theory of Justice*. Rawls asks us to imagine a hypothetical "original position" in which a group of people are formulating the basic principles for a social order. These persons are initially equal, and no one knows what his or her status will be in the society that is to be established. In agreeing on a "social contract," each person acts from rational self-interest; but impartiality in formulating the rules is guaranteed because these individuals do not know what their own social positions will be. Such a hypothetical situation can help us establish principles for the fair distribution of scarce resources. It is similar to a situation in which the child who cuts the pie does not know which piece he or she will get.

Rawls maintains that persons in such an "original position" would accept two basic principles for the social order:

1. Each person is to have an equal right to the most extensive total system of equal basic liberties compatible with a similar system of liberty for all.
2. Social and economic inequalities are to be arranged so that they are both:
   (a) to the greatest benefit of the least advantaged, consistent with the just savings principle, and
   (b) attached to offices and positions open to all under conditions of fair equality of opportunity.[17]

As Rawls develops it, the second principle requires equality in the distribution of all the primary social goods (income, wealth, power, and self-respect), with the two exceptions noted. Inequalities are allowed only if they maximize benefits to the least advantaged or are attached to offices open to all. Attention to the impact on *the least advantaged* is a product not of altruism but of the rational self-interest of people in the "original position," any of whom might end up in that worst-off status. Rawls suggests that if the least advantaged benefit, it is likely that most other social groups will benefit also. But he rejects the utilitarian view that a loss to some people can be justified by greater gains to others. The "just savings principle" deals with obligations to future generations, which will be discussed in the next chapter.

Rawls holds that rational contractors would insist that, once a minimal level of material well-being had been reached, *liberty* should have priority over *equality*. The first principle is thereafter to be fulfilled before and independently of the second. Liberty—especially liberty of conscience and political liberty (equal participation in government)—is not to be exchanged for any other benefits, including greater equality. Neither freedom nor justice is subject to trade-offs with other benefits. Political rights should not be sacrificed for the sake of economic gains, except under conditions of extreme scarcity.

Not surprisingly, Rawls has been attacked from the right for being *too egalitarian*. Defenders of free-enterprise capitalism say that the degree of equality that Rawls seeks would not provide incentives for the most able persons, and would protect the indolence of the least able. The enforcement of equality, it is claimed, would require coercive measures and would violate property rights and the acquisition of wealth by legitimate means. If one is really dedicated to freedom, one must set strict limits on the powers of the state—including its power to redistribute legitimately acquired property.[18] But Rawls is attacked from the left for *not being egalitarian enough*. Marxist and socialist critics insist that political equality, which the first principle endorses, is jeopardized by the degree of economic inequality that the second principle allows. For economic power becomes political power in capitalist societies, and inequalities perpetuate themselves.[19]

Rawls' hypothetical *"original position"* also has been criticized. If one starts from separate, autonomous individuals, can an adequate concept of community ever emerge? If one starts by abstracting the individual from all political and historical contexts, can one obtain principles relevant to actual choices in the real world? People also vary greatly in their willingness to gamble; some would risk a small loss for the sake of a possible large gain, and therefore would accept greater inequalities than Rawls. Despite such limitations, I see the "original position" as a useful analytic device for asking what would be a fair distribution of resources. It is one of the few ways of dealing with justice between generations—simply by asking you to imagine what policies you would recommend if you did not know to which generation you would belong. In later chapters I will make use of Rawls' second principle (maximizing benefits to the least advantaged) in discussing policies for pollution control, energy conservation, food production, and industrial growth.

The Concept of Freedom

Freedom, like justice, is a complex concept. Three forms of freedom may be distinguished. The first, which I will call *negative freedom*, is the absence of external constraints. Negative freedom is freedom *from* coercion or direct interference imposed by other persons or institutions. Locke, Mill, and the British tradition of liberal political philosophy interpreted freedom primarily as the absence of interference by the state. They were concerned to protect the individual against the power of government, and to allow unfettered scope for individual initiative in economic affairs and the use of private property. This view was influential among the authors of the U.S. Constitution, and was reinforced by the American experience of the frontier, abundant resources, and the assumption that this was a land of unlimited opportunities for everyone.

Second, *positive freedom* is the presence of opportunities for choice. It is not freedom *from* constraints, but freedom *to* choose among genuine alternatives, which requires a range of real options and the power to act to further the alternative chosen. Persons in poverty are not free to go to Hawaii for a vacation if they lack the means to carry out such a choice, even though no one prevents them from going. Even in the absence of external constraints, unequal power results in unequal opportunity for choice. Some degree of personal autonomy is an essential component of human dignity. Many of the conditions for the exercise of choice are *internal*. People vary widely in their awareness of alternatives, ability to make deliberate choices, and personal initiative and self-direction. But in dealing with public policy, we are concerned mainly about the *external* conditions, the social structures within which people can have some control over their own futures.[20]

Third, *political freedom* encompasses both civil liberties and democratic forms of government. Civil liberties—such as freedom of worship and the freedoms of speech, assembly, and the press—can be defended both as basic human rights and as preconditions of democracy. Political freedom entails some kinds of negative freedom (limits to the powers of government, such as censorship and arbitrary arrest), and provisions for due process and the right to a fair trial (procedural justice). It also requires institutions for political self-determination through which each person can have a voice in political decision-making processes.

There inevitably are *trade-offs* among these forms of freedom. If we try to maximize negative freedom while there are great inequalities of economic power, the weak will have little protection from domination by the strong. In a complex society, the actions of one person can greatly affect the choices open to other people. Limitations on the actions of some persons are necessary if other persons are to be able to exercise choice. *Whose* freedom is more important: the freedom of a family in seeking housing, or the freedom of landowners to sell or rent to whomever they please? Positive freedom to achieve desired outcomes exists only within an orderly society. We accept traffic lights that are sometimes red because we can go without interference when they are green. The state is an instrument of order and law, but it is also an instrument of positive freedom when it restricts some actions in order to make other actions possible.[21]

A negative and individualistic interpretation of freedom is inadequate in a *free-market industrial society*. The liberal tradition insisted on limited governmental powers in order to protect the individual from interference by the state. But only a strong government can protect the individual from the actions of such powerful private institutions as industrial corporations, and only public authorities can regulate the development of technologies. Moreover, we are more aware today of our *ecological interdependence*. "No man is an island," and many actions have far-reaching indirect repercussions. Many uses of private property have significant public consequences. Such common resources as air and water can be regulated only by collective action. The right of governments to intervene to protect health, safety, and welfare is now relevant to a wide range of policy decisions.

I suggest that freedom in an interdependent world should be understood as *participation in the decisions that affect our lives*. This definition emphasizes positive freedom within a social context, but it also includes political freedom. Participation in decisions can occur through voluntary associations, especially at the local level. Cooperatives, small-scale businesses, and community organizations provide opportunities for personal participation and local control. Public interest groups offer other channels for citizen involvement in local and regional issues, but many decisions concerning environmental preservation and resource conservation can be made only at the national level. Strict regulations are needed, yet there are pervasive dangers in bureaucracy and big government. Regulatory agencies often have become allies, if not captives, of the industries they were supposed to regulate. There is thus a continuing challenge to make government agencies more accountable to the public.

In a democratic society there is a variety of *political mechanisms* through which individuals can share in significant decisions at local, state, and national levels. The right to vote for legislative representatives is one such mechanism. Legislative hearings, regulatory hearings, court challenges, and referendum initiatives provide other avenues for citizen input. But there are four dilemmas that arise in technological and environmental decisions.

1. Citizen Participation in Technical Decisions

Choices today often involve complex scientific issues with which the public feels incompetent to deal. The risks to human health and safety from nuclear reactors and toxic substances, for instance, are very difficult to evaluate. Yet such decisions should not be left to technical experts alone, since they require comparison of diverse risks and benefits, and assessment of alternative policies, which are not purely scientific questions. An industry or a government agency that has an interest in promoting a technology usually has far more extensive legal and scientific resources than those opposing it. Technology, we have noted, is itself a major force in the concentration of economic and political power. In later chapters some procedures that can facilitate the democratic control of technology will be examined.

2. The Free Market and Government Regulation

Some types of regulation lead to much greater intervention in the market economy than others. For example, a tax on the discharge of pollutants would rely on economic incentives and would allow a greater variety of responses than strategies that mandated specific abatement technologies or set absolute standards for emissions or effluents. Energy conservation proposals range from voluntary restraint, through economic incentives, to mandatory efficiency standards and gasoline rationing. In the past, land has been viewed as private property to do with as one pleases; but land use has such widespread environmental and social impacts that zoning and siting controls are increasingly necessary. In each of these cases the consequences of insufficient regulation must be weighed against the dangers of excessive bureaucracy and possible losses in efficiency and private initiative.

3. Decentralization Versus Centralization

There is more opportunity for citizen participation in local than in national government. There is more possibility for individual or community control of intermediate-scale technology than of large-scale technology. For example, nuclear energy is complex and centralized; it demands huge capital investments and entails unusual risks that require an exceptional degree of government regulation and strict security measures. By contrast, many forms of solar energy are decentralized; equipment can be locally installed and managed. Decentralization counteracts the concentration of economic and political power and contributes to diversity and local control. Yet in many cases the centralization of authority is inescapable. Air and water pollution crosses jurisdictional boundaries. State and local governments have not been effective in controlling pollution, partly because they are dependent on industrial growth for new tax revenues. An increasing federal role in environmental regulation and resource conservation is unavoidable, but citizen participation is more difficult at the national level.

4. Authoritarian Responses to Scarcity

Is democracy doomed in a world of severe resource scarcities? It has been argued that only authoritarian governments will be able to deal with impending crises of food, energy, and population. If survival is at stake, freedom appears to be a luxury and coercion seems justifiable. If voluntary family planning fails to curb the population explosion, should overpopulated nations resort to compulsory sterilization and abortion? Or are there ways of changing behavior, such as economic incentives or social reforms, which would be both effective and morally preferable? Here as elsewhere there is a spectrum from education and rational persuasion, through economic and social measures, to physical coercion. The key questions are whether sufficiently stringent regulations can be adopted through democratic processes, and whether adequate safeguards can be provided against the abuse of new governmental powers and the erosion of civil liberties. We also must ask whether a system in which representatives are elected for short terms is compatible with the need for long-range planning. Our main concern, then, will be with freedom understood as participation in the decisions that affect our lives.

III. VALUES AND RELIGION

According to a recent Gallup poll, 94% of U.S. citizens believe in God, 86% consider religion "very important" or "fairly important," 71% are members of churches or synagogues, and 40% attend services weekly.[22] There are themes in the main American religious traditions that could make a distinctive contribution to the issues of this book. Their potential influence could be substantial because of the number of people involved.

The record of *religious institutions* in practicing the ideals they profess has been very mixed. They often have been on the side of the *status quo*, resisting changes in the social order. To victims of injustice they have sometimes offered only resigned acceptance and the consolation of a future life. Their ethical teachings frequently have been confined to the sphere of family life and personal relationships. Charity has been a substitute for justice. Yet they also have nurtured prophetic leaders who have been in the forefront of social reforms—in hospitals and prisons and the abolition of slavery in the last century, for instance, or in the civil rights and antiwar movements of the 1960s. Religious groups have an excellent record of response to immediate human need, such as famine relief; only recently have they begun to show comparable concern for the long-range causes of hunger, such as overpopulation, environmental damage, and economic exploitation. Religion thus has been a force for change as well as a conservative force. What creative role might it play in a world of technology and resource scarcities?

Values in Judaism and Christianity

Let us start by asking how the Western religious traditions have viewed the human values that we have been discussing. It should be noted at the outset that they have not minimized the importance of the *material values* of health, welfare, and employment. The laws of Deuteronomy and the message of the Hebrew prophets call for action to alleviate physical need, especially among the underprivileged. In the New Testament, love is not mere sentiment but is active caring for persons and response to the needs of the neighbor for food, clothing, and health. "For I was hungry and you gave me food . . . I was naked and you clothed me, I was sick and you visited me."[23] Agricultural and medical technology today can be seen as instruments of such caring in response to hunger and sickness. But the Bible also holds that too much concern about one's own material welfare can lead to callousness about the needs of others; the evils of excessive wealth and luxury are pointed out in many parables and teachings. In this chapter, however, we are looking primarily at four social values:

1. Distributive Justice

Both the laws of the Torah and the message of the prophets articulated the demands of justice that have been central in Judaism throughout its history.

"Let justice roll down like waters and righteousness like an ever-flowing stream."[24] "What does the Lord require of you but to do justice, and to love kindness, and to walk humbly with your God?"[25] Jesus opened his ministry with a quotation from Isaiah: "The spirit of the Lord is upon me, because he has anointed me to preach good news to the poor. He has sent me to proclaim release to the captives and recovery of sight to the blind, to set at liberty those who are oppressed."[26] Justice is a demand of the biblical law, but it is also a personal response to a God of justice and righteousness. The Hebrew prophets condemned the oppression of the poor as both a violation of the covenant with God and a violation of human relationships within the community. As in the philosophical discussion cited earlier, the central biblical meaning of justice is equal treatment. Distributive inequalities are harshly judged in the light of the fundamental equality of all persons before God. As with Rawls, the treatment of the dispossessed serves as a test case for justice in society, though the prophets go further in their impassioned concern for the specific victims of injustice.

2. Participatory Freedom

The Exodus story of liberation from slavery in Egypt has always symbolized the importance of freedom to the Jewish community. In its subsequent history as an oppressed minority it frequently has been on the side of freedom against institutions of political and religious power. There is considerable latitude for variations in individual belief within Judaism, and there is a dedication to religious freedom. But its outlook is not individualistic; persons are portrayed as social beings, interrelated selves in a community. Jewish life always has tried to balance the rights of the individual against the good of society. Throughout its history and in the modern state of Israel, it has given expression to a concept of freedom as self-determination and individual responsibility.[27]

The record of Christianity in relation to freedom has been more mixed. At times the institutional church has worked for the liberation of the oppressed; more often it has itself become an instrument of oppression, acquiring a vested interest in the structures of political and economic power. In its own internal life it often has been highly authoritarian; through most of its history, religious intolerance has prevailed. Yet the Protestant reformers defended individual conscience, and advocates of religious freedom and the separation of church and state have been prominent in all Christian traditions since the eighteenth century. The biblical insistence on the dignity of the individual and the fundamental equality of all persons was one of the roots of U.S. democracy (though the exclusion of blacks and women indicates how far practice was from theory). At the local level, one can point to the contribution of the free churches and Puritanism to participatory self-government.[28] I believe that the biblical view of personhood and community supports the idea of participatory democracy, but that it will be difficult to recover in an industrial society whose main institutions are designed to promote efficiency and material progress.

3. Interpersonal Community

Judaism has had much to say about the quality of personal interactions within communities. The God of Israel is interested in the fabric of the community's life; in His name the prophets attacked the barriers that separate people. Judaism holds up the vision of a future society of brotherhood as well as justice, a more humane social order, and a kingdom where *shalom* (peace, harmony, happiness) will reign between all persons and their neighbors. The New Testament speaks of fellowship and reconciliation between persons, and of the church as a community of mutual support. The sharing of possessions in the early church was not unlike the ideals of certain communes today. Christian writers often have defended the integrity of family and community life, and the goal of cooperation rather than competition. Interpersonal community assumes new significance today as an alternative to both individualistic isolation and the impersonal collectivism to which technological societies are prone. Local groups and voluntary organizations, including church and synagogue, provide a setting for significant personal interaction and participation.

4. Personal Fulfillment

Judaism has upheld a broad view of personhood and the many dimensions of human fulfillment. It has affirmed the goodness of life; monasticism and world-renouncing asceticism have been very rare in its history. A person's duty is to sanctify daily life. Judaism encourages a proper self-respect; the individual is free to choose the good, to turn anew, and to work for God's kingdom on earth. Study and learning—both of the Torah and of worldly knowledge—are held in high regard. Personal fulfillment can occur only in the context of the community and through the realization of harmonious relationships.

In Christianity, too, the good life is identified with personal existence in community rather than with material possessions. Fulfillment consists of right relationship to God and neighbor. The New Testament acknowledges the dangers of both poverty and affluence, but it is vehement in its criticism of wealth. In Chapter 13 I will suggest that today, in a high-consumption society, we need to recover both this critical attack on materialism and the positive witness to the quality of the life of individuals in community. Simpler life-styles are the products of new visions of the good life based on sources of satisfaction that are less resource consumptive.

Religion in a Technological Society

The contemporary crises of our technological society have stimulated interest in *Asian religions*, especially among U.S. youth. As noted in Chapter 2, the theme of harmony with nature, which has been ignored in much of the history of Christianity, is prominent in Taoism and Zen Buddhism. The Eastern traditions also have practiced meditative disciplines and fostered the inner life, to which U.S. churches have given much less attention. For the Hindu mystics, for example, both personal peace and harmony with the natural world derive from the experience of the underlying unity of all things. Buddhist meditation is a path to insight, self-realization, and wholeness. These goals contrast sharply with those of a technological, resource-consumptive society.

The West has much to learn from these Eastern traditions, especially about the inward journey and harmony with nature. But there also are neglected aspects of *the Western heritage* that could contribute to the fundamental changes in value priorities that will be required for a transition to a sustainable society. There is a long heritage of Jewish and Christian mysticism, and disciplines of individual and corporate prayer and meditation have been practiced in every age. There are Western examples of the celebration of nature and respect for the earth, as indicated in Chapter 2. There also has been a concern for social justice, stronger than in Eastern religions, which has been expressed continually though intermittently practiced. Today it is seen in liberation theology in the Roman Catholic church, and in statements of the World Council of Churches, in which it is combined with ecological themes.[29] A renewal and reorientation of religious life in the United States will not be easily achieved, but I believe it is at present more promising to draw on basic ideas in the traditions of the West than to turn to less familiar writings and practices from the East.

There are aspects of the Western heritage that are particularly relevant for a technological society. First, the Judeo-Christian tradition is a resource in defending *personal existence* against the depersonalizing tendencies of industrialism. Synagogues and churches can evoke awareness of dimensions of human experience not accessible to technical reason, and they can nourish human compassion and tenderness. They can cultivate the arts and music in their liturgical celebration. In awareness of the sacred and recognition of mystery and human limits, there are antidotes to the arrogant assumption of technological omnipotence. In receptivity and acknowledgment of grace there are attitudes that contrast with control and manipulation. Our religious traditions invite reflection on the ends of life and the nature of genuine well-being. Their images of human fulfillment go beyond consumption; a person's life cannot be measured by material possessions, nor a nation's by its GNP. Respect for human dignity today includes sensitivity to the effects of technology on people, resistance to attempts at manipulation and control, and concern for the quality of human relationships. The church and synagogue could exert a unique moral leadership in their own educational programs and in the wider society by articulating such convictions.[30]

Second, the biblical view of *human nature* combines realism with idealism. Human sinfulness refers not only to the actions of individuals, but also to social institutions. Every group tends to rationalize its own self-interest. The recognition of human fallibility and the abuse of power should make us hesitant to turn over social decisions to technical experts, however well-intentioned. It also might make us more cautious about large-scale systems in which human error or institutional self-interest could have catastrophic consequences. When technology gives us the power to destroy ourselves, humility is a survival need. But the biblical tradition is also idealistic in its vision of creative human potentialities and the possibility of a more just social order. It sees political processes as a vehicle for both the restraint of power and the fight for social justice. It would lead us to try to redirect technology rather than simply to reject it.

Third, religion can be a source of *individual and social transformation.* At the personal level, the biblical message holds out the possibility of release from guilt and anxiety, liberation from self-centeredness, and a life of genuine relatedness and openness when the power of love breaks into our encapsulated lives. Healing, wholeness, and reconciliation can take place between persons, in communities of mutual acceptance, and in the relations between groups. The biblical imagination also has looked to a future kingdom of peace and brotherhood on earth. Such images and their expression in myth and ritual can be a more powerful influence on cultural change than abstract values or philosophical principles. Today, visions of alternative futures can be a source of hope and renewal. Such a future would involve new definitions of what is necessary to sustain a good life, social patterns in which cooperation replaces competition, and life-styles that avoid the compulsion to consume. The ideal of simplicity can be recovered, not in a spirit of ascetic self-denial, but because global resources are limited—and because there are positive values in a simpler life. Some of these possibilities will be explored in Chapter 13.

While I see real hope that a revitalized biblical faith could contribute to a new consciousness and to new patterns of behavior appropriate to a sustainable society, I am aware that in the institutional church there is great resistance to change. Moreover, we live in *a pluralistic culture* in which public policies cannot be built on the assumptions of any one religious tradition. Therefore, in Parts Two and Three I will be examining policy choices in terms of general principles to which persons with diverse philosophical and religious perspectives could subscribe. I will combine a concern for individual rights that avoids absolutism, and a concern for the social good in which consequences are analyzed in a framework broader than utilitarianism. I will look at policy decisions from the standpoint of social as well as material and environmental values.

I believe that in a world of technological power, scarce resources, and increasing gaps between rich and poor, *distributive justice* is the crucial value in policy decisions today. I will employ Rawls' criterion of benefit to the least advantaged, which is consistent with the biblical demand for social justice. The basic needs of millions of people are today unmet; these are the lowest levels of Maslow's hierarchy, which are a precondition for full human development. Justice in the distribution of food and access to health care should therefore be our highest priorities. But I also will refer frequently to participatory freedom, community, and a new vision of personal fulfillment—Maslow's higher levels—which I consider essential to human dignity in a sustainable society.

NOTES

1. Kurt Baier and Nicholas Rescher, eds., *Values and the Future* (New York: Free Press, 1969), chaps. 1 and 2. See also Nicholas Rescher, *Introduction to Value Theory* (Englewood Cliffs, N.J.: Prentice-Hall, 1969); and William Frankena, "Value and Valuation," in *Encyclopedia of Philosophy*, ed. P. Edwards (New York: Macmillan, 1967).

2. See Robin M. Williams, "The Concept of Values," *International Encyclopedia of the Social Sciences* (New York: Macmillan, 1968).

3. Clyde Kluckhohn, "The Study of Values," in *Values in America*, ed. D. N. Barrett (Notre Dame, Ind.: University of Notre Dame Press, 1961); also idem, "Values and Value Orientation in the Theory of Action," in *Towards a General Theory of Action*, ed. T. Parsons and E. Shils (Cambridge, Mass.: Harvard University Press, 1951).

4. Robin M. Williams, *American Society*, 3d ed. (New York: Knopf, 1970), chap. 11; also idem, "Individual and Group Values," *Annals of Amer. Acad. of Pol. and Soc. Sci.* 371 (1967): 20.

5. Milton Rokeach, *The Nature of Human Values* (New York: Free Press, 1973). See also Norman T. Feather, *Values in Education and Society* (New York: Free Press, 1975).

6. Dennis Pirages and Paul Ehrlich, *Ark II: Social Responses to Environmental Imperatives* (San Francisco: W. H. Freeman, 1974), p. 43.

7. Willis Harman, *An Incomplete Guide to the Future* (San Francisco: San Francisco Book Co., 1976); see also George Cabot Lodge, *The New American Ideology* (New York: Knopf, 1975).

8. Abraham Maslow, *Toward a Psychology of Being*, 2d ed. (Princeton, N.J.: Van Nostrand, 1968); idem, *Religions, Values and Peak Experiences* (Columbus: Ohio State University Press, 1964); and idem, *Motivation and Personality* (New York: Harper & Row, 1970).

9. J. S. Mill, *Utilitarianism* (1863; reprinted, New York: E. P. Dutton, 1914). For a recent discussion, see J. J. C. Smart and Bernard Williams, *Utilitarianism: For and Against* (New York: Cambridge University Press, 1973).

10. See Amartya K. Sen, *Collective Choice and Social Welfare* (San Francisco: Holden-Day, 1970).

11. William Frankena, *Ethics*, 2d ed. (Englewood Cliffs, N.J.: Prentice-Hall, 1973), chaps. 2 and 3; Nicholas Rescher, *Distributive Justice* (Indianapolis: Bobbs-Merrill, 1966), pp. 40 and 47–48; and Dan Brock, "Recent Work in Utilitarianism," *American Philosophical Quarterly* 10 (1973): 241–76.

12. Frankena, op. cit., chap. 2; and Carl Wellman, *Morals and Ethics* (Glenview, Ill.: Scott, Foresman, 1975), chaps. 2 and 4.

13. W. D. Ross, *The Right and the Good* (Oxford: Clarendon Press, 1930).

14. On the concept of justice and its relation to equality, see Rescher, *Distributive Justice*, op. cit.; Hugo A. Bedau, ed., *Justice and Equality* (Englewood Cliffs, N.J.: Prentice-Hall, 1971); and Richard Brandt, ed., *Social Justice* (Englewood Cliffs, N.J.: Prentice-Hall, 1962).

15. Arthur Okun, *Equality and Efficiency* (Washington, D.C.: Brookings Institution, 1975).

16. John McHale and Magda McHale, *Basic Human Needs: A Framework for Action* (Houston, Tex.: University of Houston, 1977).

17. John Rawls, *A Theory of Justice* (Cambridge, Mass.: Harvard University Press, 1971), p. 302. See also his "Justice as Fairness," *Philosophical Review* 67 (1958): 164–94.

18. Robert Nozick, *Anarchy, State and Utopia* (New York: Basic Books, 1974); and Robert Nisbet, "The Pursuit of Equality," *The Public Interest* (Spring 1974): 102.

19. Brian Barry, *The Liberal Theory of Justice* (Oxford: Oxford University Press, 1973); also essays by Milton Fisk and Norman Daniels in *Reading Rawls: Critical Studies in a Theory of Justice*, ed. N. Daniels (New York: Basic Books, 1974).

20. On the concept of freedom, see Joel Feinberg, *Social Philosophy* (Englewood Cliffs, N.J.: Prentice-Hall, 1973), chap. 1; and P. H. Partridge, "Freedom," in Edwards, ed., op. cit.

21. See Norman Bowie and Robert Simon, *The Individual and the Political Order* (Englewood Cliffs, N.J.: Prentice-Hall, 1977), chap. 6.

22. Minneapolis *Tribune*, August 2, 1976, p. 5A.

23. Matt. 25:35.

24. Amos 5:24.

25. Mic. 6:8.

26. Luke 4:18.

27. Leo Baeck, *The Essence of Judaism* (New York: Schocken, 1948), pp. 190-225; and Abraham J. Heschel, *Who Is Man?* (Stanford, Calif.: Stanford University Press, 1965).

28. George Thomas, *Christian Ethics and Moral Philosophy* (New York: Scribner's, 1955); and L. Harold DeWolf, *Responsible Freedom* (New York: Harper & Row, 1971).

29. Gustavo Gutierrez, *A Theology of Liberation* (Maryknoll, N.Y.: Orbis, 1973); Paul Abrecht, ed., *Science, Faith and the Future* (Philadelphia: Fortress Press, 1979); and Roger Shinn, ed., *Faith and Science in an Unjust World* (Philadelphia: Fortress Press, 1980).

30. Ian G. Barbour, *Science and Secularity: The Ethics of Technology* (New York: Harper & Row, 1970), chap. 3.

5
ENVIRONMENTAL VALUES

In Chapter 2, three broad cultural attitudes toward nature were described: domination, unity, and stewardship. We now must look more specifically at some environmental values that are relevant to policy decisions: environmental preservation, resource sustainability, and ecosystem integrity. On what grounds can these values be defended? Can ethics be derived from ecology, or from the requirements for survival on a finite planet? How can environmental values be weighed against such human values as employment or justice when they conflict? Finally, how can human and environmental values be brought together, not just as ethical principles, but in policy decisions, political strategies, and individual life-styles?

I. ENVIRONMENTAL ETHICS

There are three grounds for environmental ethics: short-run human benefits from the environment, duties to future generations, and duties toward nonhuman beings. Let us consider each in turn, successively broadening the scope of ethics.

Human Benefits from the Environment

Many of the measures advocated by environmentalists can be justified by their contributions to human life during our own lifetime. An anthropocentric position need not lead to exploitative domination if *human dependence on nonhuman nature* is recognized. The environment is important to us biologically, economically, and aesthetically. Human welfare and fulfillment are dependent on the integrity of the biosphere. Enlightened self-interest provides strong arguments for action to preserve the environment.

A number of philosophers have argued that what is needed today is not a new environmental ethic but a careful application of traditional forms of ethics in ways consistent with our new *ecological knowledge*. John Passmore, for example, advocates a scientifically informed utilitarianism. He starts from human interests alone, but recognizes that if we damage the biosphere we end by injuring ourselves. The criterion of human usefulness, broadly conceived, leads him to support cautious intervention; we should "humanize and perfect nature" but acknowledge its vulnerability.[1] William Frankena similarly holds that commitment to the well-being of persons is sufficient justification for environmental action. The classical ethical principles are not at fault; it is we who have failed to live up to them and apply them intelligently. Frankena would like to see concern for the suffering of animals included in the sphere of ethics, but he can find no basis for duties toward nonsentient creatures except as they affect human welfare or aesthetic enjoyment.[2] Similarly, a religious concern for the neighbor's needs can encompass preservation of the environment, once the neighbor's dependence on the environment is fully acknowledged.[3]

Pollution of air, water, and land clearly is detrimental to human life, health, and material welfare. The correlation of air pollution with respiratory diseases (and the high incidence of both among low-income families) is amply documented (see Chapter 7). Water pollution is a health hazard, an obstacle to recreation, and a threat to the aquatic life from which people benefit. Pesticides, radioactive elements, and toxic wastes (particularly the heavy metals) have been shown to produce cancer or injure the human nervous system, and often their concentrations are greater in the higher levels of food chains from which humans draw. Whatever harms human life-support systems harms humanity. In short, on both ethical grounds and in the interest of a practical political strategy, environmental values can be defended because they are essential to the realization of human values.

Resource sustainability likewise can be defended in relation to human life, health, and material welfare. Conservation for wise human use sets limits to the exploitation of nature. Even though the goals of the conservation movement, such as multiple use and sustained yield, often were unrealistic in neglecting conflicts among uses and were distorted by pressures from particular economic interests, they did provide for considerable protection of public lands and natural resources. Human well-being requires the conservation of nonrenewable resources by recycling, reducing waste, and lowering consumption; such measures would at the same time reduce pollution and environmental damage.

Even the *preservation of wilderness* can be justified by its contribution to human life, though many of these benefits are less tangible. The most obvious benefit is recreation— camping, hunting, fishing, canoeing, backpacking, and so on—which in some areas has become so popular that permit systems have been instituted to prevent damage from excessive use. The enjoyment of natural beauty is another human benefit. As noted earlier, nineteenth-century Romanticism appreciated beauty in wild nature (rather than in the cultivated landscapes that had been extolled from antiquity to the Enlightenment), and wilderness has been widely celebrated in U.S. art and literature. As urbanization and the stresses of a technological society have increased, and wilderness has become scarcer, these recreational and aesthetic values have become more important for many persons.[4]

The positive influence of wilderness on *human character* is another intangible benefit. The solitude of woods and mountains is a source of serenity and strength, an opportunity to live in harmony with the natural order. Wilderness can teach us moral lessons; we can learn humility and gratitude, but we also can gain self-reliance, independence, and courage in facing the challenge of the wild.[5] Wilderness has molded us as a nation; Daniel Boone, the Oregon Trail, and the Western frontier are all part of our national heritage. It has been argued that we should preserve wilderness because it has come to symbolize human qualities that we respect, such as freedom, innocence, and courage. Destroying wild rivers and forests would be an attack on the human values that these natural phenomena represent in our cultural tradition.[6] These wilderness ideals can be affirmed even if one does not accept the romantic view that human nature in a natural setting, without the corrupting influence of technology and the city, is intrinsically good.

Others have found in wilderness a *spiritual significance* that goes beyond aesthetic or moral values. Thoreau, Muir, and their successors have held that the wilderness experience is a source of inspiration, an opportunity to acknowledge the divine in or beyond nature. The majesty of forests and wind-swept mountains evokes wonder, awe, and a sense of mystery. In quiet and solitude we learn reverence, which is an antidote to the spirit of technological omnipotence and the attempt to conquer nature. In wilderness we discover that we are spiritual beings and not just consumers; our imagination and primeval intuitive capacities, long suppressed by calculative rationality, may be awakened and expressed. Wilderness is a "sacred space" in which the experience of the "wholly other" can be powerful.[7] A recent study found that the majority of Sierra Club members considered aesthetic and spiritual experience more important than recreation or resource conservation as motives for wilderness preservation.[8]

A final reason for preserving wilderness is its use in *scientific research.* Ecology is a young science and needs "natural laboratories" in which to investigate the complex interactions among the members of biotic communities. No point on earth is totally beyond the reach of human influence, but there are relatively untouched areas that can provide a data base from which to study the effects of our intervention. There is much to be learned from what David Brower has called "the vanishing remnant of a world that can run itself." The disappearance of wilderness is essentially irreversible, whereas preservation leaves future options open. Wilderness constitutes an important habitat for wildlife, and in some cases for endangered species. This reservoir of *genetic diversity* is a resource whose loss would indeed by irreplaceable. Who can say what genetic strains may be useful in future plant breeding, or what clues important in agricultural and medical research may be found in unexpected places? Moreover, genetic diversity enhances the stability of the ecosystems on which we depend.[9]

Wilderness preservation, like pollution abatement and resource conservation, thus can be defended by reference to its benefits to *humanity alone*. I will suggest shortly that beyond this instrumental view of nature, there are grounds for duties toward nonhuman life apart from its usefulness to us. Such duties, as well as the intangible human benefits above, cannot be quantified for inclusion in cost-benefit analyses. Yet the preservation of wilderness and of endangered species, like pollution abatement measures, involves inescapable trade-offs with jobs, economic growth, and regional development; these have been the subjects of intense political battles that will be discussed in Chapter 7.

Duties to Future Generations

Many human environmental benefits and costs are cumulative or long-run. By depleting resources and despoiling the environment, we are increasingly jeopardizing the welfare of our descendants. Population growth, soil erosion, and the use of nonrenewable energy sources will affect human life for many generations to come. After 100,000 years, the radioactive wastes from the plutonium cycle still will be dangerous to anyone exposed to them. The alteration of human genes or the depletion of the ozone layer might affect the human race permanently. What are the grounds for ethical obligations to future generations in such policy issues?

A number of *utilitarians* have made at least passing reference to posterity. John Stuart Mill spoke of "the general interest of the human race," and pictured "all generations of man indissolubly united into a single image"; but he actually applied the principle of the greatest good to the greatest number only to living persons.[10] Among contemporary utilitarians, J. J. C. Smart holds that in principle all generations should count equally, but in practice the more remote effects of our actions usually are either too small (dissipating like the ripples from a stone in a pond) or too uncertain to be taken into account. But he says that if long-term effects might well be catastrophic (as in the case of nuclear war or the release of a potentially lethal virus), they should count heavily in our calculations.[11]

But can there be *obligations to people who do not exist*? Nonexistent persons do not have rights, though living persons or the state might assert claims on their behalf. Some utilitarians maintain that we do not have an obligation to be benevolent toward persons as such, but rather an obligation to produce what is good in itself. The obligation to increase the balance of good over evil does not require the specification of individuals because it is not owed to anyone. Other authors hold that we do have obligations to future persons as potential members of our moral community, but that these obligations diminish for more distant generations because our social ideas are less relevant to their lives under conditions that may be very different from ours.[12] It would seem, however, that such basic human needs as food and health are unlikely to change; we would at least have an obligation to avoid actions (such as ecological destruction or the massive pro= duction of long-lasting radiation) that we know will be harmful or might jeopardize the conditions of future human life.

Utilitarians usually *discount the future* because the more remote consequences of our actions are more uncertain. Events of low probability or great uncertainty are assigned low weights in utilitarian calculations. The economic system also gives priority to short-term benefits. A natural resource worth $100 ten years hence is only worth $48 to me now if I have the alternative of investing my money at 7½% interest. Democratic politics reinforces this tendency to discount the future in policy decisions. A political system with elected terms of two or four years is geared to immediate results and to current voters, since unborn beneficiaries cast no votes. Another factor in discounting the future is the prevalent confidence that technological advances will provide solutions to environmental and resource problems. In later chapters I will urge that potential long-run effects should be more heavily weighted than in current practice. New technologies will be developed, but there also are likely to be unexpected costs and risks.

Classical utilitarianism also runs into difficulties in dealing with *future population size*. In theory, it might be possible to maximize *total* happiness by having a very large population at a low level of happiness—though in a world of resource constraints it is more likely that a very large population, if it could be sustained at all, would experience considerable suffering (negative levels of happiness). To avoid this theoretical difficulty, some utilitarians have proposed that the criterion should be the highest *average* happiness, rather than the highest total.[13] But this goal could be achieved by a very small population with very high happiness (or even by killing off the least happy people to improve the average). Other formulas have been devised that combine total and average happiness.[14] But I am dubious that an "optimum population size" can be derived from any utilitarian formula, even if happiness really could be quantified.

I suggest that duties to posterity can be derived more plausibly from the idea of *justice between generations*. Several authors have developed Rawls' idea (see Chapter 4 above) of contractors in a hypothetical "original position" who do not know to which generation they will belong.[15] What policies concerning population growth and environmental degradation would you recommend if you did not know in which generation you would be born? What policies would be adopted by an assembly of representatives of all generations? Compared to future generations, the present generation may be the least advantaged with respect to economic growth and technology, but the most advantaged with repect to resources and the environment. As a minimum, each generation should leave the environment no more polluted than when it arrived (which would be comparable to the rule that a campsite should be left no worse than you found it). Only in the poorest nations today is the present generation probably the least advantaged overall, so that transferring resources from future to present generations would be justified.[16]

There also are distinctive *religious grounds* for obligations to posterity. In the Jewish tradition, individuals have a strong sense of belonging to a family and a people who will continue beyond their own life. There is a sense of solidarity in time as well as space, a covenant from generation to generation "to you and your descendants forever." Both Judaism and Christianity have expressed a univeralistic vision of the unity of humankind embracing generations yet to come. In the biblical view, stewardship requires consideration of the future because God's plan includes the future. The land, in particular, is held as a trust for future generations. This long time-perspective derives from a sense of history and an orientation toward the future, as well as accountability to a God who spans the generations. Contemporary theologians have urged that present affluence must not be allowed to rob unborn generations of basic necessities.[17]

In practice it is impossible to consider an indefinitely long series of future generations in policy decisions. But in the case of *renewable* resources, such as fish, wood products, and crops, the same result can be obtained by aiming for the *maximum sustainable yield*. Once the sustainable yield is exceeded by excessive fishing, cutting, overgrazing, and soil erosion, the productivity of oceans, forests, grasslands, and croplands is rapidly reduced. People are then consuming productive biological "capital," rather than living on the "interest" that could continue indefinitely. A fair distribution over the generations (assuming population stabilization) can be achieved by keeping within sustainable yields, which can be calculated from knowledge of ecosystems today. Future sustainable yields may be somewhat higher (due to technological advances) or lower (due to environmental damage); but use levels could be readjusted in the light of new information.[18]

How does the *discount rate* affect resource use? Consumers have a time preference favoring the present; if they put money into savings, they expect interest as a reward for postponing gratification. Producers borrow to invest in productive equipment which will bring a future return great enough for repayment with interest. In theory, the discount rate is determined by the economy's productivity, and resources are allocated to their most productive uses. But renewable resources which grow at less than the discount rate will be rapidly exhausted; unless they are protected, slowly growing biological assets will be wiped out to obtain short-term returns. Some resources such as whales are doubly vulnerable because they grow slowly and are also common-property assets. Some resource uses do contribute to capital accumulation which benefits future generations, but 90% of resources go into short-lived consumer goods. In the past, we have encouraged the extraction of minerals and other nonrenewable resources that appeared limitless. We gave away mineral rights on public lands to stimulate mining. To foster economic growth we subsidized fast extraction by granting depletion allowances. Today such policies are a threat to the future.

The goal should be *to keep the resource base intact* from one generation to the next, thereby preserving it for all generations. In the past, technology has extended the resource base at a rate which has more than offset depletion. New extraction technologies have made lower-grade ores economical, discoveries have expanded reserve estimates, and substitutes have been found for scarce materials. As a result, the price of most raw materials (in constant dollars) stayed constant or fell from 1900 to 1970. But during the 1970s shortages and rising prices affected several materials (notably oil and timber). In addition, the indirect costs of technology in the form of toxic wastes, pollution and environmental degradation are placing an increasing burden of risks on future generations. Justice between generations requires that the real costs of the main categories of essential raw materials should be kept approximately constant. We should not allow the resource base to be depleted more rapidly than it can safely be extended by technology, for it belongs equally to all generations.[19]

For *nonrenewable materials*, a severance tax is one way to ensure a more equitable distribution of resources among generations. We use income taxes to mitigate distributional inequities created by market forces within the present generation. In a similar way, a tax on mineral extraction would contribute to intergenerational justice by slowing resource depletion and waste production and it would provide a stimulus for recycling technologies. Less resource consumptive technologies and life-styles in affluent nations would also greatly reduce depletion rates. Another mechanism for giving greater institutional protection to the unborn would be to appoint proxies for them, as we do for infants and the mentally retarded and others who cannot speak for themselves; an attorney for the unborn might be paid to testify at Nuclear Regulatory Commission hearings, for example. The basic needs of future generations should have priority over the luxuries of the present generation. Planning for long-term sustainability and the preservation of a livable environment requires a wider time horizon than we have adopted in the past. As we become aware of the indirect consequences of our actions, our moral community is expanded in both space and time. The transition to a sustainable society must start now; a far greater toll in human suffering will occur if the change is forced on our descendants (see Chapter 12).

Some discounting of *the distant future* is justified because we know so little about it. Urgent current needs have priority over uncertain future ones, as long as we do not jeopardize the possibility for future generations to satisfy their basic needs. However, our present political and economic institutions discount the future so heavily that consequences more distant than a decade or two are essentially ignored. We have been incredibly shortsighted in giving massive subsidies to nonrenewable energy sources—first oil and then uranium—while until recently we gave negligible support to solar and renewable energy research. I suggest that we should think in terms of a 50-year time span—a couple of generations—which is as far as we can project consequences with any confidence. We can visualize the lives of our children and grandchildren, and we have a special interest in them that can be broadened to include the interests of our contemporaries. If we keep the resource base intact for our grandchildren they can do the same for subsequent generations.

Duties to Nonhuman Beings

Are prudential considerations enough to motivate decisive action to preserve the environment? Does human self-interest, even on a long time-scale, provide adequate grounds for the protection of other species? Are other creatures valuable in their own right, apart from their usefulness to us? Do we have duties with respect to other forms of life that are not simply derivative from the consequences for humanity?

Aristotle, Aquinas, and Kant, along with most major Western philosophers, said that we have no direct *duties toward animals* because they cannot reason. Rationality was the criterion for a sharp separation of the human and the nonhuman, and for the delimitation of the sphere of moral responsibility. Some utilitarians, on the other hand, have interpreted the greatest total happiness as including other forms of life. Bentham and Mill held that we do have duties to animals because they can experience suffering; sentience, not rationality, determines whether a being can be the object of our moral obligations.[20] Utilitarians were active in the nineteenth-century movement for the humane treatment of animals. Smart speaks for many contemporary utilitarians when he defends the goal of "maximizing the happiness of all sentient beings."[21]

Peter Singer's *Animal Liberation* is an extended defense of this duty to minimize the suffering of animals. He attacks the "human chauvinism" of restricting ethical concern to the members of our own species. Our moral responsibilities extend to any organism that can experience pleasure or pain (he draws the line "somewhere between a shrimp and an oyster"). Singer documents in detail the suffering that is common in the raising, transporting, and slaughtering of cattle and poultry, and he ends by advocating vegetarianism. He maintains that if we have a duty not to treat animals cruelly, then animals have rights (at least the right not to be treated cruelly). Others have argued that animals have interests, and that these form the basis for assigning rights, even though animals are not themselves moral agents. Such rights are, of course, not absolute, but must be weighed against the competing rights of other beings. Note that in these views we have no duties in regard to the nonsentient portions of nature.[22]

Christopher Stone proposes the extension of legal rights to *trees, rivers, and natural objects*. In the context of law, he says, to have a right means to have standing to sue in order to prevent injury or recover damages. A lawyer can present such claims on behalf of a corporation, a university, an infant, or a senile person. Stone advocates the appointment of legal guardians to defend the interests of a mountain threatened by strip-mining, a forest threatened by development, a polluted river, or an endangered species.[23] Such rights could be overridden by the rights of human beings, but at least they would have to be considered. Stone seems to view the assignment of rights as more than a legal fiction designed to ensure that environmental preservation is seriously considered in decision making. He calls for a planetary vision of the earth as one organism, but he does not develop such a wider framework within which the attribution of rights might be defended. Moreover, the idea of rights serves mainly as a protection of individual beings; there is no explicit reference to the functioning of the ecosystem of which the individual is a member.[24]

The biblical idea of stewardship provides a religious justification for *respect toward all creatures.* The doctrine of creation affirms the goodness of the world and all its inhabitants. In Chapter 2 I suggested that the biblical outlook is neither anthropocentric nor biocentric, but theocentric. God is said to delight in the manifold diversity of nature and to value it quite apart from humanity. His covenant with Noah includes both nonhuman life and posterity: "This is a sign of the covenant which I make between me and you and every living creature that is with you, for all future generations."[25] In some of the psalms, nature itself is represented as praising God. "The earth is the Lord's," and we are only caretakers and trustees, responsible to the Creator for the way we treat it. A number of biblical laws forbid cruel treatment of animals. The theme of stewardship for all creatures was compromised in later history when an absolute line was drawn between the human and the nonhuman, and it was negated in practice by institutions devoted to economic gain and technological progress. But stewardship remains a part of our cultural heritage that can be reclaimed.[26]

Among modern theologians, H. Richard Niebuhr has expressed a strong sense of *loyalty to the community of life* of which we are a part. He stresses the social character of selfhood and our involvement in the natural world. Life is a gift mediated by a long history and a continuing web of interdependencies; each of us is constituted partly by our relationships to our fellow creatures. For Niebuhr, the injunction to love one's neighbor must include the whole community of present and future beings:

> Who, finally is my *neighbor*, the companion whom I have been commanded to love as myself? . . . He is the near one and the far one, the one removed from me by distances in time and space . . . the unborn generations who will bear the consequences of our failures, future persons for whom we are administering the entrusted wealth of nature and other greater common gifts. He is man and he is angel and he is animal and inorganic being, all that participates in being.[27]

Recent documents of the National Council of Churches similarly have urged that, in an interdependent world, the "neighbor" affected by our actions must be taken to include nonhuman creatures as well as future generations.[28]

In Eastern religions, respect for the nonhuman is based on *the underlying unity of all beings.* If the divine spirit permeates all things, a rock can be as eloquent as a flower or a tree in witnessing to the universal harmony. In Taoism and

Zen Buddhism, this experience of cosmic unity is invoked as a reason for not harming other forms of life. In Hinduism; the principle of nonviolence is reflected in the widespread practice of vegetarianism. The Jains in India hold that all life is sacred; the ideal priest will sweep the path ahead as he walks to avoid stepping on an insect. "This is the quintessence of wisdom: not to kill anything."[29] In the West, the best-known modern exponent of the sacredness of all life is Albert Schweitzer. The central principle of his philosophy of ethical mysticism is "Reverence for Life." Every form of life is a manifestation of the Will-to-Live. The good is whatever preserves and promotes life; evil is whatever destroys or injures life. As he looks at a disease germ through his microscope, Schweitzer reflects that as a doctor he will have to sacrifice it to save a human life. These choices, he says, are inescapable but arbitrary, and he feels responsibility and guilt for any destruction of life.[30]

Such convictions of *the sacredness of all life*, and such gentleness in practice toward other creatures, are surely preferable to the insensitivity and violence characteristic of the modern West. But is there not a middle ground that preserves a respect for all life, yet provides for a principle of choice when the needs of diverse beings conflict? Schweitzer refuses to discriminate between higher and lower forms of life because he believes it would lead to the idea that some forms of life are worthless. But does that conclusion really follow?

The *process philosophy* of Alfred North Whitehead and his followers seems to me particularly helpful in this regard because it provides a rationale for a hierarchy of duties toward human and nonhuman beings. The organismic view of reality as a network of interacting events, rather than a collection of externally related objects, has much in common with the ecological perspective, as was seen in Chapter 2. Whitehead's stress on temporality and change, and his interpretation of the world as a creative process, were strongly indebted to evolutionary biology. There is no radical gulf between the human and the nonhuman, either in evolutionary history or in present status. Process philosophy describes all entities by the same set of basic categories. All beings are at least rudimentary centers of experience; no sharp lines can be drawn among creatures of varying degrees of complexity. But there are immense differences in the intensity, breadth, and originality of experience at various levels. A human being is capable of a far more varied and unified experience than an insect.

In process thought, every being is capable of enjoying experience, which constitutes the only intrinsic good. Every being also can contribute to the experience of other beings, which constitutes an instrumental good. In furthering the good, then, we must consider the *intrinsic good of other creatures*, and not simply their instrumental contribution to humankind. According to John Cobb and David Griffin, "The belief that all levels of actuality can enjoy some degree of experience provides the basis for a feeling of responsibility directly to them."[31] Whereas Kant said that we should treat other persons as ends in themselves and not simply as means to our own ends, the Whiteheadian tradition holds that we should treat all creatures as ends in themselves. Moreover, all types of entities are valued by God and contribute to His experience. In traditional Christianity, the worth of persons derives in part from their worth in God's sight; in process theology, all entities similarly derive significance from their participation in the divine life. In short, every creature is valuable to itself, to other beings, and to God.

But clearly there are great *differences among beings* in both the intrinsic and the instrumental good they can achieve. A rock is a mere aggregate, an object with no integrating center. A plant has no center of unified experience higher than that of individual cells, though the life of the cells is dependent on the total life of the plant. More complex beings are capable of both greater intrinsic good and a greater contribution to the experience of other beings. There may be elementary feeling but no consciousness, much less self-consciousness, in simple forms of life. In this framework it is entirely justified to destroy cancer cells or malarial mosquitoes to save human lives. If protein for starving children could be obtained only at the price of some suffering on the part of animals, the latter would be justified. (Actually, feeding grain to cattle in feedlots is detrimental both to the human food supply and to animal well-being, so we do not have to choose in this case.) All living things are valuable, but they are not equally valuable.

Because all creatures are valuable, *the preservation of endangered species* is significant not just when it benefits the human species, but even at the cost of considerable inconvenience to us. Each type of creature makes a distinctive instrumental contribution to the good of other members of the ecosystem, the diverse web of life in which it participates. Moreover, God's life is enriched by the diversity of the created order. As we become aware that we, too, participate in this wider community of being, we realize that our interests are not only dependent on but also merge with those of a larger whole. The process viewpoint thus leads to respect for the nonhuman world, commitment to the distinctive values of human life, and a basis for judging their relative importance when they do conflict.[32]

II. ECOLOGY AND SURVIVAL

Does an ecological outlook require not only an extension of traditional philosophical or religious principles, but also a radically different set of ethical assumptions? Should we start, not from individual beings, human and nonhuman, but from a holistic understanding of ecosystems and the conditions for their preservation? Can we derive ethical norms from the requirements for survival on a finite planet?

The Biotic Community

In Chapter 2, four themes in ecology were summarized: 1) the *ecosystem* concept, which expresses the interdependence of the members of biotic communities; 2) the importance of *diversity* for ecological balance and stability; 3) the *finite limits* of populations, natural resources, and the capacity of environments to absorb wastes; and 4) concern for *long-term effects* because the repercussions of our actions are far-reaching in space and time. Can ethical principles for policy choice be derived directly from these findings of ecology?

Barry Commoner's *The Closing Circle* starts from what he called the *four laws of ecology*: "Everything is connected to everything else," "Everything must go somewhere," "Nature knows best," and "There is no such thing as a free lunch."[33] It might appear that these laws, especially the last two, would allow us to derive prescriptive ethical norms from descriptive scientific statements. "Nature knows best" seems to imply "Therefore we ought not to interfere with nature." But in subsequent chapters Commoner makes it clear that he is not advocating nonintervention, but intelligent intervention based on the lessons we can learn from nature (such as the use of biodegradable materials). It may be true that there is "no free lunch," but we still have to decide how good the lunch is, how high the price is, and whether it is worth the price. Commoner insists that the crucial policy choices concerning environmental pollution, nuclear power, and so forth are value judgments rather than scientific ones. "These are matters of morality, of social and political judgment. In a democracy they belong not in the hands of 'experts' but in the hands of the people and their elected representatives."[34] For Commoner, moreover, these value judgments remain in the sphere of enlightened human self-interest.

Aldo Leopold's *A Sand County Almanac* goes further in formulating an ecological ethic that is not anthropocentric. Written in the 1930s, when even the word "ecology" was virtually unknown, these essays reflect Leopold's understanding as a scientist, his work as a forestry and wildlife manager, his firsthand experience as a naturalist, and his poetic sensitivity as a person. In "The Land Ethic," he spells out some ethical implications of an ecological outlook. The scope of a person's ethics, he suggests, is determined by the inclusiveness of the community with which he identifies himself. Slaves once were treated as property toward which the owner had no ethical obligations. The history of ethics can be read as the extension of the boundaries of community to include the tribe, the nation, and then all people. The next stage, Leopold urges, is the inclusion of the whole land community, of which people will see themselves not as conquerors but as members and citizens.

Leopold's presentation of *the biotic community* elaborates the ideas of food chains, the biotic pyramid, the carrying capacity of the land, the interactions of predators and prey, and the destructiveness of human impacts. But it is not enough, he suggests, to appeal to enlightened self-interest. Most species have no economic value—yet they have a right to exist. What is required is a new stage in the evolution of ethics:

> The "key-log" which must be moved to release the evolutionary process for an ethic is simply this: quit thinking about decent land-

> use as solely an economic problem. Examine each question in terms of what is ethically and esthetically right, as well as what is economically expedient. A thing is right when it tends to preserve the integrity, stability, and beauty of the biotic community. It is wrong when it tends otherwise.[35]

Right action is defined here in terms of the consequences for "the integrity, stability and beauty of the biotic community." The criterion seems to be the good of the total ecosystem. Integrity, stability, and beauty are holistic concepts; they are attributes of the whole system rather than aggregate functions obtained by summing up the benefits to individual members.

Leopold seems to make *the ecosystem itself* the locus of value, rather than the individual members. Yet one might argue that in a truly ecological perspective, the welfare of the members is so totally dependent on the welfare of the whole that one cannot separate them. If you really identify yourself with a larger community, your interests become those of the extended life in which you participate. Thus Paul Shepard maintains that human good and ecosystem good are the same thing differently described, since the boundary between the individual and the ecosystem is diffuse. The self is viewed as extended into the system and vice versa; self and system are interpenetrating. The world is my body, says Shepard. Some people find personal fulfillment in the welfare of other persons with whom they identify themselves, and the distinction between egoism and altruism breaks down; an even wider self-identification is encouraged by ecology.[36]

But in Leopold's case, this holism derives from his outlook *as poet and as person* and not simply as ecologist. "That land is a community is the basic concept of ecology, but that land is to be loved and respected is an extension of ethics."[37] He points to the importance of new attitudes and values, and therefore of educational and religious institutions, as well as scientific knowledge. He has much in common with Thoreau, Muir, and the romantic poets, though his orientation is more practical. For Leopold, the ground of unity with nature is more scientific than religious, but the personal and experiential aspects of his holism should not be overlooked. The qualities that he seeks in ecosystems—beauty, balance, and harmony—are judged by aesthetic and intuitive as well as biological criteria.

Can ecosystem ethics avoid making *the individual* totally subservient to *the system*? In its day, evolutionary ethics could justify any sacrifices of individuals that aided the survival of the species. There is a similar danger that in thinking about populations and ecosystems the individual is ignored in a new collectivism. Leopold himself avoids this danger, partly because his keen observation as a naturalist and his sensitive response as a person are directed to individual creatures. Does the ecologist subsume the *human* under the *biological*, however, to the neglect of distinctive features of human existence? Here I would argue that in stressing the common features of human and nonhuman life, ecology tends to ignore their differences. Consequently, ecological ethics provides no grounds for decision when the interests of diverse members of the biotic community conflict, as they sometimes do.

I would view the integrity of the ecosystem as *instrumental to the welfare of individuals*, human and nonhuman. I find in process philosophy a recognition of interdependence and the fact that individuals are constituted by their relationships, coupled with an insistence that only individuals are capable of experience—and that there are varying grades of experience. In such an interpretation, ecological integrity is a precondition of life and therefore of other values, but cannot serve as a definition of all value. The goals we seek must be *compatible* with the health of the ecosystem, but they need not be limited to the latter. A person or society interested only in satisfying biological needs would miss the most distinctive potentialities of human existence.[38]

The Ethics of Survival

Many biologists hold that survival is the supreme value and that it supersedes all other values when it is threatened. For Garrett Hardin, the main criterion for policy decisions is *long-term human survival.* He has articulated powerfully the lessons of ecology concerning finite limits and the carrying capacity of the environment. Survival is more important than freedom, he says; therefore compulsory sterilization and abortion are justified to prevent overpopulation that would endanger the human species. "Freedom to breed will bring ruin to all."[39] Similarly, Hardin holds that survival makes questions of justice irrelevant. Sending food to starving people will lead only to further population growth and a greater catastrophe later. Like people in a crowded lifeboat, we all will go down together if we try to help people whose lives are now in jeopardy. "The criterion is survival. Injustice is preferable to total ruin."[40]

I find the *lifeboat analogy* very misleading. Affluent nations are not on lifeboat rations, and they could help the starving without any great sacrifice. Their consumption habits contribute to starvation. The carrying capacity of poor countries can be considerably increased by assistance for agricultural development, and the prospects for population stabilization without coercive measures are not as dismal as Hardin assumes. Even if we were in a lifeboat situation, survival would not make other values irrelevant. If it were clear that resources were not sufficient for everyone, the decision as to *who should survive* would still be subject to criteria of justice. No human group should be excluded from the class of potential survivors. Comparable sacrifices should be expected from all nations; clearly it would be unjust to allow some to starve while others lived in luxury. One might want particularly to question the impartiality of any policy that favors the survival of one's own group or nation.[41] Other aspects of "lifeboat ethics" are discussed in Chapter 11.

"Human survival" is frequently invoked by biologists, and the phrase legitimately underscores human dependence on the ecosystem and the importance of an extended time-frame. But the *survival of the species* does not appear to be at stake in current decisions, though it might be in decisions a generation or two hence if past trends continue. Conceivably a global nuclear war or the unforeseen destruction of a major component of the biosphere (such as life in the oceans) might result in irreversible damage to the life-support system. But even the worst scenarios in *Limits to Growth* do not endanger the human species, though a large fraction of humankind would not survive the projected catastrophes.[42] We should talk, therefore, about the *survival of individuals*—of whom millions are threatened today, and billions will be if environmental degradation continues at increasing rates. What place should individual survival have in a scheme of values?

In Maslow's hierarchy (see Chapter 4), *survival and physical security* are the most urgent needs when they are threatened. In situations of extreme scarcity they preoccupy a person's attention. They are a precondition of all other values. At the same time, they are the lowest values in the hierarchy; people who have to devote all their efforts to biological survival are hardly living a human life. One can always ask: for what is a person surviving? Social relationships are as essential for a distinctively human existence as food and shelter. Maslow's highest levels—self-respect and self-actualization—are the most desirable. I will argue that if global resources are wisely and equitably used, they are indeed adequate for the survival of all, on the time-scale I have projected. Policies therefore should be designed to take into account all the levels of Maslow's hierarchy. Survival is important as a precondition of other values, not as a substitute for them.[43]

Above the level of bare survival, what relative priority should be given to justice, freedom, and material welfare? I noted that Rawls gives the highest priority to freedom—except under conditions of *radical scarcity*, in which he considers the abridgment of civil liberties justifiable in order to provide for basic necessities. Under conditions of extreme deprivation, the opportunity to exercise individual rights is at best very limited; authoritarian measures that are efficient in meeting basic needs can raise living standards and establish conditions in which freedom can flourish later. But Rawls insists that the principle of justice holds even when material levels are very low.[44] Other authors also have maintained that at low levels of development, efficiency in meeting physical needs is more important than freedom. As living standards rise, there are more choices in employment and consumption, and political liberty would contribute more to human dignity and self-respect than additional affluence would; freedom should have increasing priority as material levels rise.[45] When there is extreme economic inequality, some people have great power over others, whose freedom is thereby limited; justice is thus a precondition of freedom.

Donella Meadows points out that there are distinctive value priorities implicit in *alternative models of society*. The dominant U.S. model stresses material welfare and freedom; these values are evident in assumptions about economic and industrial growth, the free market, and political democracy. The socialist model emphasizes material welfare and justice; economic and industrial growth along with distributional equity are promoted through the powers of the state. In the environmental model, by contrast, policies are evaluated mainly in relation to long-term survival; the control of birth rates, resource consumption, and environmental degradation may require sacrifices in freedom, justice, and industrial growth. In forming a composite model, Meadows draws from all three models, but she relies most heavily on the last.[46]

Meadows defends the following value priorities. 1) *Survival* is a precondition for all other values. Any possible threats to environmental integrity must be taken very seriously. The short-term sacrifice of other values is justified to avoid long-term survival risks, including the uncertainties of heavy dependence on unproven technologies. 2) *Material welfare* up to the level of basic human needs comes next, but progress beyond this level goes to the bottom of the list. Since equality in poverty has little appeal, the satisfaction of basic needs comes before justice. 3) *Justice* is to be understood as equal opportunity to realize the other values listed. 4) *Freedom* should be curtailed when it conflicts with the preceding values, but only to the extent necessary to realize them. For example, the restriction of reproductive freedom is justified because such freedom conflicts with the future achievement of material welfare for all.

I agree with these priorities, but would put greater stress on the ways in which the values are *interrelated* rather than being mutually exclusive. I have urged that even when survival is threatened we must ask who will survive and who will make decisions affecting survival; questions of justice and freedom are not irrelevant. If justice lies in maximizing the welfare of the most disadvantaged, the fulfillment of basic human needs for all is precisely what justice demands. Justice and freedom are similarly linked; wide participation in decision making is the most promising way to achieve greater justice. Authoritarianism often leads to new forms of privilege, rather than greater equality. I will suggest in Chapter 12 that sustainability and justice require extensive government powers, but not the abandonment of political democracy.

Let me summarize the position I will adopt in the remainder of this volume regarding environmental values. *Environmental preservation* and *resource sustainability* are goals that should be sought primarily for their human benefits. The ascription of primacy to the interests of humanity is justified by ethical principles, and it also is a dictate of practical politics in gaining public support for environmental measures. Control of the pollution of air, water, and land is essential for human health and welfare. Resource sustainability, however, requires the adoption of a long time-scale and the acceptance of obligations to future generations. It can be achieved only by stringent conservation, a shift to renewable resources, and restraint in resource consumption by affluent nations. In the more distant future, environmental destruction could threaten the survival of mankind as a species, but even now our decisions jeopardize the survival of millions of individuals. Since diverse costs, benefits, and risks—including risks to survival—fall unequally on various population segments, distributive justice and participatory freedom always are relevant to environmental and resource decisions.

Ecosystem integrity likewise contributes to the long-term interests of humanity. Ecology has made us more aware of our dependence on the biosphere; we have begun to acknowledge the vulnerability of ecosystems and the importance of balance and diversity. But we hardly have started to learn the lessons of finite carrying capacity and limits to growth. Wilderness preservation, I suggested, can be defended from the standpoint of its various benefits to humanity. In addition, I have urged *respect for other creatures* based on the biblical concept of stewardship of the created order (even apart from its usefulness to us), and on process philosophy (the intrinsic and instrumental value of the diverse levels of experience in all beings). Some would go further in affirming the sacredness of all life, based on the experience of an underlying unity, a spiritual reality pervading all nature; but such a view seems to provide no guidance for decisions when the welfare of various forms of life conflict.

III. FROM VALUES TO POLICIES

We now must try to bring together some of these diverse values and ask about their implications for the policy decisions that will be explored in subsequent chapters. How can commitment to human and environmental values be combined in practice as well as in theory? Does social change start from changes in values or changes in institutions?

Environment, Employment, and Justice

In many of the policy decisions we will be considering, the central conflicts are among environmental preservation, employment opportunities, and social justice. To be effective, these values must be brought together at three levels: philosophy and ethics, policy decision, and political strategy. We have been considering mainly the first level, but now we must look ahead to the other levels.

The *environmental movement* has been strongly criticized for its neglect of *social justice*. Richard Neuhaus, for example, attacks environmentalism as a cop-out from the problems of racism and poverty, a diversion from issues of human oppression and suffering. He claims that an educated, middle-class, white, suburban elite has enjoyed the luxury of escape to nature and turned its back on the poor in cities.[47] Sociological studies lend some credibility to the charge of elitism. In one study of members of several environmental organizations, 80% of the sample had some college education and 23% had a graduate degree. In a survey of Sierra Club members, only 7% were clerical or blue-collar workers.[48] Middle-income white families are heavily overrepresented among users of national parks, wilderness areas, and water-based recreational areas.

However, a succession of surveys has shown that environmental measures continue to have *broad public support*, cutting across lines of race, education, and income. A national survey in 1978, soon after the tax-cutting Proposition 13 had been adopted in California, found that support for environmental protection remained strong, despite deep concern about taxes and inflation. Of those who responded, 62% said they would be willing to pay higher prices to protect the environment, while only 18% chose lower prices and more pollution—figures that were virtually the same as those three years earlier. At all income levels, at least 60% regarded themselves as "active participants" or "sympathetic" to the environmental movement (rather than "neutral" or "unsympathetic"); responses from black and white groups were almost identical. 67% agreed with the statement: "An endangered species must be protected even at the expense of commercial activity."[49] In another 1978 opinion survey, 52% said that the federal government should spend more on environmental protection.[50]

Anyone concerned about social justice has reason for concern about the forms of environmental degradation of which *low-income families* are the main victims. The correlation of environmental quality with socioeconomic status has been established in case after case. The poor almost always have a disproportionate share of the burden of noise, air pollution, and lead poisoning. For example, levels of carbon monoxide (mainly from autos) and sulfur oxides (from stationary sources) were highly correlated with poverty in three cities studied. In Washinton, D.C., 23% of the tracts in which the median family income was less than $7,000 had particulate levels of 5 to 10 mg/cu. meter, whereas only 3% of those in the $16,000 to $20,000 bracket had particulate levels in the same range.[51] The correlation of air pollution with illness and death rates has been well documented (see Chapter 7).

Furthermore, when action is taken, the poor often carry a disproportionate share of *pollution abatement costs*. They have less economic and political power in defending their own interests when regulatory measures are introduced. Most abatement costs are passed on to the consumer in the form of higher prices; the price of basic commodities represents a higher fraction of a low-income family budget than of a high-income one.[52] Rising prices for electricity hit low-income families with particular severity, since utility bills are a large fraction of their total income. The use of sales taxes to finance public projects is another practice whose impact is regressive.

In many situations, *environmental degradation* and *social injustice* are linked together. The exploitation of nature and the exploitation of human beings are often products of the same economic and political forces. In Appalachia, the landscape and the people have suffered for the same reasons: the economic profits of coal companies and the political power that they wield in state legislatures. Strip-mined hillsides, declining land values, and inadequate mine-safety precautions are jointly products of the pursuit of financial gain and the exercise of political influence. A society pursuing affluence destroys human community along with natural resources. The manipulative mentality leads to the control of both things and people in the interest of efficiency. People find themselves alienated from nature and from each other. Technologies are geared to the interests of organizations rather than the welfare of persons or the earth.

To be sure, environmental and human values sometimes conflict, especially when it comes to *employment*. A proposed oil refinery is likely to be opposed by environmentalists as a source of pollution, but favored by labor unions as a source of jobs. Occasionally the enforcement of air or water standards will force a plant to close, depriving a whole community of employment and tax revenues. Even the threat of loss of jobs is politically potent, though it often turns out that standards can be met with added costs passed on to the consumer, or that the plant was really obsolete or of marginal efficiency and would have had to close soon anyway. Moreover, environmental legislation actually has created far more jobs than it has jeopardized. According to a 1978 report, 678,000 people were employed in new pollution abatement activities, while only 21,900 had lost their jobs because of plant closings due to environmental regulations.[53] The human impact of such plant closings could be reduced by job retraining, relocation subsidies, and adjustment assistance. Proposals for full employment or a guaranteed annual income would promote employment and social justice, and at the same time reduce the effectiveness of plant-closing threats as a form of environmental blackmail.

A task in the remainder of this volume, then, is to seek policies and political coalitions that combine environmental preservation and social justice. At the level of *policy decisions*, we always will have to ask both what impacts there will be on the environment, who will benefit, and who will pay the costs, direct and indirect. Such distributional questions seldom are raised in cost-benefit analyses or in environmental impact statements. Again, in the agenda for pollution control, highest priority should go to the forms of pollution that affect large numbers of people—especially among the disadvantaged, if we take the Rawls criterion seriously. We will find, for example, that air pollution control will benefit low-income families more than most forms of water pollution control (especially proposals aimed at preserving fishing, boating, and water recreation far from cities). When we come to examine regulatory strategies (absolute standards versus effluent taxes or subsidies, for instance), questions of distributional equity as well as economic efficiency must be raised. As economist William Baumol has recommended:

> Environmental legislation should be paired with measures explicitly designed to offset its undesired distributive consequences. . . . Coupling environmental and redistributive measures will enable us to increase their political acceptability and to deal more appropriately with social priorities. By acting simultaneously and explicitly on both issues we can have some assurance that environmental policy does not become yet another influence that makes the rich richer and the poor poorer.[54]

At the level of *political strategy*, it should be noted that in the past labor unions often have joined industry in opposing environmental regulations that might harm employment. An effective political strategy will require coalitions drawing from environmentalists, labor unions, consumer groups, community organizations, and civil rights groups. Environmentalists and labor unions have cooperated on issues in the industrial environment, occupational health, and pollution in the workplace (such as asbestos as a cause of cancer and lung disease).[55] Environmentalists have worked with urban groups on air pollution, traffic congestion, toxic waste problems, and the planning of parks and recreation areas accessible to inner city residents. A 1979 conference sponsored by the Sierra Club and the National Urban League (a major voice for city blacks) was devoted to the improvement of urban environments through land-use plans, community self-help, recycling and solar energy projects, and locally controlled, environmentally sound programs to create new jobs.[56]

While the goals of environmental quality, employment, and social justice often can be combined, there are many situations in which they *inescapably conflict*, and one value must be sacrificed for the sake of the other. The same valley cannot be preserved as a wilderness and dammed to form a reservoir for the water supply or recreation of a city population. If there is a real commitment to stewardship of the earth, a location that is one of the few habitats of an endangered species sometimes must be set aside, even though its use for a power plant site would provide local employment, county taxes, and electricity for the city. Preservation of scenic areas for future citizens can be defended in the name of intergenerational justice, and maintenance of ecosystem integrity is a precondition of long-term human welfare. But there are some decisions in which there will be significant human costs for environmental preservation. Even then, however, every effort should be made to see that these costs are distributed as equitably as possible. Some of these hard choices are discussed in Part Two.

Sources of Change: Values and Institutions

Throughout Part One we have been examining attitudes and values. But do values really make any difference in actual behavior? If we want to change behavior, should we start by trying to change institutions rather than values?

Similar questions arose in a historical context in Chapter 2. We asked whether the historical roots of the Western environmental crisis are to be sought in *values and beliefs*, such as the biblical idea of dominion, or in *social institutions*, such as industrial capitalism. I argued that economics and technology were the most powerful forces influencing the way people have treated nature in practice, but that these institutions were themselves in part the product of distinctive attitudes. Differing theories of social change have emphasized differing types of historical cause. Max Weber, especially in his earlier work, stressed the role of cultural values (the contribution of the Protestant ethic to the rise of capitalism and the growth of technology). On the other hand, Karl Marx gave priority to economic institutions, especially ownership of the means of production, as the primary sources of historical change. I myself am convinced by the contemporary sociologists who maintain that attitudes and institutions are inseparable. Peter Berger, for example, shows that social institutions shape individual world views, while individual world views shape institutions.[57] We make corporations, schools, and governments, which in turn make us what we are.

Psychologists are divided concerning the roles of *values* and *institutions* as determinants of behavior. In the previous chapter, Rokeach's empirical studies on the ranking of values were mentioned. Rokeach holds that values influence behavior, which results in institutional change. He says that values are "standards which guide conduct" and "standards that determine actions." He is impressed by the correlation of values and behavior evident in his studies.[58] Other psychologists are impressed by the discrepancies between professed values and actual behavior. Stanley Milgram, for example, has shown that experimental subjects will act in obedience to authority in violation of the values they claim to hold.[59] Daryl Bem maintains that "cognitive dissonance" between behavior and values usually is resolved by a change in values. For him, the causal sequence runs primarily from institutions to behavior; values are a subsequent rationalization of behavior. As an example he cites the changes in attitudes toward racial justice that occurred after desegregation laws were enforced. If we change institutions, he concludes, changes in attitudes and values will follow.[60]

I believe that both these views are partly valid, but that neither can be taken alone. Values influence institutions, but institutions also influence values. Each of the five main methods for changing environmental behavior involves both values and institutions.

1. Political Action

Since environmental damage is mainly the product of institutions, environmental preservation requires the regulation of institutions. Public policies conducive to resource sustainability also must be adopted on the national, state, and local levels. In Part Two I will try to show that despite the immense political power wielded by industrial interests, citizens working within the system through existing political institutions can bring about significant changes. But both the motivation for political action and the priorities for legislation will depend on the values of legislators and citizens. Without changes in values, politics perpetuates old patterns, and individuals are co-opted by the status quo.

2. Institutional Change

Existing institutions are in many respects ineffective for achieving the environmental and human goals we have been discussing. New mechanisms for decision making and new forms of participatory democracy can be created by modifications and reforms in current political structures. Some people believe that only more radical changes in political and economic institutions can effect a redistribution of wealth and power. They hold that real shifts in power will not occur without conflict, struggle, and perhaps violence and revolution. Those proposing radical social change often stress the importance of institutions, yet their own ideological positions reflect strong value commitments (to the priority of justice, for example). Other changes can be made more gradually—for instance, the development of communities organized around appropriate technologies that are less destructive of both the natural and the human environment. Decentralized control mitigates the concentration of economic power that so often has led to the exploitation of people and nature. Again, the goal of global justice can be achieved only through changes in the international order, for which a new global outlook is a prerequisite. These institutional changes, which are both products and causes of value change, are discussed in Part Three.

3. Education

Environmental education in school, church, and community should convey the scientific understanding of ecological interdependence. In addition to cognitive knowledge, it can encourage affective appreciation arising from firsthand experience. Voluntary community groups also contribute to environmental awareness and consciousness raising through programs of public education and activity. The church, the school, and the family are the main social institutions through which values are transmitted and modified. In all these settings there are challenging opportunities for alternative visions of the future, new images of human fulfillment, and new appreciation of the community of life, as we will see in later chapters.

4. Individual Life-Styles

A person's first responsibility is for his or her own life. There are many actions that can be taken without waiting for social institutions to change. Some represent attempts to simplify one's life and reduce one's resource consumption at home or work, without withdrawing from society. Other life-styles can be adopted only in distinctive new communities. New patterns of significant work and leisure may serve as examples that later will lead to wider social and institutional change. In either case, the decentralization of control and the development of less consumptive forms of satisfaction are among the goals sought. The value changes that would support such alternative life-styles are considered in the concluding chapter.

5. Crises and Disasters

Are changes from all these sources too slow to avert catastrophe? Will it take major crises to convince people that resources are finite and that the environment is vulnerable? Perhaps oil spills, oil embargoes, blackouts, pesticide poisonings, and massive famines will have to occur on a dramatic scale before people will wake up. My own conviction is that external pressures and crises often are essential for social change, but, taken alone, they may lead to undesirable changes. In the case of civil rights, the limited social change that has occurred would not have taken place without pressures from the black community. But without at least some national dedication to justice, these pressures might have led only to further repression and violence. Resource crises alone might lead to environmentally destructive technical fixes, political authoritarianism, or greater exploitation of other countries—unless, along with the crises, there are new visions of the good life and a new dedication to justice, freedom, and respect for the earth. I believe that in the combination of political action, education, and recurring crises there may indeed be hope of moving toward a just, participatory, and sustainable society.

NOTES

1. John Passmore, *Man's Responsibility for Nature* (New York: Scribner's, 1974), chap. 5. Cf. W. T. Blackstone, "Ethics and Ecology," in *Philosophy and Environmental Crisis*, ed. idem (Athens: University of Georgia Press, 1974).

2. William Frankena, "Ethics and the Environment," in *Moral Philosophy and the 21st Century*, ed. K. M. Sayre and K. E. Goodpaster (Notre Dame, Ind.: University of Notre Dame Press, 1979).

3. Thomas Derr, *Ecology and Human Need* (Philadelphia: Westminster Press, 1975).

4. Roderick Nash, *Wilderness and the American Mind*, rev. ed. (New Haven: Yale University Press, 1973), chaps. 3, 4, 5, and 8.

5. William O. Douglas, "Wilderness and Human Rights," in *Voices for the Wilderness*, ed. William Schwarz (New York: Ballantine, 1969).

6. Mark Sagoff, "On Preserving the Natural Environment," *Yale Law Journal* 84 (1974): 205-67.

7. Sigurd Olson, "The Spiritual Aspects of Wilderness," in Schwarz, op. cit.; and Linda Graber, *Wilderness as Sacred Space* (Washington, D.C.: Assoc. of Amer. Geographers, 1976).

8. Kent Gill, "Preservation and Re-creation," *Sierra Club Bulletin* 60 (February 1975): 15.

9. Bruce Kilgore, "Wilderness and the Self-Interest of Man," and Frank Fraser Darling, "Wilderness, Science and Human Ecology," in Schwarz, op. cit.; William Godfrey-Smith, "The Value of Wilderness," *Environmental Ethics* 1 (1979): 309-19; and David Ehrenfeld, "The Conservation of Non-Resources," *American Scientist* 64 (1976): 648-56.

10. See John Black, *The Dominion of Man* (Edinburgh: Edinburgh University Press, 1970), chap. 8.

11. J. J. C. Smart and Bernard Williams, *Utilitarianism: For and Against* (New York: Cambridge University Press, 1973), pp. 62-67. See also Jan Narveson, *Morality and Utility*

12. Martin Golding, "Obligations to Future Generations," *Monist* 56 (1972): 85; also his article in *Encyclopedia of Bioethics* (New York: Macmillan, 1978). Cf. J. Barton Stearns, "Ecology and the Indefinite Unborn," *Monist* 56 (1972): 612; Daniel Callahan, "What Obligations Do We Have to Future Generations?" *American Ecclesiastical Review* 164 (1971): 265; Passmore, op. cit., chap. 4; and Black, op. cit., chap. 5.

13. See chapters by Robert Scott and L. W. Sumner in *Obligations to Future Generations*, ed. R. I. Sikora and Brian Barry (Philadelphia: Temple University Press, 1978).

14. Peter Singer, "A Utilitarian Population Principle," in *Ethics and Population*, ed. Michael Bayles (Cambridge, Mass.: Schenkman, 1976).

15. John Rawls, *A Theory of Justice* (Cambridge, Mass.: Harvard University Press, 1971), pp. 284–98; Ronald Green, "Intergenerational Distributive Justice and Environmental Responsibility," *Bioscience* 27 (1977): 260–65; idem, *Population Growth and Justice* (Missoula, Mont.: Scholars Press, 1976); D. Clayton Hubin, "Justice and Future Generations," *Philosophy and Public Affairs* 6 (1976): 70–83; and Talbot Page, *Conservation and Economic Efficiency* (Baltimore: Johns Hopkins University Press, 1977), chap. 9.

16. Victor Lippit and Koichi Hamada, "Efficiency and Equity in Intergenerational Distribution," in *The Sustainable Society*, ed. Dennis Pirages (New York: Praeger, 1977); and R. Routley and V. Routley, "Nuclear Energy and Obligations to the Future," *Inquiry* 21 (1978): 133–79.

17. Derr, op. cit., chap. 5; Hugh Montefiore, *Can Man Survive* (London: Collins Fontana, 1970), pp. 43–67; Patrick Dobel, "Stewards of the Earth's Resources," *Christian Century*, October 12, 1977, pp. 906–09.

18. Mary B. Williams, "Discounting versus Maximum Sustainable Yield," in Sikora and Barry, op. cit.

19. See Page, op. cit., chaps. 7–9.

20. On animal rights, see Tom Regan and Peter Singer, eds., *Animal Rights and Human Obligations* (Englewood Cliffs, N.J.: Prentice-Hall, 1976); the whole Spring–Summer 1979 issue of *Inquiry* 22; Richard Watson, "Self-Consciousness and the Rights of Nonhuman Animals and Nature," *Environmental Ethics* 1 (1979): 99–129; and Richard Morris and Michael Fox, eds., *On the Fifth Day: Animal Rights and Human Ethics* (Washington, D.C.: Acropolis, 1978).

21. Smart, op. cit., p. 67; cf. Carl Wellman, *Morals and Ethics* (Glenview, Ill.: Scott, Foresman, 1975), p. 47.

22. Peter Singer, *Animal Liberation* (New York: New York Review, 1975); and Joel Feinberg, "The Rights of Animals and Unborn Generations," in Blackstone, op. cit.

23. Christopher Stone, *Should Trees Have Standing? Toward Legal Rights for Natural Objects* (New York: Avon, 1975).

24. See Sagoff, op. cit.; Laurence Tribe, "Ways Not to Think About Plastic Trees," in *When Values Conflict*, ed. Laurence Tribe, Corinne Schelling, and John Voss (Cambridge, Mass.: Ballinger, 1976); John Rodman, "The Liberation of Nature," *Inquiry* 20 (1977): 83–131; and Kenneth Goodpaster, "From Egoism to Environmentalism," in Sayre and Goodpaster, op. cit.

25. Gen. 9:12.

26. H. Paul Santmire, *Brother Earth* (New York: Thomas Nelson, 1970); Eric Rust, *Nature–Garden or Desert?* (Waco, Tex.: Word Books, 1971); for other references see Chapter 2.

27. H. Richard Niebuhr, *The Purpose of the Church and Its Ministry* (New York: Harper, 1956), p. 38; also idem, *The Responsible Self* (New York: Harper & Row, 1963).

28. National Council of Churches, *Energy and Ethics* (New York: National Council of Churches, 1979).

29. William Gerber, *The Mind of India* (Carbondale, Ill.: Southern Illinois University Press, 1977), p. 80.

30. Albert Schweitzer, *Out of My Life and Thought* (New York: Henry Holt, 1933), chaps. 13 and 21.

31. John Cobb and David Griffin, *Process Theology* (Philadelphia: Westminster Press, 1976), p. 44.

32. Ibid., chaps. 4 and 9; John Cobb, "Ecology, Ethics and Theology," in *Toward a Steady State Economy*, ed. Herman Daly (San Francisco: W. H. Freeman, 1973); and

Charles Hartshorne, "The Rights of the Subhuman World," *Environmental Ethics* 1 (1979): 49-60.

33. Barry Commoner, *The Closing Circle* (New York: Knopf, 1971), chap. 2.

34. Ibid., p. 196.

35. Aldo Leopold, *A Sand County Almanac* (New York: Oxford University Press, 1949), pp. 224-25.

36. Paul Shepard, "Ecology and Man—A Viewpoint," in *The Subversive Science*, ed. Paul Shepard and Daniel McKinley (Boston: Houghton Mifflin, 1969).

37. Leopold, op. cit., Foreword. See also the chapter on Leopold in Nash, op. cit.

38. Thomas Colwell, "The Balance of Nature: A Ground for Human Values," *Main Currents in Modern Thought* 26 (November 1969): 46-52; and Holmes Rolston, "Is There an Ecological Ethic?" *Ethics* 85 (1975): 93-109. See also Frankena, note 2 above.

39. Garrett Hardin, "The Tragedy of the Commons," *Science* 162 (1968): 1,243-48; and idem, *Exploring New Ethics for Survival* (Baltimore: Penguin Books, 1973).

40. Garrett Hardin, "Lifeboat Ethics: The Case against Helping the Poor," *Psychology Today*, September 1974, pp. 38-43 and 123-24; another version is in *Bioscience* 24 (1974): 561-68. See Chapter 11 for further references.

41. Drew Christiansen and Charles Wolf, "Environmental Ethics: The Problem of Growth," in *Encyclopedia of Bioethics*, op. cit.

42. Donella Meadows et al., *Limits to Growth* (New York: Universe, 1972).

43. Daniel Callahan, *The Tyranny of Survival* (New York: Macmillan, 1973), chap. 4.

44. Rawls, op. cit., pp. 541-48.

45. Norman Bowie and Robert Simon, *The Individual and the Political Order* (Englewood Cliffs, N.J.: Prentice-Hall, 1977), chap. 7.

46. Donella Meadows, "The World Food Problem: Growth Models and Nongrowth Solutions," in *Alternatives to Growth–I*, ed. Dennis Meadows (Cambridge, Mass.: Ballinger, 1977).

47. Richard Neuhaus, *In Defense of People* (New York: Macmillan, 1971); and David Sills, "The Environmental Movement and Its Critics," *Human Ecology* 3 (1975): 1-41.

48. Clem Zinger et al., *Environmental Volunteers in America* (Washington, D.C.: National Center for Voluntary Action, 1973), pp. 5-23; and Don Coombs, "The Club Looks at Itself," *Sierra Club Bulletin*, July/August 1972. See also Irving Horowitz, "The Environmental Cleavage: Social Ecology versus Political Economy," *Social Theory and Practice* 2 (1972): 125; and Allan Schnaiberg, "Politics, Participation and Pollution: The Environmental Movement," in *Cities in Change*, ed. J. Walton and D. E. Cairns (New York: Allyn & Bacon, 1973).

49. Richard Mitchell, "The Public Speaks Again: A New Environmental Survey," *Resources* (September-November 1978): 1-6. Summary in *Science* 203 (1979): 154.

50. Kathryn Utrup, "Environmental Public Opinion: Trends and Tradeoffs, 1969-1978," cited in Mitchell, op. cit.

51. Julian McCaull, "Discriminatory Air Pollution," *Environment* 18 (March 1976): 26-31; cf. Daniel Zwerdling, "Poverty and Pollution," *Progressive* 37 (January 1973): 25.

52. A. Myrick Freeman, "Income Distribution and Environmental Quality," in *Pollution, Resources and the Environment*, ed. Alain Enthoven and A. Myrick Freeman (New York: W. W. Norton, 1973); a longer version is in Allen Kneese and Blair Bower, eds., *Environmental Quality Analysis* (Baltimore: Johns Hopkins University Press, 1972).

53. *Environmental Quality 1978* (Washington, D.C.: Council on Environmental Quality, 1978), pp. 431-32.

54. William Baumol, "Environmental Protection and Income Distribution," in *Benefit-Cost and Policy Analysis 1974*, ed. Richard Zeckhauser et al. (Chicago: Aldine, 1975), p. 266.

55. James Noel Smith, ed., *Environmental Quality and Social Justice in Urban America* (Washington, D.C.: Conservation Foundation, 1974), especially chaps. 1 and 2. Cf. "Jobs and Energy" (1977) and other publications of Environmentalists for Full Employment.

56. *Proceedings from the City Care Conference: Toward a Coalition for the Urban Environment* (Washington, D.C.: Environmental Protection Agency, 1979).

57. Peter Berger, *The Sacred Canopy* (Garden City, N.Y.: Doubleday, 1969), chaps. 1 and 2.

58. Milton Rokeach, *The Nature of Human Values* (New York: Free Press, 1973), chaps. 1 and 5.

59. Stanley Milgram, *Obedience to Authority* (New York: Harper & Row, 1974).

60. Daryl Bem, *Beliefs, Attitudes and Human Affairs* (Belmont, Calif.: Wadsworth, 1970).

PART TWO: ENVIRONMENTAL POLICIES

6
POLITICAL PROCESSES

The next four chapters deal with conflicting values in political decisions about the environment. Here policy choices are related to environmental and human values, on the one hand, and the realities of politics on the other. Today, as in earlier history, technology is a crucial factor in environmental degradation (Chapter 2). If technology is an instrument of power (Chapter 3), how can it be democratically controlled? How can freedom, understood as participation in decisions affecting one's life (Chapter 4), be exercised in policy choices involving complex technical questions? How can long-range environmental values (Chapter 5) be balanced against immediate economic benefits?

Three questions run through Part Two: 1) How can environmental and human values be effectively included in decision-making processes? 2) What is the relationship between scientific judgments and value judgments in policy decisions concerning technology and the environment? 3) How can input from elected representatives, agency administrators, citizens, and technical experts be brought together in environmental policy decisions?

In Part Two we will be examining environmental policies exclusively in the context of U.S. politics. Some environmental impacts are indeed international, but most are limited geographically and are therefore essentially national problems. By contrast, the resource issues discussed in Part Three are more international in scope. Resource decisions raise questions of global justice and international politics not present in most environmental decisions.

In Part Two we will be looking at policy decisions in the context of prevailing economic and political institutions. Various institutional reforms are proposed, but we will defer until Part Three the discussion of more radical changes in cultural assumptions, social institutions, and individual life-styles that may be required for the transition to a sustainable society. The United States has taken many significant steps in responding to environmental problems, but hardly has begun to face resource scarcities.

The present chapter describes the political processes through which environmental and technological policy decisions are made. The roles of the main actors in the political arena are analyzed: the federal government, special interest groups, citizens, and technical experts. Subsequent chapters are devoted to specific policy issues within this framework, including air and water pollution, wilderness preservation, water resources, toxic substances, and environmental impact statements.

I. THE FEDERAL GOVERNMENT

Let us look first at the nature of environmental politics at the federal level. All three branches of government take part in decisions on technology and the environment, in accordance with the separation of powers in the U.S. Constitution, but the executive branch has the advantage of unified authority and greater technical expertise. Many of the basic initiatives on environmental policy and the control of technology come from the White House—through presidential decisions, cabinet-level planning, proposals to Congress, and the annual budget message. Congress must enact legislation and authorize funds for programs that the federal agencies are to administer. The agencies themselves, as part of the executive branch, have responsibility for carrying out these directives, subject to review by Congress and the courts. In environmental issues, the judicial system has played a particularly important role in decisions as to how legislation is to be implemented by the agencies.

Many of the issues concerning the relationships among the three branches of state government are similar to those at the federal level. In addition, there is a host of problems—some of which are taken up in later chapters—created by the vertical fragmentation of authority among federal, state, and local levels. The present chapter deals only with federal decision making.

Congress and the Agencies

Congress has the tasks of writing and adopting legislation, approving appropriations, and conducting oversight and investigative hearings. These tasks are carried out mainly by the committees, whose decisions are rarely reversed by the full House or Senate. Several features of the congressional committee system are important for our subsequent discussion.

First, *the fragmentation of responsibilites* among committees makes integration of their activities very difficult. At least 22 congressional committees have important environmental assignments. Various aspects of President Carter's 1977 energy plan were taken up in more than a dozen separate committees (though there was some effort at coordination in the House through an ad hoc committee). There have been many jurisdictional disputes over the assignment of bills to particular committees. The communication among committees is sporadic, and there is a long-standing rivalry between House and Senate. The fate of a bill often depends on the personality and power of a single subcommittee chairman who favors or opposes it. Chairmanships usually have been assigned by seniority; some people acquire great power simply by being in Congress for a long time. Any realistic discussion of the legislative process must recognize these changing balances of political power within the tangled web of committee responsibilities.

Second, *special interest pressures* are often strong, especially in technological and environmental decisions with important economic consequences. Members of Congress try to represent their constituencies and usually support measures that benefit major groups of voters. But despite legislation requiring the reporting of campaign contributions, there are still many ways in which groups with strong financial stakes in a bill can try to influence congressional votes. For many years the combined lobbies of the auto, oil, highway construction, and insurance industries, and the corresponding unions, effectively opposed any use of gasoline taxes for mass transit development. Some committees have tended to be closely identified with particular economic interests; for instance, for years the Interior Committee favored Western timber and mining companies.

Third, *alliances between committees and agencies* hinder the effective discharge of oversight responsibilities. There are many committees whose status and power are dependent on the strength of the agencies they are supposed to oversee; consequently, more effort goes into defending the agencies than into monitoring them. The accountability of an agency to the public through elected representatives is particularly in jeopardy when there is a "cozy triangle" of mutual support between a committee, an agency, and an industry. Such a triangle existed between the Joint Committee on Atomic Energy, the Atomic Energy Commission (AEC), and the nuclear industry. Another such triangle has linked the Armed Services Committees, the Defense Department, and defense and aerospace contractors. These alliances are strengthened by their virtual monopoly on expertise and information.[1]

Fourth, *scientific advice to Congress* has been inadequate. Congressional committees often have had to depend on the judgment of agency scientists in hearings. For instance, the claims of Defense Department experts on proposed weapons systems often have gone unchallenged; Congress funded a new system if the Pentagon said it was needed. Some progress has been made during the 1970s in providing Congress with more adequate scientific advice, though it still does not begin to match the expertise available to the executive branch. A few committees have technical experts on their staffs, but most have to rely on expert witnesses at hearings. The Office of Technology Assessment in Congress is beginning to provide committees with comprehensive and balanced reports on policy options, using a wide range of evaluation criteria (see Chapter 9).

Committee recommendations must be acted on by the full House and Senate. Here additional political pressures come into play, and in controversial cases there is protracted debate, negotiation, and amendment. Differences between House and Senate bills then must be resolved in a specially appointed

Conference Committee. Further debate may occur in authorizing and appropriating funds to support programs adopted. The votes of representatives are influenced, in varying proportions, by party loyalty, benefits to a constituency, and political ideology or individual conscience.[2] Since most members of Congress want to be reelected, they are especially responsive to well-mobilized and strongly held public opinion in their home districts, and to individual or corporate sources of campaign contributions. We will see all these forces at work in particular policy decisions in later chapters.

There are two types of federal agencies that carry out the administration of legislated policies related to technology and the environment. First, the *mission-oriented agencies* administer programs aimed at fulfilling specific national objectives. Among these are the Departments of Defense, Transportation, Agriculture, and Energy (which now includes nuclear energy, formerly under AEC). We could include here the Department of the Interior, which is responsible for public lands, national parks, fish and wildlife, and other natural resources. All of these agencies have programs for promoting, developing, or managing particular technologies or resources, and each is subject to oversight by one or more congressional committees.

The activities of every such agency are directed both to the fulfillment of its mission and to the promotion of its own organizational interests. In the *bureaucratic politics* of competition for power and funds, agencies seek support from Congress, other agencies, and client groups. Every agency must interact with private interests; the Forest Service, for example, sells timber rights on public lands to lumber companies. Agencies work closely with the primary clients who support their power base, and they tend to insulate themselves from interference by other parties. Agencies develop their own professional expertise that closely matches that of their primary clients, but this inhibits accessibility to other groups.[3] The chief vehicles for greater accountability on the part of bureaucratic officials are more critical legislative oversight, judicial review, and public participation. In each case, access to information is crucial to pluralistic participation and more open decision making.

The *Army Corps of Engineers* is a prime example of an agency that has had very cordial relationships with both Congress and a special clientele. While technically part of the army, it has been virtually autonomous and reports directly to Congress. Up to $10 million can be spent by committee resolution alone. The corps has been widely used for "pork barrel politics," in which members of Congress secure federal funds and jobs for their home districts without careful scrutiny by their colleagues. The corps has had strong support from construction companies, recreational interests, land developers, and barge and ship operators, all of whom stand to benefit from federal contracts. It has a record of professional competence in planning dams, waterways, and shoreline projects, but also of insensitivity to environmental values—especially in such massive construction projects as the nearly completed $2 billion Tennessee-Tombigbee Canal. In Chapter 8 we will examine the neglect of environmental costs in the corps' use of cost-benefit analysis to justify dam construction. There is some evidence, however, that in response to a series of court rulings, it more recently has improved its environmental impact statement process and its provisions for public participation.[4]

The second type of agency is the *regulatory agency*; we are interested in those whose task is the protection of the environment or of human health and safety. Included here are the Environmental Protection Agency (EPA), Food and Drug Administration (FDA), Occupational Safety and Health Administration (OSHA), and Nuclear Regulatory Commission (NRC) (which took over the AEC's regulatory functions). The AEC provided the most flagrant example of the difficulties that arise when the same agency has responsibility for both promoting and regulating a technology, but a similar ambivalence is present in many agencies. Furthermore, even the primarily regulatory agencies, such as FDA, tend to develop very close ties to the regulated industries. It is common for administrative officials and technical staff to go back and forth between positions in industry and the corresponding regulatory agency.

If regulation is to be effective, the relationship between a *regulated industry* and a *regulatory agency* should involve consultation and negotiation, rather than an exclusively adversary role. But the fine line between close cooperation and capitulation often has been crossed, especially if an agency has only weak and diffuse public support and faces a strong industry or coalition of industries. For example, Labor Department safety administrators and inspectors worked for years in concert with chemical companies to delay enforcement of safety measures for asbestos pollution.[5] Such government-industry coziness is less likely to occur if third parties (such as members of unions, consumer and environmental groups, universities, and community organizations) are present on advisory boards and review panels and are active at agency hearings.

The main regulatory agency for environmental pollutants is the *Environmental Protection Agency.* Created by executive order by President Nixon in 1970, it was given responsibility for air and water quality, pesticides, solid wastes, and (jointly with AEC) radiation—to which noise and toxic substances were later added. In addition to standard-setting and enforcement activities, EPA has developed a growing research and monitoring capability. It has ten regional offices that work with the states, especially on air and water plans. Early administrators took a firm stance, but by the mid-1970s prolonged litigation and technical and political obstacles resulted in many delays and postponed deadlines (see Chapter 7). On many issues, such as auto emission standards, it was caught in the crossfire from industry, environmentalists, and a Congress increasingly concerned about energy, economic recession, and inflation. But the EPA has escaped the fate of many regulatory agencies that have been captured by the industries they were supposed to regulate—partly because it deals with such a diverse array of industries, and partly because its constituency includes well-organized environmental groups.[6]

The Courts and the Environment

The judicial system has played a prominent part in the interpretation and implementation of environmental legislation. In many cases the original legislation was deliberately vague and ambiguous. Agencies often were given considerable administrative discretion because Congress lacked technical expertise, or wanted to leave some flexibility, or wanted to avoid making controversial decisions itself. Almost every major EPA action has been challenged in court, either by industry or by environmentalists, and industry has been sued on many occasions by EPA or by environmentalists. Broad rights to challenge agencies concerning the fulfullment of their legislative mandates were given to citizens under the National Environmental Policy Act (NEPA, 1969), the Clean Air Amendments (1970), and the Water Pollution Control Act (1972). Of the early environmental challenges in district courts, 50 percent were successful, and soon even the threat of litigation was a useful weapon.[7]

Another reason for the frequency of *court action* was the emergence of environmental organizations and legal firms, such as the Sierra Club and the Environmental Defense Fund, which could provide technical experts and competent lawyers. Grounds for standing to sue were broadening. The *Scenic Hudson* ruling (1965) established that a plaintiff could have standing for a court case even without an economic interest in it. In *Sierra Club* v. *Morton* (1972), the Supreme Court reaffirmed earlier decisions that "scenery, natural and historical objects, and wildlife" constitute sufficient grounds for a suit, though it ruled that the Sierra Club had not shown that the aesthetic and recreational interests of its own members would be threatened. But in its first case under NEPA, the Supreme Court ruled in 1973 that five law students, Students Challenging Regulatory Agency Procedures (SCRAP), had shown sufficient stake to challenge the Interstate Commerce Commission action in approving railroad rates that discourage the use of recyclable materials.[8]

The courts usually have tried to base their rulings on whether the agencies have complied with the *procedural requirements* laid down by Congress. In a number of cases, environmental impact statements, which were required under NEPA, were declared inadequate to fulfill the purposes of the act. In other cases, courts have ruled that EPA did not provide adequate evidence to support a regulatory standard, or had not adequately taken into account the economic costs of compliance as required in the legislation. The courts do not have the expertise to decide technical issues, and they often remand cases to agencies for reconsideration. But they have inevitably become involved in technical judgments. For example, in *International Harverster* v. *Ruckelshaus* (1973), the court ruled that EPA had not presented convincing evidence that the auto industry could develop the technology to reduce auto emissions to the required levels within the time allotted; EPA was forced to postpone the deadlines and weaken the proposed standards (see Chapter 7).

In many cases Congress has instructed an agency to balance the economic *costs* of a proposed regulation against environmental and health *benefits*. Usually it has not specified how such trade-offs between conflicting values are to be made, and often there has been considerable uncertainty in the scientific data on which estimates of health risks are based. In judging the adequacy of such balancing decisions, the courts have had a role in the formulation of standards, and industrial and environmental groups have been able to influence regulatory policy through court hearings. Congress can eventually override judicial decisions in such cases. Thus all three branches of government, as well as a diversity of special interest groups and technical experts, have been actively involved in the evolution of environmental policies.

Two Examples: DDT and SST

The complex interactions among these many actors on the political stage can be illustrated in two landmark cases from the early 1970s. The ban on the pesticide DDT was the first decision in which environmentalist forces achieved a major victory against a powerful alliance of a government agency, a major industry, and several congressional committees. The defeat of the supersonic transport (SST) was the first case in which the nation decided to forego the benefits of a promising technology because of its potential environmental and human costs.

The ten-year story of the banning of DDT starts in 1962 with the publication of Rachel Carson's *Silent Spring*. The book was forcefully written and cited extensive evidence of the concentration of DDT in higher members of food chains, and the resulting decline in the population of several bird species. Members of the President's Scientific Advisory Committee (PSAC) were sufficiently impressed by her argument to appoint a study panel, which concluded that the hazard was indeed serious; the panel urged that most uses of DDT be curtailed immediately. President Kennedy ordered the implementation of these recommendations, but they were opposed by the U.S. Department of Agriculture (USDA). The chairmen of the crucial subcommittees of both the Agriculture and Appropriations Committees, in both House and Senate, were all from Cotton Belt states, and cotton accounted for two-thirds of the use of DDT. One of these chairmen, Jamie Whitten of Mississippi, wrote, as a rebuttal to *Silent Spring*, a book that was subsidized by three pesticide companies. The pesticide industry lobbied strenuously against any controls on DDT use. USDA did discontinue massive use of DDT in its own area insect-eradication programs, but endorsed its continued use in agriculture.[9]

In 1968 the General Accounting Office, which serves as a watchdog for Congress, issued a report highly critical of the USDA record in regulating pesticides. USDA had cited hundreds of repeated major violations of the law, but in 13 years not a single case had been referred to the Justice Department for prosecution. Wisconsin and then several other states banned DDT after extensive hearings, debates in state legislatures, and national media coverage. Fish kills in rivers and lakes, new evidence that DDT caused cancer in mice, and the appearance of insect strains resistant to DDT further dramatized the issue. The Environmental Defense Fund, which had led the court challenges in several states, filed suit against USDA and then against EPA, to which jurisdiction on pesticide regulation was transferred in 1970. Whitten's subcommittee also had authority over EPA's budget. In 1971 the Court of Appeals ordered EPA to cancel the registration of all DDT uses. The EPA administration delayed, pending a study on the availability of substitutes, but finally announced the ban in 1972. In this case the combination of a well-written book, a scientific panel, an environmental organization active in state and federal court challenges, media coverage, and an aroused public opinion prevailed over a powerful alliance among an agency, an industry, and four congressional committees.

In the second example, the proponents of the *supersonic transport* in the late 1960s were the Department of Transportation and the aircraft industry. Advocates claimed that building a fleet of SSTs would strengthen the aerospace industry and maintain U.S. prestige and balance of payments in international competition with the French and British (who were developing their own supersonic plane, the Concorde). The agency initially estimated that 75 planes would be ordered by airlines; as the debate progressed the estimates went steadily up to 1,000. Nixon asked for two reviews of the SST proposal. The first, conducted by PSAC, was highly critical and concluded that high fuel consumption, low payloads, and small demand for the plane would make it very expensive to operate. The second, by a panel of high-level officials from several federal agencies, also concluded that the SST would be uneconomical, and that the sonic booms it produces (which sound like very loud claps of thunder) would be unacceptable in populated areas. But both reports were kept secret.

A physicist, William Shurcliff, started a newsletter and wrote a book in opposition to the SST, and founded the Citizens League Against the Sonic Boom. But a report by a panel of the National Academy of Sciences, issued under contract with the Department of Transportation, said that the damage from sonic booms would be small. Nixon decided to disregard the advice of his two review panels and approved the SST project. In Congress the battle for the SST was led by Senators Warren Magnuson and Henry Jackson, both chairmen of important committees and both from the state of Washington, home of Boeing, the prime contractor. At the hearings, agency officials dramatized the time saved by flying at three times the speed of conventional jets, and they gave optimistic estimates of the demand. In 1969, the appropriations bill easily passed in both houses.[10]

But during 1970 a full-scale national debate ensued. Russell Train, chairman of the Council on Environmental Quality (CEQ), established under NEPA, was the first administration official to speak out on the seriousness of the sonic boom and the possibility that pollutants from SST exhaust might affect the ozone layer in the stratosphere, which in turn might increase the incidence of skin cancer. Environmental groups publicized these uncertain but possibly disastrous consequences, on which only sketchy scientific evidence was available.

The SST drew national attention in 1970 election campaigns. Some opponents argued that the same money spent on mass transit would create as many jobs and a more socially useful product. In 1971 Congress terminated the project, mainly on economic but also on environmental grounds, after having spent $1 billion on research and development.

Subsequent events have confirmed the wisdom of this decision. The development, production, and operating costs of the British and French Concorde SSTs were far over the original estimates. Three billion dollars was spent on R & D, and each plane cost $65 million; by 1977 only nine had been ordered (all by British and French national airlines) and production was being shut down. With fuel costs rising and fuel consumption per passenger twice that of conventional jets, the project has become an economic white elephant. Only 100 passengers can be carried (at 20 percent more than first-class fare) and the range is limited. In 1976 the U.S. Secretary of Transportation granted the Concorde landing rights for two flights daily to Washington and four to New York, but only subsonic speeds are allowed over land. Local protests and court challenges delayed the New York landing rights, which were finally upheld by the Supreme Court in 1977.[11]

The SST seems to have been the first case in U.S. history in which a major technological advance was renounced because the economic, environmental, and human costs were seen to be too high in comparison with the benefits. Proponents saw it as the dawn of the supersonic age and said, "You can't stop technological progress." Opponents saw it as a symbol of misplaced priorities when budgets were tight and urgent human needs were unmet. Distributional justice also played a part in the decision. There were strong objections to large expenditures of public funds to subsidize a project of which the main beneficiaries would be the aerospace industries and a few rich travelers who could afford to pay an exorbitant price to save a few hours.

The SST decision, like that in the DDT case, was made initially by technical experts in industry and federal agencies, with the support of sympathetic congressional committees. In both cases there was a subsequent broadening of scientific input and a wider participation by the public and other elected representatives. In each case, the decision criteria were broadened significantly during the debate to include environmental and human values that were not initially taken into account. Both decisions occurred in the early 1970s when environmental concern was widespread and the Vietnam War and domestic discord were undermining faith in the omnipotence of U.S. technology. In subsequent chapters we will see these same political forces at work in more recent debates within a somewhat different climate of public opinion.

II. CITIZENS AND SPECIAL INTEREST GROUPS

In Chapter 4 it was argued that the most important form of freedom today is participation in the decisions that affect one's life. But individual citizens seem helpless in a world of huge government bureaucracies and industrial corporations. Does the concentration of economic and political power, to which technology contributes, subvert the democratic process? Can the public even understand the complex technical issues that arise in environmental and technological policy decisions?

In a democracy, the citizen can vote for *elected representatives*, and the importance of the right to vote should not be minimized. In some election campaigns there really are significant policy issues, and not merely different personalities. Working within a party, a citizen can help to select its candidates. In many states there are provisions for a referendum on specific issues. Nevertheless, elections are infrequent and offer limited opportunity to influence specific legislative policies. Citizens can write to their representatives, and an outpouring of mail is taken seriously by legislators. But these individual expressions of opinion are relatively ineffective compared to the influence of organized groups.

There are two kinds of special interest groups in politics. *Private interest groups* represent the interests (primarily economic) of particular industries, businesses, or occupations. Trade associations, labor unions, farmers' organizations, and commercial associations are active in lobbying for legislation favorable to the institutions or population segments for which they speak. We will look particularly at the political activities of industry, which is the main sponsor of technology and a major source of environmental degradation and risks to health and safety; agriculture, land development, and domestic wastes also harm the environment, but not on the same scale as industrial technology.

Public interest groups, on the other hand, claim to represent the interests (mainly noneconomic) of the wider public, such as civil liberties, consumer safety, or environmental quality. Critics maintain that no one speaks for the public, and that some of these groups have rather small memberships—mostly a white, middle-class, college-educated constituency. Even if these nonprofit citizens' organizations do not represent the public, they do operate in independence from the main centers of economic power, and they often defend environmental and human values neglected by private interest groups. Let us examine the way these two types of interest groups operate in the sphere of politics.

Industry and Private Interest Groups

The benefits of environmental protection are typically long-term, intangible, and distributed over a wide public. The economic costs are more immediate and most of them fall initially on particular industries (though they eventually may be passed on to the consumer). The CEQ estimates that the *costs of complying* with environmental regulations over the ten-year period from 1977 through 1986 will total $361 billion (of which $212 billion is for air pollution control). Of this total, $220 billion will be paid by industry (including $74 billion by electric utilities). Nineteen percent of the steel industry's capital expenditures in 1977 were for pollution control; metal processing, paper, and chemical companies also faced high abatement costs.[12] It is not surprising that industry has been very active in opposing environmental regulations.

The *political influence* of large industries is enormous. There are various ways in which economic power is translated into political power. The most potent weapon is the threat that a plant might close, depriving a community of jobs and taxes. Local and state governments are highly responsive to pressures from industries on which the local economy is dependent. Campaign contributions are another important political tool; expansion of the public financing of campaigns would be a major step in making elections truly democratic. Media advertising and extensive lobbying in support of specific legislation favorable to an industry also have been common. Industry is active in promoting its interests in regulatory agencies at any points where there is administrative discretion (including standard-setting, compliance deadlines, and enforcement action), or where judgments have to be made about scientific uncertainty, technical feasibility, or the economic impact of regulations. Finally, industry has the financial resources and the legal and technical expertise for protracted court battles.[13]

Some examples can illustrate *the political activities of industry*. In the DDT case, the pesticide industry launched a public relations campaign to discredit Rachel Carson and *Silent Spring*, calling her "fanatical," "hysterical," and "misinformed." It sponsored publications by scientists who defended the importance of DDT to agriculture and minimized its dangers. Three billion pounds of pesticides were being used annually in agriculture, and the companies felt that high profits were justified as a return on their heavy investment in research, development, and production facilities. They supported USDA in advocating massive area programs aimed at insect eradication (rather than control). In the SST case, there was very active lobbying by Boeing and other contractors, who argued that government subsidy was justified because a strong aeronautics industry is important to the nation. In trying to sell the SST to Congress, they repeatedly overstated the advantages and ignored or played down the design problems, economic costs, and environmental impacts.[14]

No industry has a longer record of political activity to further its own interests at the expense of environmental and human values than *the coal industry*. Under "broad form" deeds, mining rights in Appalachia were bought up for as little as 10 to 50 cents per acre, with clauses ensuring immunity from damage suits. Both deep-mine wastes and extensive strip-mining resulted in low land values, polluted streams, and the marring of areas of great natural beauty.[15] The industry has a long history of bitter labor conflicts and strikes and of company opposition to safety standards to reduce the heavy toll from mine accidents and black-lung disease. The power of the industry over state legislatures, and the threat to move to other states if stricter standards were introduced, blocked effective state legislation and enforcement.[16] There was strenuous industry opposition to federal safety legislation, but the Coal Mine Health and Safety Act was finally passed in 1969, though enforcement has been very uneven.

The coal industry's *opposition to strip-mining legislation* continued through most of the 1970s. The National Coal Association (NCA) fought land reclamation standards under consideration by Congress from 1972 to 1977, and financed intense lobbying campaigns in 1974 and 1975. The NCA claimed that the added cost of reclamation would slow the expansion of coal production to meet the energy crisis. The bill passed by Congress in 1974 received a pocket veto from President Ford too late to be overridden; passed and vetoed again in 1975, there were too few votes to override. But western states were becoming alarmed about the environmental destruction, water scarcity, and human impacts that expanded strip-mining would bring. The bill finally enacted in 1977 imposes a reclamation tax, sets moderately strict reclamation standards, and requires that highwalls at the edges of stripped areas be back-filled.[17] The NCA's lobbying efforts then were directed against proposed regulations to require the installation of scrubbers to reduce the sulfur dioxide emissions from coal-burning plants (see Chapter 7).

A *coalition of private interest groups* can form a formidable political force. The alliance of auto-related interests opposing public transportation already has been mentioned. Another convergence of interests has effectively opposed reusable bottle ("ban the can") legislation. Steel and aluminum companies, can and bottle manufacturers, beer and soft drink industries, retailing and marketing firms, and several unions all have a stake in perpetuating throwaway beverage containers–which environmentalists oppose because they use 2 percent of all industrial energy and add to litter and solid waste. From 1970 to 1977, seven state and local referenda failed, following intensive lobbying by the beverage interests.[18] By 1979, however, throwaway beverage containers had been banned in seven states. Similarly, saccharin manufacturers and the soft drink industry combined forces to oppose the proposed ban on saccharin (see Chapter 8). The utilities joined the coal companies in opposing clean air regulations. In the next chapter we will analyze the role of the auto industry and unions in attacking auto emission standards.

Citizens and Public Interest Groups

In the face of such powerful private interest groups, citizens concerned about environmental and human values can do little as individuals; but often they can do a lot if they are organized. In the DDT case, public opposition was mobilized first at the state level in New York, Wisconsin, and Michigan. There was press coverage of protests against massive DDT-spraying programs on Long Island. In Wisconsin there were six months of hearings that received extensive coverage by the media. The hearing examiner concluded that DDT is harmful, and the state assembly voted 90 to 0 to ban it in all but emergency use. Seven states banned DDT before there was federal action. As is frequently the case, public opinion on a national issue could be mobilized when it was related to local issues.[19]

The *mobilization of citizens* can occur in a variety of ways. A dramatic event can focus public attention on a continuing issue. Mercury poisoning in Japan, the London smog that killed 4,000 people in 1952, the oil spill off Santa Barbara in 1969, and the Three Mile Island nuclear accident in 1979 all received extensive media coverage and aroused widespread public concern. Initially there may be only a small group of persons committed to a cause, but a broader movement can be formed if there is sufficient public response. In this way successive environmental questions have become political issues in election campaigns and legislative debates.[20] The formation of coalitions can broaden the base of popular support. Environmental and consumer groups, for example, have worked together on such issues as the regulation of chemicals in the environment.

The *environmental organizations* have provided channels for many of these forms of voluntary citizen action. Some organizations work primarily through the legal system—for example, the Environmental Defense Fund, with a staff of lawyers assisted by scientists and other experts willing to testify in court. Other groups, such as Environmental Action, concentrate on legislative lobbying. Still others work mainly with the public in programs of education and publication, including the Wilderness Society, the National Audubon Society, and the National Wildlife Federation (the largest, with 3.5 million members). Some, such as the Sierra Club (160,000 members), combine a broad spectrum of legal, lobbying, and educational activities. The student-financed Public Interest Research Groups combine education and lobbying at the state level. In addition, temporary coalitions are formed around such specific issues as the SST and the Alaska pipeline. Thousands of local groups are organized around local issues and engage in educational activities, community projects, and protest demonstrations.[21]

Political strategies and styles of activity also have varied widely. The rallies, demonstrations, and protests of the civil rights and antiwar movements of the 1960s were continued in the environmental movement of the early 1970s. By 1975, the more traditional forms of education, lobbying, and litigation were more common. But confrontational politics, exemplified by demonstrations at the nuclear plant site in Seabrook, New Hampshire in 1977 and 1978, continue among those who feel that their viewpoint is not represented in existing decision-making processes. In an earlier chapter, I noted the sense of powerlessness and alienation that a technological society engenders. Impersonal aggregations of power, whether in industry or government, seem to acquire a momentum toward organizational goals over which citizens feel they have little control. Public hearings are confined all too often to details of policy implementation, with little opportunity to question fundamental assumptions. Those who try to influence decisions through advisory committees often find themselves co-opted by the bureaucratic framework. Confrontational politics are motivated by the desire to regain control over these basic decisions, as well as by the hope of mobilizing public support for strongly held beliefs that run counter to prevailing institutional assumptions.

Public Participation

I would suggest, however, that opportunities for citizens to participate directly in decisions by *administrative agencies* (rather than by working through legislatures) have improved, and further improvements are possible. The National Environmental Policy Act, the Freedom of Information Act (and its 1974 amendments), the Administrative Procedure Act, and the Advisory Committee Act all allow citizens to obtain a vast array of records, reports, transcripts, and minutes that otherwise would be inaccessible in agency files. NEPA also requires that a draft environmental impact statement (EIS) must be circulated to citizens' organizations for comments, which are to be included in the final EIS. Many legislative acts and agency regulations require public hearings in connection with rule making, standard setting, the issuing of licenses and permits, and other agency actions.

Public participation in agency decisions runs the gamut from empty formalities, through varying degrees of consultation and interaction, to the delegation of at least limited decision-making power to participants.[22] As an example of participation with virtually no influence on decisions, consider the *public hearings on nuclear plant sites* that were held by the AEC. The hearings were held after the AEC had approved the preliminary safety analysis report; government and industry were by then allies against citizens who raised questions about safety or environmental impact, and the expertise and power were all on their side. According to one study, the hearings "gave only lip service to citizen participation . . . agency arrogance and stacked-deck proceedings were typical."[23] The biases of the agency and the industry were prominent in the EIS reports of the AEC, according to this study. The procedural rules which hinder effective public participation have continued under the Nuclear Regulatory Commision, as we shall see.

It has been proposed that such public hearings should be conducted by *independent hearing officers*, rather than by agency staff members whose objectivity might be questioned. Such officers could make their own reports, along with a verbatim transcript of the hearings. Hearings also would be more helpful early in the planning while options were really open. Because the balance of technical expertise and financial resources has been so one-sided, there have been several proposals for providing nonprofit voluntary organizations with better technical support (for witnesses at public hearings, preparation of critiques of draft EIS, public interest representation on advisory committees, and so forth).[24]

Workshops have been used effectively by some agencies in the planning of local or regional projects. Attempts were made to obtain a cross section of community viewpoints by inviting persons with diverse interests. In some cases the group would meet for a weekend or two early in the planning process to have some voice in the determination of objectives and the exploration of alternatives, and then meet again several times during the development of specific plans.

The participants became quite well-informed on the issues, and took part in working sessions with the project staff. In other cases, community organizations were encouraged to submit alternative plans and comment on early drafts of planning documents. Such community involvement was used with some success in a Susquehanna River Basin project.[25] The Corps of Engineers has conducted experiments with "fishbowl planning" and "iterative open planning," involving extensive public participation and repeated revision of plans. Citizen involvement is advocated in the Corps' new planning guidelines.[26]

Citizen participation in *the management of public lands and forests* also has been tried. In the past, a common goal was "harmonious multiple use," and it was assumed that the achievement of this goal was a matter of expert management. But now it is evident that uses often conflict. Part of the agency's task, consequently, is conflict management, which can be furthered by open planning, public involvement, and the weighing of trade-offs. But it must be acknowledged that citizen participation frequently leads to confrontation and polarization, and that even a partial consensus is reached only with patience, time, and real openness. Yet the integration of input from citizens, specialists, and agency officials sometimes has been effectively achieved in face-to-face interaction in such local and regional projects.[27]

Finally, let us note some of *the objections* that have been raised against most of these types of public involvement. Many administrators still see citizen action as a threat to orderly government, an interference with their work as duly appointed officials. They believe that an agency acts in the public interest when it follows the policies and regulations established by democratically elected legislative bodies; if citizens want to express their opinions, let them write their representatives, or vote for a candidate with more acceptable views at the next election. I would reply that working through elected representatives is indeed central in any democratic government, but there also are other channels through which citizenship should be expressed. Active participation at a variety of levels contributes to the vitality of public debate and self-government. It often has been the vigilance of citizens that has increased the accountability of elected and appointed officials who were devoted to narrowly conceived missions or subject to pressures from economic interests. As one study of citizen involvement in environmental decisions concludes, "Voluntary citizen action has been a positive and constructive force for more responsible government."[28]

Another objection is that the citizens who take part in environmental issues are *not representative of the views of the public*. Members of environmental groups are predominantly middle-class, well-educated professionals, as pointed out earlier. The witnesses at public hearings usually are either spokesmen for economic interests, or a small, self-selected group of highly motivated, articulate citizens. The vast majority feel too uninformed to participate, or simply do not want to get involved—even though their welfare may be affected.[29] There is considerable validity to this objection. My reply would be that citizen activists do not represent everyone, but they do speak for some people and some values that otherwise might be neglected. An additional answer is that greater effort should be made to seek out participants from various segments of the population affected by a decision.

One further objection is the *delay and obstructionism* to which public involvement so often leads. This is particularly evident in protests over the siting of airports, power plants, and other large facilities. A small group of local citizens and dedicated activists can block a project from which a much wider public might benefit. No major new airport has been started since the 1960s, partly because of citizen protests. Airport siting was once seen as a technical decision for experts. Today public hearings are required, and local residents can protest the increasing liabilities (especially noise) that they would have to bear.[30] Perhaps obstructionism would be reduced if there were mechanisms whereby the beneficiaries of siting decisions were required to provide more substantial compensation to those adversely affected. Delays also might be reduced if public participation occurred early in the exploration of alternatives, rather than in opposition to already formulated plans. People are more likely to accept a decision made through a process that they consider legitimate.

The *effectiveness of government* can indeed be jeopardized when there is extensive citizen participation in a fragmented society. During the 1960s, diverse population segments (such as blacks, women, youth, and environmentalists) were more self-conscious, more active politically, and placed greater demands on the government. By the 1970s, more and more groups were pressing conflicting claims on the overloaded federal budget, creating what some authors have called "a crisis of democracy."[31] Special interest organizations and protest movements, unwilling to compromise for common goals, often exercised a virtual veto power that led to stalemate and impasse. Following Vietnam, Watergate, and rampant inflation, the public had less confidence in political leaders, and the authority of the presidency declined. Political parties, which formerly served to aggregate diverse interests and to build coalitions, were themselves fragmented. In the absence of social consensus or agreement on fair principles to adjudicate competing claims, Congress avoided hard decisions about the allocation of limited resources. In 1978 and 1979, intense conflicts between regions and between various economic interests prevented the adoption of effective energy legislation.

Democracy does indeed face serious challenges in times of social conflict. The answer, I suggest, lies in the development of *common social purposes*, not in the reduction of participation or increased reliance on government authority and technical expertise. Only when there are common loyalties to more inclusive social goals are people willing to compromise private ends in the public interest, as occurs during time of war or threats to the nation. As the United States finally awakens to the reality of resource limits, it is possible that greater consensus on priorities will develop, along with more agreement on principles of fair allocation. The value changes explored in Part Three would encourage commitment to such common goals. But in the face of scarce resources, the conflicts among social groups are likely to increase. In subsequent chapters we will return to these issues: the trade-offs between administrative efficiency and pluralistic participation, and the search for procedures for integrating expert judgment, representative government, and citizen involvement.

III. THE ROLE OF TECHNICAL EXPERTS

In most policy decisions about technology and the environment there are technical questions that legislators and citizens are incompetent to answer. Yet policy choices involve value judgments that should not be left in the hands of experts alone. How can technical experts contribute most effectively to policy choice? How can decisions be made democratically in a technological society? Are democracy and expertise compatible? What should be done when experts disagree?

The Scientist as Adviser and Witness

There are various positions inside and outside governmental institutions from which scientists (along with engineers and other technical experts) can take part in policy decisions.[32] At one extreme are the "insiders" who are full-time staff members of federal agencies or (in a few cases) congressional committees or the White House. Staff members are part of a team to which they are expected to be loyal and with which they enjoy confidential relationships. When advisory panels hold closed meetings and issue confidential reports, they have a primarily "inside" role. The insiders can have considerable influence when they are trusted and share the assumptions of an administration. They are less effective in opposing administration positions or challenging basic assumptions. The PSAC report and the interagency review that opposed the SST were kept secret during the debate and ignored by Nixon. A summary of the interagency panel's conclusions, given by the Undersecretary of Transportation, was so misleading that panel members finally protested. But it was mainly outsiders rather than insiders who introduced important new issues into the public debate.

Consider next the scientists on *government advisory panels*. Political leaders and agency officials often have used advisers to legitimate decisions already made on political grounds. They wanted not advice but ammunition to defend their positions. Agency officials can choose experts favorable to their own view, loading the panel membership to secure the desired outcome. A consulting firm preparing a report is inclined to tell the contracting agency what it wants to hear, and thereby encourage future contracts. The National Academy of Sciences (NAS) is autonomous, but members of its review panels often have had close associations with the agencies that commissioned the reviews. Many of the NAS panel on DDT, for instance, were pesticide enthusiasts who had previous ties to USDA or pesticide manufacturers. In the SST case, officials from the agency that paid for the NAS study took part in the selection of panel members and the drafting of the final report.[33]

The impartiality of an NAS panel is suspect when most of its members have had very close *associations with industries or agencies* promoting the technologies under scrutiny. All the members of a panel on radioactive wastes, for instance, had previous ties to the AEC. The chairman of a committee on dog and cat food standards was an official of Ralston Purina, a major pet food producer. The Aeronautics and Space Engineering Board was dominated by aerospace executives and government officials.[34] But NAS recently has made efforts to secure greater diversity of viewpoints and disciplines. A "disinterested" panel, of course, would be an ignorant one, but a greater variety of interests and at least some representation of more independent "outsiders" is now sought. There also are new provisions for review committees, drawn from experts in other fields, to scrutinize early drafts of reports, looking for biases and self-serving conclusions. More diverse criteria are being employed now, as we will see in the 1979 NAS energy study.

The public, understandably, is confused by *disagreement among experts* on advisory panels, at legislative or regulatory hearings and court cases, or in public policy statements. One group of Nobel laureates was strongly opposed to the breeder reactor, but an equally eminent group put out a statement in favor of it. Scientists can be found on opposite sides of almost every controversial environmental or technological issue. How can qualified experts disagree?

1. Uncertainties in Data

Many recent controversies have been about new technologies with which there had been little previous experience, and environmental phenomena of which there was very limited knowledge. Little was known about several of the relationships linking the SST to ozone depletion, climate changes, ultraviolet radiation, and skin cancer—and estimates varied widely. In that case the uncertainties could be reduced by further research, but in other cases this is not possible. There is no possibility for the direct testing of nuclear reactor accidents, and tests on separate components may neglect interactive or very improbable failure modes—to say nothing of human errors. In discussing risk-benefit analysis (Chapter 8), we will examine the particular difficulties that arise in studying low-probability events and low-level exposure to radiation or chemicals.[35]

2. Formulation of Issues

The way the issues are defined in the context of policy debate will determine what data are relevant. The time-scale employed may emphasize short-term or long-term effects. The boundaries of analysis in space and time, and the types of impacts considered, will affect one's conclusions. In the debate over nuclear energy, the issues are so complex that very diverse kinds of questions can be raised. In such disputes the central issues for one side may be peripheral or irrelevant for the other, so that the opponents talk right past each other. The most important decision may be the way a problem is conceptualized and bounded.

3. Value Judgments

In the 1960s scientists were prominent as advisers in areas such as space and defense in which the goals were clear, the problems were mainly technical, and distributional questions did not arise. Means could be debated separately from ends. But in the 1970s scientists more often were involved in choices in which there were conflicting goals, major social problems, and issues of distributional equity.[36] When options have multiple objectives and multiple consequences, the separation of means and ends is more difficult. The weighing of incommensurable costs and benefits involves trade-offs among diverse values. Judgments of "acceptable risk" are never purely scientific. The SST debate involved priorities among very disparate impacts: employment, the balance of payments, national prestige, passenger convenience, noise, and possible environmental damage. "Policy-relevant studies and advice can never be value-free, even when carried out by scientists and engineers."[37]

4. Professional Biases

In predicting the consequences of alternative policies, experts use conceptual models that selectively represent those causal relationships they consider most significant. Specialists tend to think that the variables with which they can deal are the most important ones; yet policy issues are inescapably interdisciplinary. Disciplinary training also may encourage certain general attitudes toward problems. An engineer and an ecologist are likely to bring divergent assumptions to the appraisal of technical fixes for environmental problems. People who have devoted most of their lives to a particular technology are predisposed to promote it. The presence of such professional biases is one of the reasons for insisting that technology assessment teams (Chapter 9) should be genuinely interdisciplinary.

5. Institutional Biases

At one point during the congressional hearings on the antiballistic missile (ABM), it was pointed out to Secretary of Defense Packard that almost all of the scientists he had consulted had been Pentagon employees. He replied: "I do not consider that when you are involved with scientific matters it is important whether you have people outside the Defense Department or not. Scientists, to me, are objective about such matters."[38] But a subsequent study casts doubt on the claim of objectivity. Of the scientists who took active pro-ABM stands in the debate, 79 percent had received research funds from the Defense Department, whereas only 18 percent of the anti-ABM scientists had.[39] As Jeremy Stone puts it, "Where you stand depends on where you sit."[40] Another author analyzes the influence of institutional perspectives in the outlook of scientists and concludes: "Research on science advice has set aside the myth of objectivity."[41]

John Kenneth Galbraith has shown that most people either agree with the goals of their employer when they take a job, or adjust their goals to those of the organization for which they work.[42] As mentioned earlier, every institution directs its activities to the perpetuation of its own power and to a narrow range of objectives or missions. Information is selectively filtered—deliberately or unconsciously—as it is transmitted within organizations. People tend to tell their superiors what they want to hear as was evident in the misleading "body counts" in the Vietnam War, from the platoon level to the Secretary of Defense. Institutional assumptions influence perceptions as well as value judgments.

6. Conflicts of Interest

Occasionally witnesses or advisers have direct financial interests in decisions on which they are testifying. More commonly they stand to benefit personally in indirect ways, including future employment opportunities, grants, and contracts. A university scientist testifying on the NASA budget said, "I have no vested interest in the space program." Later, asked if he had any government contracts, he replied, "Oh yes, indeed, I do receive funds from NASA for part of my work."[43] Few people are impartial when their own advantage is at stake. Daniel Greenberg proposes a first law for science policy advice: "Never ask a barber whether you need a haircut."

How can these problems in conflicting expertise be avoided? The last is the easiest to confront. Advisers to government agencies are required to disclose any potential *conflicts of interest*, and a similar declaration is called for in some hearings. It is important for witnesses to recognize honestly their own *professional* and *institutional biases*; unconscious biases are more likely to result in distortions of evidence than those openly acknowledged. Moreover, in selecting witnesses or advisory board members, diversity in professional and institutional affiliation should be sought. No panel of experts can provide a purely objective analysis, but people who are less directly involved in the outcome of decisions generally will have a more independent viewpoint.[44]

Scientists have no greater wisdom than other people concerning conflicting *value judgments*; these should be resolved primarily through democratic political processes. Scientists offering testimony inevitably will express their own value commitments, but should try to make them explicit and distinguish them from scientific evidence.[45] Similarly, the *assumptions* used in the interpretation of data should be made clear and the presence of *uncertainties* acknowledged. The credibility of scientific witnesses is suspect when they claim a greater certainty than the often-ambiguous evidence warrants. Scientific integrity requires both careful documentation and appropriate tentativeness in stating conclusions.

Adversary Procedures

Adversary procedures can be useful in bringing uncertainties, assumptions, value judgments, and hidden biases to light. They can help to sort out points of agreement and disagreement. In the courts there is provision for cross-examination of witnesses, within accepted rules of legal procedure, which often have helped to clarify controversial policy issues. There have been proposals for a "science court" in which scientist-judges would weigh the evidence presented by scientist-advocates and give impartial rulings on the scientific facts alone. In theory this would allow "the separation of scientific fact from value-laden issues."[46] I am dubious about the proposal because the crucial questions are so often the conceptualization of the problem and the definition of its scope; these will determine which facts are relevant to a policy choice. The really controversial disagreements between scientists almost always reflect political and social assumptions that should be debated in the context of political institutions.[47]

One way to do this would be to extend opportunities for cross-examination. In *congressional hearings* there has been no opportunity for witnesses to respond to each other and no direct confrontation of opposing views. Skillful questions by members of Congress or committee staff occasionally are fruitful in getting at the reasons behind disagreements, but usually witnesses talk past each other and their assumptions remain unexamined. One study of such hearings concludes: "The best way of keeping scientists honest and relevant would be to have witnesses confront peers of equal competence."[48] A panel of technical consultants could be hired for the duration of certain hearings to submit questions on disputed issues to the witnesses and receive written answers for the record.

The effectiveness of an adversary format has been demonstrated in a number of *regulatory agency hearings*. An interesting example occurred in 1972 when Henry Kendall, an MIT physicist, and other intervenors were able to challenge the AEC's nuclear reactor safety standards. The issue was the emergency core cooling system, designed to spray water on the reactor core of a nuclear power plant in case the normal water circulation system fails (as occurred subsequently in the Three Mile Island accident). No full-scale working test of the system had ever been made, and critics questioned several key assumptions in the AEC's theoritical calculations of its operation under emergency conditions. The intervenors presented an 80-page critique of AEC standards for the system. Under the Freedom of Information Act they obtained AEC documents showing considerable uncertainty about its effectiveness, as well as evidence of intimidation of dissenting employees. The intervenors' testimony stood up well to two weeks of cross-examination by the AEC team, whereas the AEC experts could provide little evidence to support some of their assumptions. The hearings led to modifications in the safety standards and the appointment of a new safety director.[49]

There are a number of problems and limitations in the adversary format, but they do not appear to be insuperable. *Unequal access to technical expertise* is typical of recent controversies. Industry and government agencies have their own in-house experts and vast financial resources to hire consultants to provide the witnesses and the technical analysis to support the positions they favor. The opposing side often has to rely on volunteers testifying "on their own time," with limited research support from hard-pressed environmental or consumer groups. However, the technical base of citizens' groups has been improving and has received limited support from several government programs.[50]

A more serious shortcoming of adversary procedures is their tendency to foster a premature *polarization of viewpoints*. When a specific legislative or regulative proposal is being debated, it is natural for testimony to be bipolar, for and against. But when policy is being formulated, or when a proposal is being amended, there are seldom just two opposing sides that should be represented. In adversary situations it also is tempting to present extreme positions and exaggerated or one-sided claims in order to counterbalance the opposition's claims. Uncertainties are neglected, ad hominem arguments are used, persuasive rhetoric and attempts to discredit the opposition's integrity may replace careful analysis of issues. These dangers can be reduced by provision for cross-examination and by careful staff planning to avoid oversimplified pro-and-con confrontations, but they cannot be totally eliminated.[51]

Despite these limitations, the introduction of structured adversary procedures would contribute to the *informed debate* that is central in a democracy. Lakoff has maintained that technical issues arise in so many current decisions that "the participation of scientists as adversaries in a variety of political processes could prove essential to the survival of government by discussion."[52] He proposes that the Office of Technology Assessment could play a role in helping congressional committees organize hearings that would illuminate the technical and political issues at stake in policy choice, and could provide funds for the participation of more scientists from outside the government. A series of standing national commissions also has been proposed as a forum for discussion of long-term policy issues with technical components. The goal would not be the resolution of such disputes, but the provision of "the diversity of opinion that would foster informed public debate on genuinely controversial issues."[53] Experimentation with a variety of adversary formats might indicate their appropriateness for various kinds of controversy.

Scientists and the Public

We have seen a number of cases in which the warnings of inside scientists were ignored, but outside scientists were able to bring neglected risks to the attention of the public. Recall Rachel Carson in the DDT case, or Shurcliff's efforts against the SST. Many such independent scientists have contributed to public debate on policy issues related to science—through courts, congressional or agency hearings, state and local government, or the public media. Joel Primack and Frank von Hippel have examined a number of such case histories and conclude that the effectiveness of these scientists depended on four factors: the timeliness and local impact of the issues, an appropriate forum (such as a court or hearing), the credibility of the scientists (including the documentation they supplied) and news media coverage.[54] Independent scientists have enhanced significantly the public debate on national issues in which speakers for industry or government would otherwise have had a virtual monopoly of scientific expertise[55]

Scientists also have made *efforts to inform the public* of the social implications of their own fields by writing articles and giving talks. Others have been active in local citizens' organizations and environmental groups. At the national level, the Federation of American Scientists has lobbied for legislation on the environment and the control of nuclear weapons. Several organizations and periodicals have tried to combine information and advocacy. The *Bulletin of the Atomic Scientists* deals with a broad range of issues on science and public policy. In the early 1960s the St. Louis Committee on Nuclear Information publicized studies on radioactive fallout from nuclear testing, and later did research and publicity on other threats to health and the environment. From these local efforts a national organization was born, the Scientists' Institute for Public Information, and a journal, *Environment*. Such journals have published extensive scientific information, but also have provided a forum for policy debate on issues in which technical and political aspects are intertwined.

The scientific *professional societies*, whose activities traditionally have been limited to the communication of knowledge and the advance of the profession, have begun to offer some encouragement for scientists to take part in public policy issues. The American Physical Society (APS), for example, sponsors a Newsletter and a Forum on Physics and Society, in which current policy questions are discussed. It finances congressional fellowships that allow physicists to spend a year on the staff of members of Congress or congressional committees. It also gives an annual Award for Physics in the Public Interest, acknowledging that such activities are forms of professional responsibility and deserve recognition. The APS also has prepared reports on such policy issues as reactor safety and energy efficiency. Many scientific and engineering societies have started similar programs.[56] The American Association for the Advancement of Science has devoted a substantial fraction of its annual meetings and its weekly journal, *Science*, to problems in the relation of science and public policy.

Scientists and engineers working in industry and government are under more severe constraints than those in universities in *speaking out on controversial policy issues*. Whereas the work of academic scientists is judged mainly by the scientific community, industrial and governmental employees are more dependent on the judgment of their immediate superiors. They are part of a team and serve the purposes of an organization as well as those of a profession. They have multiple loyalties to their employers, their professions, and the public. They will hestitate to speak out when the interests of their employers and those of the public conflict. In a survey of 1,100 members of one engineering society, 38% said they felt restrained from writing letters to the editor, 48% from taking part in political affairs, and 44% from criticizing their employers' activities or products.[57]

In the course of work, an industrial or governmental scientist or engineer occasionally may become aware of *a serious potential hazard* to people or to the environment. If all channels for action within the organization have been exhausted, he or she may feel that the only way to protect the interests of society is to go public, calling the danger to the attention of regulatory agencies, legislative committees, or news media.[58] For example, three engineers working on plans for the Bay Area Rapid Transit system in California pointed out defects in its computer control design. They were fired for speaking out—though later the defects proved to be so serious that the control system had to be abandoned.[59]

In another case, a number of industrial scientists were aware of evidence that vinyl chloride is cancer producing; they kept quiet though thousands of workers were exposed to dangerous concentrations of the chemical.[60]

Scientists and engineers in government and industry who *"blow the whistle"* on their employers are very vulnerable to retaliation. Doctors and lawyers have considerable autonomy and professional independence, partly because they have many clients. Scientists in universities have some protection—through tenure regulations and the traditions of academic freedom—when they take controversial stands. Even industrial workers have some support from their unions against arbitrary dismissal. In the case of scientists and engineers in government and industry who have none of these safeguards, the professional societies could offer some protection by instituting procedures for airing grievances and holding hearings. A few professional societies have provided legal assistance to members who have been fired or transferred for speaking out in the public interest.[61] These are, of course, exceptional cases; ordinarily scientists or engineers will participate as citizens in the processes of public decision making, or try to influence from within the policies of the organizations for which they work.

My own conclusion is that *democratic political processes* can work in environmental policy decisions, despite the severe problems introduced by 1) the concentration of economic and political power to which technology contributes, and 2) the presence of technical questions on which the opinions of experts must be sought. Input from citizens, individually and through public interest organizations, has occurred in all branches of government—judicial, administrative, and above all, legislative—and has helped to bring about significant changes in environmental policy during the 1970s. I have suggested the importance of continuing efforts to make public debate more informed, legislatures more responsive, and administrators more accountable. Outside advisers, adversary procedures, and efforts by scientists to inform the public can help to resolve the tensions between democracy and expertise.

By the end of the 1970s, however, there were *reactions against citizen participation* in judicial and administrative processes. There had been many projects in which opposition by small minorities had caused lengthy delays. Well-organized single-interest groups exercised a virtual veto power over some programs. As energy production began to assume higher priority than environmental quality, there were moves to "cut through the red tape" and streamline permit and license procedures. As inflation continued, the taxpayers' revolt against government spending threatened environmental programs, and some people feared that the economic costs of pollution control would hinder economic growth. Conservative voices urged less governmental regulation and greater reliance on market forces. These trends are traced in subsequent chapters.

There is an even deeper question: Can democracy cope with *resource scarcities*? Political as well as economic processes tend to discount the future. Elected representatives seek measures that pay off before the next election, and citizens vote in the light of short-term benefits, deferring costs to the future. In the face of scarcities, people seem to put material values ahead of social and environmental values. In Part Three it will be suggested that the transition to a just, participatory, and sustainable society will require significant changes in our basic values.

In the final analysis, it is citizens who must defend the environmental and human values that are the concern of this volume. Industrial and bureaucratic interests wield immense power in the political arena and have great advantages in technical expertise, but within a democratic society an informed and aroused electorate can have the last word. But the citizenry can do little if it is uninformed or apathetic—or deeply divided over conflicting values.

NOTES

1. Sanford Lakoff, "Congress and National Science," *Pol. Sci. Quarterly* 89 (1974): 589-611; and Clarence and Barbara Davies, *The Politics of Pollution*, 2d ed. (Indianapolis: Pegasus, 1975), chap 3.

2. Robert Bernstein and William Anthony, "The ABM Issue in the Senate," *Amer. Political Science Rev.* 68 (1974): 1198-206.

3. Cynthia Enloe, *The Politics of Pollution in Comparative Perspective* (New York: David McKay, 1975), chap. 3; and Walter Rosenbaum, *The Politics of Environmental Concern*, 2d ed. (New York: Praeger, 1977), chap. 4.

4. Rosenbaum, op. cit., chap. 5; and Richard Andrews, *Environmental Policy and Administrative Change* (Lexington, Mass.: Lexington, 1976), chap. 4.

5. Paul Brodeur, *The Expendable Americans* (New York: Viking, 1974).

6. Davies, op. cit., chap. 5; and Charles Jones, *Clean Air* (Pittsburgh: University of Pittsburgh Press, 1975). For other references, see Chapter 7.

7. Davies, op. cit., chap. 6; and Bill Shaw, *Environmental Law* (St. Paul, Minn.: West, 1976).

8. Frederick Anderson, *NEPA and the Courts* (Baltimore: Johns Hopkins University Press, 1973); and *Sierra Club* v. *Morton*, 405 U.S. 727 (1972).

9. Frank Graham, Jr., *Since Silent Spring* (New York: Fawcett World, 1970); and Joel Primack and Frank von Hippel, *Advice and Dissent* (New York: Basic Books, 1974), chaps. 3 and 10.

10. Primack and von Hippel, op. cit., chaps. 2 and 4; Ian Clark, "Expert Advice in the Controversy about Supersonic Transport in the U.S.," *Minerva* 12 (1974); 416-32; Luther Carter, "SST: Commercial Race or Technology Experiment?" *Science* 169 (1970); 352-55; and William Shurcliff, *SST/Sonic Boom Hand Book* (New York: Ballantine, 1970).

11. Thomas Means, "The Concorde Calculus," *George Washington Law Rev.* 45 (1977): 1037-65; and Mary Ames, *Outcome Uncertain: Science and the Political Process* (Washington, D.C.: Communications Press, 1978), chap. 2.

12. Council on Environmental Quality, *Environmental Quality 1978* (Washington, D.C.: CEQ, 1978), 428 and 437.

13. Cf. Davies, op. cit., chap. 4; and Rosenbaum, op. cit., chap. 5.

14. See notes 9, 10, and 11.

15. "Strip Mining," *Congressional Quarterly*, February 1, 1973, p. 33.

16. Harry M. Candill, *My Land is Dying* (New York: Dutton, 1971); and David Davis, *Energy Politics*, 2d ed. (New York: St. Martin's Press, 1978).

17. "Strip Mining," *Congressional Quarterly*, July, 1977, 1495-1500.

18. Rosenbaum, op. cit., chap. 9.

19. Primack and von Hippel, op. cit., chap. 10.

20. Enloe, op. cit., chap. 2.

21. Odum Fanning, *Man and His Environment: Citizen Action* (New York: Harper & Row, 1975); and Lynton Caldwell, Lynton Hayes, and Isabel MacWhirter, *Citizens and*

the Environment: Case Studies in Popular Action (Bloomington: University of Indiana Press, 1976).

22. Sherry Arnstein, "A Ladder of Citizen Participation," *J. Amer. Inst. of Planners*, (July 1969):216.

23. Steven Ebbin and Raphael Kasper, *Citizen Groups and the Nuclear Power Controversy* (Cambridge, Mass.: MIT Press, 1974), chap. 1.; and Dorothy Nelkin and Susan Fallows, "The Evolution of the Nuclear Debate: The Role of Public Participation," *Annual Review of Energy* 3 (1978): 274-312.

24. Ibid., chap. 5; and M. Rupert Cutler and Daniel Bronstein, "Public Involvement in Government Decisions," *Alternatives: Perspectives on Society and Environment* 4 (1974): 11-13.

25. Thomas Heberlein, "Some Observations on Alternative Mechanisms for Public Involvement," *Natural Resources J.* 16 (1976): 197-212.

26. L. Ortolano, "A Process for Federal Water Planning at the Field Level," *Water Resources Bull.* 10 (1974): 766-78; T.P. Wagner and L. Ortolano, "Testing an Iterative Open Process for Water Resources Planning" (Ft. Belvoir, Va.: U.S. Army Engineers Institute for Water Resources, 1976); and "Planning Process: Multiobjective Planning Framework" (Washington D.C.: Office of the Chief of Engineers, U.S. Army, 1975).

27. Lloyd Irland, "Citizen Participation: A Tool for Conflict Management on the Public Lands," *Public Admin. Rev.* 35 (1975): 263-69. See articles by Norman Wengert and by W. Derrick Sewell and Timothy O'Rirodan in the "Symposium on Public Participation in Resource Planning," *Natural Resources J.* 16 (1976): 1-236.

28. Caldwell et al., op. cit., p. xxix.

29. William Burch, "Who Participates: A Sociological Interpretation of Natural Resources Decisions," *Natural Resources J.* 16 (1976): 41-54; and James A. Riedel, "Citizen Participation: Myths and Realities," *Public Admin. Rev.* 32 (1972): 211-19.

30. Jerome Milch, "Feasible and Prudent Alternatives: Airport Development in the Age of Public Protest," *Public Policy* 24 (1976): 81-109.

31. Daniel Bell, *The Cultural Contradictions of Capitalism* (New York: Basic Books, 1976), chap. 6; Michel Crozier et al., *The Crisis of Democracy* (New York: New York University Press, 1975), chaps. 3 and 5; and Theodore Lowi, *The End of Liberalism*, 2d ed. (New York: Norton, 1979).

32. Harvey Brooks, *The Government of Science* (Cambridge, Mass.: MIT Press, 1968); and Daniel Greenberg, *The Politics of Pure Science*, rev. ed. (New York: New American Library, 1971).

33. Primack and von Hippel, op. cit., p. 50.

34. Philip Boffey, *The Brain Bank of America* (New York: McGraw-Hill, 1975); and Hugh Folk, "The Role of Technology in Public Policy," in *Technology and Man's Future*, 2d ed., ed. Albert Teich, (New York: St. Martin's Press, 1977).

35. Alvin Weinberg, "Science and Trans-Science," *Minerva* 10 (1972): 209-22; reply by Harvey Brooks, *Minerva* 10 (1972): 484-86; and Allan Mazur, "Disputes between Experts," *Minerva* 11 (1973): 243-62.

36. Dean Schooler, *Science, Scientists, and Public Policy* (New York: Free Press, 1971); and Dorothy Nelkin, "The Political Impact of Technical Expertise," *Social Studies of Science* 5 (1975); 35-54.

37. Eugene Skolnikoff and Harvey Brooks, "Science Advice in the White House?" *Science* 187 (1975): 41; see also Harvey Brooks, "Expertise and Politics—Problems and Tensions," *Proc. Amer. Philosophical Society* 119 (1975): 257-61.

38. *Strategic and Foreign Policy Implications of ABM Systems* (Subcommettee on International Organization and Disarmament Affairs, Committee on Foreign Relations, U.S. Senate, 1969), part I, p. 308.

39. Anne Cahn, "American Scientists and the ABM," in *Scientists and Public Affairs*, ed. Albert Teich (Cambridge, Mass.: MIT Press, 1973): 86.

40. Cited in ibid., p. 112.

41. Albert Teich, "Objectivity and Advocacy in Scientific Advice," (paper given to the American Association for the Advancement of Science, February 22, 1976).

42. John Kenneth Galbraith, *The New Industrial State,* rev. ed. (Boston: Houghton Mifflin, 1972), chap. 12.

43. Amitai Etzioni, "When Scientists Testify," *Bull. Atomic Scientists* (October 1964): 24.

44. Skolnikoff and Brooks, op. cit.

45. Daniel D. McCracken, *Public Policy and the Expert: Ethical Problems of the Witness* (New York: Council on Religion and International Affairs, 1971), chap. 7.

46. Arthur Kantrowitz, "Controlling Technology Democratically," *American Scientist* 63 (1975): 505-09; idem, "The Science Court Experiment: Criticisms and Responses," *Bull. Atomic Scientist* 33 (April 1977): 44-50; and Task Force of Presidential Advisory Group on Anticipated Advances in Science and Technology, "The Science Court Experiment: An Interim Report," *Science* 193 (1976): 653-56.

47. Barry Casper, "Technology Policy and Democracy," *Science* 194 (1976): 29-35; Dorothy Nelkin, "Thoughts on the Proposed Science Court," *Newsletter on Science, Technology, and Human Values,* no. 18 (January 1977): 20-31; and Allan Mazur, "Science Courts," *Minerva* 15 (1977): 1-14.

48. Paul Doty, "Can Investigations Improve Scientific Advice? The Case of the ABM," *Minerva* 10 (1972): 280-94.

49. Primack and von Hippel, op. cit., chap. 15. See the series of articles on "Nuclear Reactor Safety" by Robert Gillette in *Science* 172 (1971): 918; 176 (1972): 498; 177 (1972): 771, 867, and 970; and 178 (1972): 482.

50. "Public Service Science Residencies and Internships" (National Science Foundation, 1977); "NSF Authorization Coming Up," *Science* 196 (1977): 1423; and Tersh Boasberg et al., *Implications of NSF Assistance to Nonprofit Citizen Organizations* (Washington, D.C.: National Science Foundation, 1977).

51. Mazur, "Disputes between Experts," op. cit.

52. Lakoff, op. cit.

53. Nelkin, op. cit., p. 28.

54. Primack and von Hippel, op. cit., chap. 16.

55. Articles by James Sullivan, and by Martin Perl et al., in *Physics Today* (June 1974): 23-37; and A. Robert Smith, "The New Scientist-Advocates," *Bull. Atomic Scientists* 31 (February 1975): 16-18.

56. Barry Casper, "Physics and Public Policy: The Forum and the APS," *Physics Today* (May 1974): 31-37; idem, "Scientists on the Hill," *Bull. Atomic Scientists* 33 (November 1977): 8-15; and *Scientists in the Public Interest: The Role of Professional Societies* (Alta, Utah: College of Engineering and Department of Physics, University of Utah, 1973).

57. James Olson, "Engineer Attitudes toward Profession, Employment and Social Responsibility," *Professional Engineer* (August 1972): 30-32.

58. Ralph Nader, Peter Petkas, and Kate Blackwell, eds., *Whistle Blowing* (New York: Bantam, 1972).

59. John Edsall, *Scientific Freedom and Responsibility* (Washington, D.C.: Amer. Assoc. for the Advancement of Science, 1975), pp. 37-39; and S.H. Unger, "The BART Case: Ethics and the Employed Engineer," *Newsletter*, no. 4 (September 1973), Committee on Social Implications of Technology (IEEE).

60. John Edsall, "Scientific Freedom and Responsibility," *Science* 188 (1975): 690.

61. Ibid. Stephen Unger, "Engineering Societies and the Responsible Engineer," *Annals of the N. Y. Academy of Sciences* 196 (1973): 433-37; and Committee on Scientific Freedom and Responsibility, *1978 Annual Report* (Amer. Assoc. for the Advancement of Science).

7
POLLUTION AND LAND USE

Of all the environmental policy decisions in which conflicting values are at stake, none has been more controversial than pollution and land-use legislation. Here all the political pressures and diverse interests discussed in Chapter 6 can be seen in operation. Section I deals with the politics of pollution and the setting of air and water standards by Congress and the Environmental Protection Agency (EPA). In Section II the environmental and human values relevant to pollution control decisions are taken up and some economic and ethical issues in alternative regulatory strategies are considered. The trade-offs among environmental quality, economic growth, distributive justice, and individual freedom also are analyzed. The final section looks at ethical questions in land-use policies, especially solid waste disposal, zoning ordinances, and the preservation of wilderness, wildlife, and endangered species.

I. AIR AND WATER POLLUTION

The high tide of public environmental concern in the period from 1970 to 1972 produced the Clean Air Amendments and the Clean Water Act. While considerable progress was made in reducing air and water pollution during the 1970s, a series of technological, economic, and political obstacles prevented the attainment of the goals set forth in this legislation. We will examine these difficulties in the cases of auto emissions, stationary sources, and water pollutants.

There are five *air pollutants* for which primary standards have been established: carbon monoxide and photochemical oxidants (mainly from autos); sulfur oxides and suspended particulates (from stationary sources); and nitrogen dioxide (about equally from autos and stationary sources). Carbon monoxide is directly emitted by autos, whereas photochemical oxidants (ozone and "smog") are products of hydrocarbon auto emissions and nitrogen oxides reacting in sunlight. The levels of these auto-related pollutants have improved steadily since 1973, but still violate standards in some cities. The effects of auto emission controls are evident, since levels would have been much higher without them. By the late 1970s, however, nitrogen oxides had shown no improvement. Sulfur oxides and particulate levels from stationary sources showed some improvement but violated standards in many areas, especially in the region from Ohio to the northeast.[1]

Auto Emissions

The 1970 Clean Air Amendments were adopted unanimously by Congress. For new autos, the legislation required a 90% reduction in hydrocarbon (HC) and carbon monoxide (CO) emissions by 1975, and in nitrogen oxides (NOx) by 1976. In deriving this figure, the Muskie subcommittee started from 1967 data for the highest ambient levels of these pollutants in any U.S. city (averaged over eight-hour periods), doubled them to allow for projected growth during two decades, and then calculated the percentage reduction that would be needed to reach the ambient levels below which no study had found *any adverse health effects.* Critics said that this calculation was based on the worst cities, that it assumed that all of these pollutants had come from autos, and that it implied that public health should be protected no matter what the cost. Defenders of the act replied that when health risks are uncertain but potentially serious, it is wise to make assumptions that are on the safe side.[2]

Congress believed that the auto industry previously had put little effort into *emmision-control technology,* and that tight standards and deadlines would force technological development. But achievement of the new standards proved more costly and difficult than anticipated, and the industry had many incentives for seeking delay. In *International Harvester* v. *Ruckelshaus* (1973), four auto companies sued EPA for refusing to extend deadlines; they insisted that necessary technology was not currently available. The Court of Appeals ruled that probable technological advances were a legitimate basis for setting standards, but that the EPA had not adequately shown that the standards could be met. Moreover, said the court, the costs to the economy and to workers would be very high if EPA's judgment were wrong and the industry had to shut down. The court remanded the case to EPA with the suggestion that it establish less demanding interim standards, which EPA did. Since the case involved technical factors, the court wanted to rule on procedural rather than substantive issues, and said the EPA's decision had not been supported by sufficient evidence. The judges became, in effect, umpires concerning a disputed decision, even though they stopped short of making a standard-setting decision themselves.[3]

A succession of other factors provided *reasons for further delays.* In 1974 it was the energy crisis that prompted Congress to extend the deadlines; exhaust controls were expected to reduce fuel efficiency by 5%. In 1975 another delay was granted because of fears (which turned out to be exaggerated) that catalytic converters might generate serious sulfate emmissions. In 1977 Congress extended for three more years the interim 1977 standards that set NOx emissions at 2.0 grams per mile rather than 0.4 gpm as originally projected.[4]

Catalytic converters are the simplest add-on devices for reducing emissions. The catalyst cuts down greatly on HC and CO; NOx is somewhat reduced by recirculating air. The catalytic converters available on 1975 models to meet the stricter standards in California achieved a 75% reduction in emissions. Volvo developed an almost pollution-free auto with 10% better gas mileage than its regular models; a three-way catalyst reduced pollutants well below the requirements of the 1970 act, but at considerable expense. An NAS study in 1975 concluded that there was no technical reason to delay further in enforcing the strict standards. Other studies emphasized that lighter cars would cut down on both emissions and fuel use; by the late 1970s Detroit finally was investing more heavily in small car production.[5]

Transportation controls are an effective way of limiting air pollution in urban areas, but they have been strongly resisted by cities and states. Even with 90% emission reduction, the ten largest cities would be unable to meet federal air quality standards without restricting auto use. But in 1974 Congress rejected an EPA plan that would have required states to control auto use in central areas (by parking restrictions, surcharges, vehicle-free zones, and so on). By 1977 EPA was proposing only voluntary strategies to reduce traffic, such as car pools and bus lanes. It also was urging a mandatory vehicle inspection program that would force owners to maintain motor tune-up; poorly tuned motors produce far more emissions than the prototypes tested in Detroit. As we will see in Chapter 10, mass transit systems offer the greatest opportunity for conserving fuel and reducing pollution, but they are strongly resisted by a population that has become dependent on the convenience of the auto.

Alternative designs for auto engines also hold considerable promise for reduced pollution. The stratified-charge engine is a modification of the present internal-combusion engine; a dual carburetor supplies a rich fuel mixture around the spark plug and a leaner mixture to continue combustion, achieving both lower emissions and improved fuel economy. As early as 1973, the Honda with stratified-charge could meet the original 1975 standards, even after 50,000 miles. Diesel engines also are economical and low in emissions, though they are noisier and less powerful for a given motor weight. More radical departures from the internal-combustion engine, such as turbine or steam engines or electric motors, offer prospects for even greater emission reduction, but are still at the experimental stage.[6]

There is some question as to whether the benefits of achieving the original *NOx emission standards* would be worth the cost. A 1974 NAS study estimated the total benefits of achieving all the original auto emission standards at $5 billion per year (including property damage, health costs, and deaths at $200,000 per person). The control costs were estimated at $8 billion per year, but this would drop to $4.7 billion if the interim NOx standard of 2.0 grams per mile were used, instead of the original 0.4 gpm.[7] The additional NOx removal would be costly and the health benefits appear small compared to the effects of other pollutants, especially those from stationary sources. Some authors thus urged a revision of the original standard, despite the great uncertainty in data on synergistic effects among the pollutants. The 1977 amendments did in fact relax the long-term NOx standard (for 1981 and beyond) to 1.0 gpm, while tightening the standards for HC and CO. Another cost-saving option would be a "double standard" or "two-car policy" that would require higher standards for cars registered in metropolitan areas than in other areas.

Stationary Sources

Under the 1970 act, EPA was authorized to establish *primary air quality standards* "to protect public health . . . allowing an adequate margin of safety." The states were to formulate implementation plans and issue permits and schedules for reducing emissions to achieve these quality standards by 1975. Emission limits for existing sources could vary among regions, but new sources were to conform to uniform emission limits established nationally for various categories of industry. While air quality standards were to be based solely on health, emission limits were to be determined in part by expected abatement costs. EPA thus has had wide discretionary power, and there has been prolonged negotiation and litigation with particular industries.[8]

Once again, the record has shown considerable progress but many delays. By 1976 emitters accounting for 80% of pollution from stationary sources were in compliance or on a schedule to attain it.[9] The energy crisis created strong pressures to relax plant emission standards. Following the oil embargo, Congress permitted plants switching from oil to coal to delay until 1979 their compliance with 1977 standards, and in 1977 the deadline was extended another two years, but attempts to weaken the standards were defeated. The costs of air pollution control are very large, especially for heavily polluting industries such as electricity generation, steel mills, and copper smelters. With rising unemployment and continuing recession, these economic costs were a potent weapon in arguing for delay.

Under such conditions, the *political influence* of the affected industries has been strong. With pollution controls, as with many other regulations, public benefits are diffuse and often poorly represented, while costs fall initially on a few industries that have a lot at stake (even though most costs are eventually passed on to consumers in the form of higher prices). State and local governments want to attract industry to increase their tax base and provide local jobs, and so they are very sensitive to any threat of plant closings. At the federal level, the polluting industries exert great political power through lobbying and campaign contributions. Typical corporate responses include: "The problem is exaggerated," "The health data are uncertain," "The technology is unproven," "The costs are too high," and "Regulation would retard economic growth."[10]

The most serious air pollution problems arise from *sulfur emissions*. Several studies have shown that it is not sulfur dioxide (SO_2) itself that is the health hazard, but the sulfates formed from it in the atmosphere. Asthma attacks increase above a daily average sulfate concentration of 6 to 10 micrograms per cubic meter, and children exposed to 13 micrograms for several years show increases in acute lower respiratory diseases. Heart and lung diseases are aggravated, especially in elderly persons, and daily mortality rates rise rapidly above 25 micrograms. This may represent a threshold for sulfate-induced mortality, but there is some evidence that even at lower levels there is a small addition to "natural" death rates that would be significant in large populations. Current daily sulfate levels in several cities are in the 15 to 20 microgram range, which provides little if any margin of safety. Furthermore, the rate of formation of sulfates from SO_2 varies widely with temperature, humidity, and particulate levels.[11]

Removal of SO_2 from plant stack emissions can be accomplished by *limestone scrubbers*, but the process is expensive and only 90 to 95% effective. Scrubbers would add 20% to the incremental cost of bulk electricity from a new plant; this is equivalent to a 14% increase in the consumer's bill, which includes transmission and distribution costs.[12] Adding scrubbers to old plants would be even more expensive. Scrubbers also produce vast quantities of sludge that must be disposed of. Power companies claim that they are unnecessary with low-sulfur coal and tall stacks, and that intermittent controls during periods of poor air quality would be adequate. In 1979 EPA announced strict emission standards for new plants, but older plants can continue to operate without scrubbers if they use low-sulfur coal.[13] We will return to this issue in Chapter 10.

The conflict between industrial growth and air quality is particularly acute in *nonattainment areas* where primary air standards have not been met. In 1977 Congress approved EPA regulations for new plants in these critical areas. Such plants must install the best available control equipment, regardless of cost. A company also must offset any new emissions by a reduction in emissions from existing plants (its own or those of other companies). For example, in order to obtain a permit for a proposed tanker terminal in California, Standard Oil agreed to spend $65 million for scrubbing equipment for an oil-burning plant owned by a utility company. In 1977 Congress also acted to prevent *significant deterioration* in areas that currently have high air quality. No deterioration is to be allowed in large national parks and wilderness areas, and permits are to be issued governing increments in pollution in other areas that states can designate.[14] We will analyze these regulatory strategies after looking at the clean water legislation.

Water Pollution

The Water Pollution Control Act Amendments of 1972 (commonly called the Clean Water Act) set extremely ambitious goals, to be achieved in three stages. The *"best practicable technology"* was to be installed by 1977. EPA was authorized to issue permits specifying maximum amounts of waste effluents, based on judgments of practicability, including the cost of control. All municipal waste treatment plants were to have at least secondary treatment, and $18 billion was authorized for federal grants for plant construction. In the second stage, the "best available technology" was to be installed by 1983, with the objective of "fishable and swimmable waters." Extensions were to be allowed only if a company could show that economic costs would be unreasonable for the pollution reduction achieved. The final goal was zero discharge of pollutants by 1985.

Congress left great discretionary latitude to EPA in *interpreting and administering* the act. Since no water quality standards were established, EPA had to make difficult judgments about the "practicability" of particular technologies. It set uniform effluent limits for whole classes of industry, despite the great variations in the age of individual plants and the condition of the rivers or lakes into which they discharge. But exemptions in meeting the 1977 deadlines were granted to 400 major polluters, often because of possible economic impacts.[15] For example, the steel plants in the Mahoning Valley in Ohio were marginally profitable and their closing would create serious local unemployment. EPA exempted the plants from 1977 water standards until 1983, but it was unwilling to relax air standards. One plant (Youngstown Sheet and Tube) did close, laying off 4,100 workers in 1977. Overall, some 80 to 90% of the nation's industries met the 1977 water deadlines, and pollution in the Great Lakes and the largest rivers has been reduced.

The *high cost* of achieving "fishable and swimmable waters" has been increasingly evident. For example, even before the 1972 act, the Delaware River Basin Commission initiated measures to clean up the river between Philadelphia and Wilmington. It concentrated on the oxygen demand of municipal and industrial effluents, ignoring toxic chemicals and nonpoint sources. 95% of the projected benefits were recreational (mainly boating and fishing, which are predominantly middle-class activities). But, according to one study, these benefits were "grossly overstated"; the recreational potential of a heavily industrialized stretch of river is limited at best, and rain runoff from agricultural and urban areas (nonpoint sources) probably would prevent the achievement of these goals. There are coastal and urban recreational areas, and also less polluted rivers, where much greater benefits could be obtained for much smaller costs than on the lower Delaware.[16] For the country as a whole, the National Commission on Water Quality concluded in 1976 that the "best available technology" would cost $43 billion, and recommended postponing second-stage deadlines for five to ten years.[17]

The *Clean Water Amendments of 1977* represented a considerable retreat from the second-stage goals of the 1972 act, but they dealt with several problems that had been neglected in the interim, including toxic pollutants and nonpoint sources. The Senate wanted to retain the requirement of "best available technology" by 1983; the House was willing to settle for stricter enforcement of the "best practicable technology" requirement. The compromise agreed on was "best conventional technology," which EPA was to determine after considering costs, energy use, and equipment age, balanced against expected benefits, with deadlines extended to 1984. No exemptions are to be allowed, however, in the case of toxic pollutants, starting with a list of 65 substances to which EPA can add new chemicals found to be hazardous. The municipal treatment program was continued, with $25 billion in grants over a five-year period. Special aid was earmarked for alternative treatment methods. Spraying wastewater on fields, for instance, can provide inexpensive treatment as well as nutrients for crops if toxic substances are controlled.[18]

Nonpoint sources contribute half of the nation's water pollution. The two main forms are agricultural drainage (sediment, fertilizer, pesticides, and feedlot wastes) and urban runoff (storm sewers carrying street debris directly into rivers). Section 208 of the 1972 act specified control of nonpoint sources, but EPA gave them lower priority than domestic and industrial wastes. Thousands of separate sources, including mining wastes, are harder to control than effluent pipes. The 1977 amendments authorized grants to farmers for management of agricultural runoff, and EPA launched a program to deal with urban runoff. State plans under Sec. 208 must be approved, and applications for any water treatment grants must include plans for "best management practices" for nonpoint sources. Pollution from storm sewers, for example, can be reduced by better street cleaning, by use of storage lagoons, and partial treatment of the first half inch of rainwater runoff.[19]

While *the EPA* often has extended deadlines and withdrawn proposed regulations, it has not capitulated to regulated industries as frequently as have other regulatory agencies in the past. The EPA has been under intense pressure from both environmentalists and industry because it has considerable administrative discretion delegated by Congress. In particular, it often must decide whether a given type of control is technologically reliable and economically feasible. It often must balance substantial economic costs against uncertain benefits to health and the environment. EPA's research capability and its expertise concerning health hazards and abatement technologies have grown slowly, but often the scientific data used in setting standards are very limited.

II. ECONOMIC AND ETHICAL ISSUES

The regulation of pollution has proved more difficult and the technology more costly than had been anticipated. Which of the regulatory strategies used so far has been most effective? What ethical issues arise in attempting to balance a clean environment against employment, distributive justice, and individual freedom?

Regulatory Strategies

In theory, *the marketplace* provides an efficient allocation mechanism in which both consumer and producer have maximum freedom of choice without government interference. Producers choose their own ways to minimize costs in a competitive market. In practice, competition may be restricted by monopolies or price agreements and the consumer's choice may be limited by inadequate information, but for many commodities the price system works fairly well. However, a serious market failure occurs when there are "externalities"—benefits and costs not included in profit-and-loss calculations. Common-property resources to which private-property rights are not attached will tend to be overused, leading to what Garrett Hardin has called "the tragedy of the commons." A "commons" is a pasture available at no charge to anyone who wants to use it; at least in the short run, individuals gain more than they lose by using the commons more heavily.

If air and water are viewed as free resources, their use will create *indirect costs* for other parties. A city that dumps untreated sewage in a river may jeopardize the water supply of communities downstream. Such externalities result in both inefficiency (higher total costs to society) and injustice (some people benefit while others bear the costs). Yet a competitive market system may force firms to take advantage of the air and water as waste disposal systems that are free to them, though costly to society. The nonpolluting firm will have higher costs and a competitive disadvantage that could drive it out of business.

Political institutions can correct some of these shortcomings in economic institutions. Legislation can reduce the imposition of indirect costs on other parties and cause some of the externalities to be internalized in economic decisions. There are four types of regulatory strategy, each of which has distinctive advantages and problems.

First, emission regulations can be based on *quality standards* alone. This was the approach of the 1970 Clean Air Act, which required EPA to establish national ambient air standards to "protect public health . . . allowing an adequate margin of safety." Standards were to be based on health data, without consideration of economic costs. Emissions from both moving and stationary sources were to be reduced to the extent necessary to achieve these standards. This approach makes sense if it is assumed that 1) there are thresholds for the concentration of each pollutant, below which there are no hazards to health, and 2) technologies can be developed to keep pollution levels below these thresholds at a reasonable cost. Health data on high-level, short-time exposures to air pollutants did suggest the existence of thresholds, but the assumption now appears dubious, especially for low-level, long-time exposures.

Lester Lave and Eugene Seskin have summarized evidence for the correlation of *mortality rates* and *air pollution levels* even below supposedly "safe" thresholds. If there is no threshold, and if the cost of removing a pollutant increases rapidly as 100% removal is approached, there will be very large costs in achieving very small additional benefits in the pursuit of zero pollution and absolute safety. Some weighing of costs and benefits must then enter the setting of standards. Lave and Seskin concluded from statistical studies of health data and pollution levels in major cities that a 50% reduction of sulfate and particulate levels resulted in an increase in average life expectancy by nine months. They calculated that in 1979 the annual cost of achieving the ambient standards for SO_2 and particulates would be $9.5 billion (in 1973 dollars) while the annual health benefits would be $16 billion (plus lower cleaning costs and crop losses, and aesthetic and recreational gains).[20] While cost-benefit analysis is subject to severe limitations (see Chapter 8), it appears that enforcement of SO_2 and particulate standards at least as strict as those of the Clean Air Act can be justified by comparing costs and benefits.

The use of emission regulations as a way of *forcing the development of control technology* seems to have had very mixed results. In the case of auto emissions, the pressure of imminent deadlines encouraged short-term, high-cost solutions, such as catalytic converters, rather than more fundamental changes that might be cheaper in the long run, such as stratified-charge engines. The enforcement procedures were not effective. The act provided fines of $10,000 for each new car not in compliance with statutory standards; since such fines would have forced the auto industry to shut down, they did not constitute a credible sanction. The industry had few incentives for compliance, and a succession of reasons for seeking delay. Moreover, regulations setting emission limits provide no incentive to reduce emissions below the permitted limits.[21]

A second alternative is the use of *technology-based effluent limits.* This was the approach taken by Congress in the Clean Water Act, which required best practicable technology (BPT) by 1977 and best available technology (BAT) by 1983. BPT effluent standards for a given type of industry were to be uniform nationally, despite the variations in the assimilative capacity of the river into which a particular plant would discharge. The standards are inflexible and not directly related to measures of river water quality. BAT would give some effluents much more treatment than would be needed to achieve a "swimmable, fishable river." But for other rivers, especially if there is extensive nonpoint runoff, not even BAT would achieve that goal. In between there are cases where a very clean effluent could be obtained with high costs and small benefits. The National Council on Water Quality estimated that the same water quality improvement could be achieved at 30 to 35% lower cost if effluent limits were varied according to local river quality and control costs.[22]

With technology-based standards, an industry has *little incentive to improve control technology*, since such improvements would result only in tighter standards. When a particular technology has been approved by EPA, an industry may be hesitant to look into cheaper alternatives whose approval is uncertain. Because the criteria were ambiguous, many EPA wastewater regulations have been challenged in court. If Congress is too ambiguous and gives too much discretion to an agency, there are long delays from judicial review of administrative interpretation. But if Congress is too specific, there is little flexibility in dealing with individual cases and changing circumstances. In the Clean Water Act, the key terms were not well defined, and Congress gave no guidance on the determination of "practicable technology" and "reasonable cost"; the goal of "fishable, swimmable rivers" was not translated into water quality standards that along with costs could be used to set effluent limits.

A third strategy is the provision of *subsidies for waste treatment*, as in the federal program for construction of municipal sewage plants. This is a popular policy politically, since it brings funds to the home districts of members of Congress; it is now the nation's largest public works program. The program gives some subsidy to industries whose effluents are municipally treated, though there are provisions for municipalities to recover part of the treatment costs, and industries have some incentives to reduce these wastes. However, this strategy, like the previous one, deals mainly with treatment of wastes after they are formed, rather than with processes that might generate less waste in the first place. The uniform requirement of secondary treatment leads to overtreatment of some wastewaters and undertreatment of others. Only recently have novel methods of treatment (storage ponding, on-land treatment, and so on) been funded. The grant program biases local choices toward systems with high capital costs and low operating costs, since federal subsidy applies only to the former. The policy also encourages communities to overbuild, with excess capacity for future growth, which weakens the incentives to control growth.

A fourth strategy, *effluent taxes or charges*, has not been attempted in the United States, but is advocated by many economists. The tax would be proportional to the quantity of a pollutant emitted. Polluters could decide their own type and level of control, but the tax rate would be steep enough that average emissions would be brought down to the desired levels. For example, the tax per pound of SO_2 emitted might be very high in regions of poor air quality; industries would have incentives for diverse responses in fuel use, technological research, plant location, and so on. In the case of autos, an emission rating for each vehicle, determined by periodic inspections, would determine its tax rate for gas at the pump. Consumers could reduce their "smog tax" payments by buying cleaner cars (or adding control equipment), or by tune-up maintenance, or by driving less. Auto manufacturers would have strong market incentives to produce cleaner cars.

Effluent taxes offer a number of advantages. They rely on *economic incentives* that allow greater freedom of choice for both producers and consumers. Total costs would be lower; plants with low control costs would install more control than those with high costs. Regional variations could be easily introduced to cut down on costly overtreatment, such as to take advantage of the natural assimilative capacity of a river. Administrative costs probably would be lower; once the tax level is set, the main regulatory task is checking the monitoring equipment installed by each industry. In contrast to subsidies, taxes would provide revenue, which could be used for pollution control research. Above all, there would be incentives for industrial research on a variety of technologies and for other pollution-reducing responses. Incentives would be graduated; heavy polluters would be heavily penalized, especially in critical regions, but there would be some incentive to reduce pollutants below present standards and in currently "clean" regions.[23]

Effluent taxes do have some *disadvantages*, however. The setting of tax rates would be difficult and controversial. The Netherlands tried such a system and found that fee setting required extensive information, much of which was difficult to get. Estimates must be made of probable average costs and benefits, which require data on technology and health effects.[24] But it would be relatively easy to adjust the tax rate in the light of further data, since these standards would not be tied to specific technologies. The main obstacles to effluent taxes are political. The present system, for all its shortcomings, has a sizable constituency in Congress and federal and state agencies, and major changes would involve further uncertainty and delay. Effluent taxes would be highly visible and would be viewed by some people as a license to pollute; elected officials prefer policies with hidden costs that can be passed on to the unwary or to those with least political power. The rich easily could afford a tax on auto emissions, but it would be a heavy burden on the poor who drive older cars.

While major reforms are thus unlikely, *modifications in existing policies* could introduce greater economic incentives. One hybrid strategy would be the replacement of fixed noncompliance fines with a system of graduated noncompliance fees, based on the quantity of effluents above permit limits and on the cost of compliance. The 1977 Clean Air and Clean Water Amendments both specify that EPA penalties are to be at least equal to a firm's economic gains from noncompliance. Another proposal would allow permits for air emissions from new plants to be bought and sold; this would be a logical extension of the present procedure by which a permit for a new plant in a nonattainment area can be obtained only if the company agrees to offset new emissions by a reduction in existing emissions. A state could introduce a SO_2 tax or a water effluent charge as part of its implementation plans.[25] But in general we can expect a continuation of quality-based and technology-based standards, in which better health-effect data and some weighing of costs and benefits will play an increasingly important part. The chief arguments for relaxing standards will probably continue to be based on energy needs and economic costs.

Justice, Freedom, and Pollution Control

Closely related to these economic issues are several ethical issues in decisions about pollution control policy.

1. Environment and Employment

Do environmental regulations lead to widespread plant closings and unemployment? We noted earlier that from 1971 through 1977, 21,900 persons lost their jobs because of 118 plant closings due at least in part to air and water standards. But in this same period 678,000 new jobs were created in pollution abatement activities (including equipment manufacturing and operation and sewage plant construction). Two out of three threatened plant closings did not occur, and most of the plants that did close were marginally profitable or had obsolescent equipment, so that pollution control was only one of several factors involved.[26] A study by Chase Econometrics concluded that environmental regulations would lead to an increase in employment until 1983, followed by a small decrease thereafter from higher prices and a slight slowdown in the growth of real GNP, which would have a very minor national impact.[27]

Even though the loss of jobs from plant closings is more than compensated by new jobs nationally, there are *severe local impacts*, especially when a threatened plant provides a high percentage of the employment in a community (as in the Reserve Mining case discussed in Chapter 8). The threat of closing sometimes is an excuse to avoid compliance, but in other cases it is seriously contemplated and may be legitimate ground for postponing deadlines (as in the Mahoning Valley steel plants). Relocation assistance or job-retraining programs should be expanded so that when a plant does have to shut down, one group of people will not have to bear the brunt of costs for the sake of widely distributed benefits. Similar special assistance has been provided for workers whose jobs have been eliminated by the lowering of protective tariffs.[28] In the last analysis, legislation for full employment or a guaranteed annual wage would be the most effective way to prevent fear of unemployment from undermining environmental goals.

2. Environment and Inflation

The Chase Econometrics model predicts that from 1970 to 1983 federal pollution regulations would add only 0.3 to 0.4% to the Consumer Price Index. Moreover, pollution control brings benefits not reflected in the Index, such as public health, recreation, and a more aesthetic environment. Today, resource scarcities are one of the main causes of worldwide inflation, and resource stewardship is a wise long-term investment. In many cases, delays in pollution control will result in higher cleanup costs later or else in the deterioration of human health and agricultural productivity. There often are high initial costs, but the stream of benefits increases with time as less-polluting equipment and methods are gradually introduced.[29]

3. Distribution of Benefits and Costs

The *benefits* of pollution control fall very unevenly. Inner-city low-income families are the primary beneficiaries of air pollution control. The poor are more likely to live downwind from a polluting factory than the rich, who can escape to cleaner suburbs. In Chapter 5 some figures were cited on the correlation of air pollution levels with average income in various parts of Washington, D.C. and other cities. By one estimate, 30% of the national benefits of clean air would accrue to five metropolitan areas with 8% of the U.S. population.[30] Water pollution control, by contrast, is weighted toward more affluent beneficiaries. 70% of the benefits of improved water quality are recreational, and upper- and middle-class families are the major users of water-based recreation.[31]

The *costs* of pollution controls also fall unevenly, though not as unevenly as the benefits. Many air control costs are paid initially by industry but are passed on to the consumer as higher prices, such as utility rates and auto prices. Increased prices for basic commodities are regressive in impact, since they take a larger fraction of the budget of a low-income family. Auto emission control costs are high and quite regressive.[32] The subsidies for municipal water treatment plants, on the other hand, come from federal income taxes that are more progressive than either state or local taxes.

Taking both *benefits and costs* into account, stationary source air controls yield a progressive net benefit, automotive emission controls a regressive net cost, and water treatment a progressive net cost.[33] The economic costs of air and water controls together were somewhat regressive in 1977 and were expected to be more strongly regressive by 1980, totalling 2% of an annual family income of $5,000, but only 1% of a $40,000 income.[34] Of all current programs, air controls for stationary sources appear to have the most progressive distributional impact and the most favorable cost-benefit ratio (though cost-benefit analysis as a tool for policy analysis has severe limitations, which are discussed in the next chapter).

4. *Centralization of Authority*

I have said that, in general, decentralization facilitates freedom, understood as participation in decision making. Yet the centralization of pollution control authority in the federal government, which has occurred since 1970, seems inescapable since both air and water pollutants cross jurisdictional boundaries. Moreover, state and local governments have been ineffective against the power of major industrial polluters with extensive financial, scientific, and legal resources. State governments want to attract industry to broaden their tax base and to provide jobs; industries vital to a state's economy can exert great political power to block effective legislation.

The *federal government* thus has had to assume the principal role in setting environmental standards, though it has tried to leave as much as possible to the states in formulating implementation plans. In the setting of standards there are provisions within EPA for input from citizens, state officials, and spokespeople for industry and environmental groups. Considerable authority has been given to EPA regional offices in reviewing state plans. But the average citizen has little opportunity to influence basic environmental policy except through congressional representatives.

There also is a tension between democratic political processes and the need for *long-term comprehensive planning*, as was noted earlier. Authority in the United States is distributed, not only among federal, state, and local governments, but among legislative, executive, and judicial branches. Elected officials have relatively short terms in office. Most policy changes are short range, piecemeal, uncoordinated, and relatively modest, the product of negotiation and compromise.[35] The political style of such incrementalism is poorly adapted for dealing with long-term, systemwide environmental consequences. For many environmental policies the costs are immediate, local, and economic—whereas the benefits are longer range, more widely distributed, and often less tangible. Pollution control also requires the integration of diverse policy components; air pollution is strongly affected by policies for transportation, plant siting, urban zoning, and energy use.[36] For some forms of pollution there are appropriate areas for regional planning, such as airsheds or river basins. But the few serious attempts of states to establish regional authorities (such as the Ohio River or the Delaware River basin) have had limited success because each state insisted on retaining veto power. Once again, an increasing federal role seems inescapable.

5. Restraint in Consumption and Growth

Most of the policies above are designed to remove pollutants after they have been formed. Policies also can be directed at reducing the quantity of pollutants formed in the first place. Since fuel combustion is the major source of air pollution from both mobile and stationary sources, energy conservation and solar energy measures will contribute to improved air quality. Recycling is aimed at the reduction of both resource use and pollution from solid wastes. The trend toward greater reliance on synthetic materials and persistent chemicals can be reversed in favor of biodegradable materials. In a later chapter I will argue that a more equitable use of world resources requires a slowdown of the escalation of consumption in affluent nations, which also would slow the generation of pollutants. While I do not accept the "no growth" position, I will defend selective growth in which consumption and production are shifted toward less energy-intensive and resource-intensive goods and services, which generally create less pollution.

III. LAND-USE POLICY

Land-use policies influence air and water pollution and almost every other form of environmental degradation. In decisions about land use, the most fundamental economic, human, and environmental values come into conflict. Yet the mechanisms for making and implementing land policies are limited and authority is fragmented. We will examine solid waste disposal, zoning and siting regulations, the ethics of land use, and the preservation of wilderness, wildlife, and endangered species.

Solid Waste Disposal

Solid waste disposal involves both pollution and land use. The largest volume of solid wastes comes from agriculture and mining, but the most hazardous are the half billion tons of municipal and industrial wastes generated each year. *Municipal solid wastes* alone would fill the New Orleans Superbowl from floor to ceiling twice every day.[37] Solid wastes are managed mainly by local governments, subject to state regulations that vary widely. The Resource Conservation and Recovery Act (1976) calls for an end to open dumping by 1983, but offers few incentives for a shift to sanitary landfills (in which garbage is covered daily with a layer of earth). Sites for landfills must be chosen carefully to minimize the risk of groundwater contamination; today they are very difficult to find in many regions, and local opposition is often intense.

10–15% of *industrial wastes* are hazardous because of pesticides, heavy metals, acids, and other reasons. Many of these chemicals can be toxic for long periods—PCBs are dangerous for decades, for instance, and cadmium forever. In New Jersey thousands of drums of hazardous chemicals have been left in vacant lots, abandoned warehouses, and industrial landfills. In the Love Canal district of Niagara Falls, N.Y., the Hooker Chemical Company used an abandoned canal as a dump for drums of wastes. The dump was closed in 1953; a school was built on the landfill, and homes constructed adjacent to it. Water seeped in and chemicals oozed out, creating hazardous levels of toxic chemicals in nearby basements. Miscarriages and birth defects occurred in neighboring blocks at three times normal rates. In 1978 the state of New York finally purchased the 235 houses most affected, and helped the occupants find new homes. The cleanup bill will exceed $30 million, and $2 billion in damage suits have been filed. Seepage from such industrial landfills into ground or surface waters are a continuing threat in many parts of the country.[38]

The Resource Conservation and Recovery Act (RCRA) of 1976 gives EPA authority to develop site criteria. To be eligible for federal grants for solid waste programs, a state must develop an enforceable program requiring compliance in not more than five years. EPA also was authorized to regulate hazardous wastes. By 1979 it had begun the immense task of getting data on existing disposal sites and developing standards for treatment and disposal in the future. The proposed regulations include: a long list of chemicals considered hazardous; permits for storage, transportation, and disposal of hazardous wastes; and a system of manifests to keep track of these chemicals "from cradle to grave." Disposal companies would have to monitor sites for 20 years after they were closed, assume liability for damages up to $10 million, and post sizable bonds to ensure proper closing and monitoring. Safer disposal requires sealing with a thick layer of impervious clay or synthetic liners, and a drainage system in case leakage does occur; several methods for neutralizing, detoxifying, or solidifying liquid wastes appear promising though costly.[39]

RCRA also authorizes grants to states to develop *resource recovery programs.* Recycling reduces waste disposal, resource depletion, and energy consumption at the same time. 20% of paper products are now recycled; if 50% were recycled it would save 3 billion trees a year. Recycling a stack of newspapers three feet high saves one tree.[40] Aluminum extraction from ore is highly energy intensive; 7 billion aluminum cans were recycled in 1978, but this was a small fraction of those used. Other metals (iron, steel, and copper) and glass also can be recovered easily if they are separately collected or taken to recycling centers. In Atlanta, magnetic separators recover 99% of tin cans from mixed garbage. Sweden has used a combination of magnets, air currents, and water flotation to separate materials in automated recovery centers. But separation at the household level is the cheapest method, even with the cost of separate collection.[41]

Energy recovery from solid wastes also appears promising. Organic wastes have been burned to produce steam (for heating or electricity generation) in Europe and in Chicago, Nashville, St. Paul, and other U.S. cities. Wastes can be mixed with fossil fuels, or used to dry sewage sludge, which itself burns when dry. Air pollution controls must be used, however, and industrial wastes containing toxic metals must be avoided. Heating garbage without oxygen (pyrolysis) produces gas and liquid fuels and avoids air pollution, but the technology is still experimental. Even small communities can efficiently use incinerators with heat recovery, and metals and glass can be recovered from the residue, according to the Office of Technology Assessment (OTA).[42] Connecticut has launched an ambitious program to use garbage for fuel and scrap iron. The use of agricultural and forest wastes, as well as municipal wastes, to produce energy is discussed in Chapter 10.

The chief obstacles to recycling are *economic* rather than technical. In the past, disposal costs have been low and they have been borne by the public rather than by individual producers or consumers. Environmental costs of resource extraction and waste disposal have not entered profit-and-loss calculations. The omission of such external costs has encouraged the generation of wastes. In addition, we have subsidized resource extraction industries by depletion allowances and by taxing virgin materials as capital gains. We have discriminated against recycled materials. Freight rates for scrap iron are still more than twice those for iron ore. As the costs of raw materials and waste disposal rise, there will be greater incentives for conservation and recovery, but legislation is also needed: the repeal of rates, taxes, and regulations that penalize recycling, and the introduction of incentives for material recovery. The Energy Act of 1978 took a small step in this direction by giving industries a 10% tax credit on the purchase of recycling equipment. Federal funding for research on recycling technologies also should be greatly expanded.[43]

Beverage containers constitute 7% of municipal solid waste. We throw away 60 billion cans annually. Aluminum production, largely for cans, uses 5% of the nation's electricity. Recycling glass and aluminum containers saves energy, but the energy saving with refillable bottles is much greater (about 75%), apart from savings in materials and pollutants.[44] Refillables are 20% cheaper and the number of jobs is almost identical with jobs from throwaways (though they are in differing industries).[45] Oregon banned throwaway beverage containers in 1971, and by 1979 similar laws had been adopted by seven states. Federal laws for mandatory bottle deposits have been discussed, but no action has been taken. 60% of municipal garbage (by weight) consists of *paper products*; a tax on packaging materials, or on all paper (levied at the mill) would provide incentives for conservation. More far-reaching proposals call for a disposal charge imposed on the manufacturer of all products (but not on recycled materials) and a tax on the extraction of virgin materials (see Chapter 12).

Zoning and Siting

Land use affects almost every environmental problem from air, water, and noise pollution to soil erosion and the loss of wildlife and natural resources. Patterns of land use are central in such contemporary urban problems as highway strip development, the deterioration of the inner city, and traffic congestion. Expanding cities absorb productive farmland, open space, and wildlife habitat. Highway construction and continuing dependence on the auto contribute to air pollution, energy use, and urban sprawl. Land has been viewed as private property, to do with as one pleases; as available land becomes scarce, speculation and pressures toward development increase. By far the largest portion of land-use decisions are made at the local level; local communities derive most of their revenue from property taxes and therefore usually have welcomed residential and industrial development.[46]

During the 1970s, many communities have become more aware of *the indirect costs of rapid growth.* The traditional assumption that growth is progress has been questioned. Environmentalists have opposed developers, and local groups have said "slow down" or "no more growth." For example, Dade County (Florida) and Suffolk County (Long Island, New York) both passed highly restrictive zoning laws and authorized public purchase of wetlands and ecologically fragile areas. Some communities have limited construction to single-family homes on large lots. But civil rights groups have charged that many zoning laws are discriminatory, designed to protect white, middle-class suburbs by excluding low-income families and racial minorities.

The *purposes of zoning ordinances* have been at the center of several important court cases. An ordinance in Sanbornton, N. H. required six-acre lots for most of the land in a new housing development. The state court ruled that ordinances can legitimately protect ecological balance, open space, scenic beauty, and even the rural character of a small town—but they cannot limit population as such, since people have to go somewhere. The Sanbornton zoning was allowed only as a stopgap measure pending formulation of long-range plans for orderly growth. Petaluma, California set maximum population limits and a five-year moratorium on subdivision, but it specified that 8 to 12% of new units were to be low-income and multiple-family housing. The Supreme Court (1974) allowed the ordinance to stand, arguing that the "general welfare" can include open space and low density, and that the town was not exclusionary since there were provisions for low-cost housing. The ruling indicates that ordinances that exclude all low-income residents would not be acceptable.[47]

Other court cases have dealt with *the taking issue.* The Fifth Amendment asserts that private property "shall not be taken for public use without just compensation." One survey found that courts usually have required compensation if more than two-thirds of the financial value of a piece of land is lost because of a regulation.[48] But courts also have said that regulations designed to prevent public harm do not require compensation; with new ecological awareness there is a greater recognition of the harmful environmental impacts of land use. In 1972 the Wisconsin Supreme Court upheld a regulation prohibiting the filling of wetlands, even though it would result in financial losses to landowners:

> An owner of land has no absolute and unlimited right to change the essential character of his land so as to use it for a purpose for which it is unsuited in its natural state and which injures the rights of others. . . . The changing of wetlands and swamps to the damage of the general public by upsetting the natural environment and the natural relationship is not a reasonable use of that land.[49]

More recently, courts have upheld uncompensated land regulations if they are the result of some sort of balancing of *public benefits* against *private interests.* In the past, such balancing has favored property owners and developers, since the economic value of land can be readily estimated, whereas its social and environmental value is unquantifiable and often long term. But courts are now more likely to accept a regulation if it has a plausible justification in relation to community welfare, orderly growth, environmental preservation, or any of a broad range of public objectives.

Another crucial issue is the division of authority between *local, state,* and *federal* governments. While land control is constitutionally part of the police powers of states, it traditionally has been delegated to municipalities and counties. However, many environmental and social impacts of land use are regional in character. Most local governments have little technical expertise, and they are inclined to promote growth because most of their revenues come from property taxes. They find it difficult to resist pressures for individual variances or exceptions to zoning ordinances. It is not surprising that state regulation has been increasing, especially for land uses that affect people outside local boundaries. In some cases a state sets standards, issues permits (for example, for wetland dredging and filling), and enforces guidelines for action by local authorities. Legislation for national land-use planning has been defeated twice in Congress after long debates.[50]

Many states have taken action to protect *coastal zones* that are of scenic, recreational, and biological importance. Half of the biological productivity of the world's oceans is dependent on coastal areas. Marshes, estuaries, and tidal wetlands are spawning grounds for fish and shellfish and habitats for wildlife; they also absorb wastes, recharge groundwater aquifers, and provide protection against flood and storm tides. In 1972 Congress authorized federal grants to cover two-thirds of the cost of state coastal zone management programs. All 34 states and territories bordering oceans or the Great Lakes have applied for grants, but by 1979 only six had developed zoning and management plans. California adopted by referendum a plan setting up six regional commissions to issue development and building permits for all land within 1,000 yards of the ocean. The commissions have denied permits for several large projects, and have set conditions for others, though the overwhelming majority of applications have been approved. Delaware prohibits industrial development within two miles of the coast.[51]

Other states have passed measures to *protect critical areas.* In Florida, for instance, voters approved a bond issue to purchase environmentally fragile lands and to protect scenic areas. To protect agricultural lands, Hawaii has since 1961 zoned all its land. To preserve its natural beauty and rural life-style against the inroads of second-home and recreational development, Vermont passed legislation in 1970 requiring permits for all subdivisions of ten or more lots and all construction above an elevation of 2,500 feet. To discourage land speculation it also adopted steep capital gains taxes—as much as 60% on gains from land held less than a year. But in 1974 and 1975 the Vermont legislature turned down a proposal for statewide zoning; by then the economic recession and the drop in tourism had intensified the opposition to further measures that might restrain growth. Other states have passed laws to protect scenic rivers and wetlands.[52]

The *siting of major facilities* often has been the subject of extended controversy and local opposition. Power plant construction usually requires state permits for land use, as well as water and air permits. Planning for airports and highways that use federal funds must include environmental impact statements and public hearings. Because of land shortages and public objections to the prospect of aircraft noise, no new major airport has been initiated in the 1970s. Highway decisions have far-reaching social as well as environmental impacts; a highway can disrupt a neighborhood and determine future patterns of community growth. Opportunities for public participation in such siting decisions are discussed in Chapter 9.

The Ethics of Land Use

Historically, land in the United States has been viewed as *private property.* John Locke and other British political philosophers had argued that people who own property have a stake in preserving and improving it; they will look after it and use it productively. The authors of the U.S. Constitution—most of whom were themselves property owners—followed the British legal tradition in defending the right to use private property for economic gain, limited only by the state's right to protect public health, safety, and welfare. Through most of our history land was plentiful, and its development and use were major national goals. In the late nineteenth and early twentieth centuries there was concern about conservation of natural resources and preservation of scenic areas by setting aside public lands, but there was virtually no regulation of private land use.

The state can influence the use of land through *economic incentives.* In order to preserve agricultural land, more than half the states give it preferential tax treatment by assessing it at its value for farming rather than for development. (The protection may be short-lived, however, and may merely subsidize speculators unless there are provisions to recover back tax breaks if the land is developed within a given time period.) State authorities can acquire land essential for public purposes by exercising the right of eminent domain, but the owners must receive fair compensation. Governments also can enact regulations (without provision for compensation) for land uses that are detrimental to the public welfare—which in the last decade has been understood to include environmental protection. Today it is evident that many people besides the owner of a piece of land have a stake in its use. The public must have a larger role in land-use decisions because environmental and social impacts are far-reaching in space and time.

Decisions about construction of *public facilities* (schools, roads, sewers, water service, and so on) are one important policy tool for influencing patterns of urban development. In large-scale housing projects ("cluster development" or "planned unit development"), developers can be required to provide open spaces and recreational areas. There also can be strict requirements for landscaping, parking space, public access to riverfronts and beaches, and integration of residential and commercial areas. Other changes can reduce the dependence of local communities on property taxes, which bias decisions toward uncontrolled growth. A larger fraction of public school support can be shifted to the state level. Regional tax sharing can lessen some of the pressures for development. Thus the Minneapolis–St. Paul Metropolitan Council redistributes (mainly on the basis of population) 40% of any growth in local tax revenues in a seven-county area. The council also plans all sewer and waste treatment systems and coordinates transportation systems for the whole region.[53]

Comprehensive land-use planning requires an even broader and more systematic approach. Some regional planners, such as Ian McHarg, start from superimposed maps showing geological, ecological, and social characteristics of various areas to determine those most suited for conservation, recreation, residence, commercialization, and so on. Other planners have designed "new towns" with high-density settlements surrounded by greenbelts. In the past, home mortgage and tax policies as well as zoning laws have favored low-density, single-family housing—which wastes land, energy, and services and leads to heavy use of autos. High-density areas with extensive open spaces and provision for multifamily and low-income housing are more amenable to mass transit systems—which both environmental and civil rights groups have supported. Recreational areas accessible to low-income families should be given high priority in such regional planning.[54]

There is some evidence that *a new land ethic* is emerging in the United States. A report to the CEQ speaks of a "quiet revolution" occurring as people realize that land is a national resource in which all citizens have an interest.[55] A citizens' task force, appointed to advise the president on land use and urban growth, writes:

> It is time to change the view that land is little more than a commodity to be exploited and traded. We need a land ethic that regards land as a resource which, improperly used, can have the same ill effects as the pollution of air and water, and which therefore warrants similar protection.[56]

The report delineates a broad range of economic, social, and environmental criteria that should enter land-use decisions, including natural beauty, social diversity and openness, strong neighborhoods and a sense of community, and the management of finite resources for future generations.

This task force also proposes that *development rights* should be separated from *land ownership*. Many increases in land value are created by society (through public expenditures for roads and sewers, new technological processes, changes in zoning, and so on). The value of a lot near a projected interstate highway exit rises enormously almost overnight; the development rights could be auctioned or sold to help pay for the highway. Great Britain has pioneered in legislation by which the general public, rather than private landowners, receives the added value created by publicly financed improvements. An alternative would be to tax heavily the increases in land value resulting from rezoning or public expenditures.[57]

Decisions about land use, like those about air and water pollution, frequently entail trade-offs between *short-term economic benefits* (profit from land development, increased tax revenues, new jobs, or markets) and *long-term environmental and social costs.* We should seek neither uncontrolled growth nor the end of growth, but planning for directed growth; preservation and development must be pursued, each in appropriate places. Because such decisions require trade-offs between unquantifiable values, as well as the reconciliation of divergent interests, they should not be made by planners or technical experts alone, but through political processes in which there is provision for public participation at both local and state levels. Regulation by state and regional authorities does diminish local independence, but there are many opportunities for citizen input in such decisions.

Wilderness and Wildlife

There is a wide range of conflicting values in the management of public lands. Federal and state lands are potential sources of a diversity of renewable resources (mainly timber and grasslands), nonrenewable resources (coal, oil, and minerals), and water resources (watersheds and water control projects). I will consider the less tangible values associated with public lands: recreation, scenic beauty, and the protection of wildlife and endangered species.

Some of these values are partially fulfilled on lands managed for multiple purposes. Thus the Multiple Use–Sustained Yield Act of 1960 specifies that *National Forests* are to be devoted to recreation, grazing, timber, watershed

tinued work on the dam, claiming that it should be exempted because it was started two years earlier. The dam would provide 4,000 construction jobs and other federal contracts in an area of unemployment, so congressmen from the state were active in its support and tried to weaken the 1973 act. In 1978 the Supreme Court upheld the ban, arguing that the absolute priority for preservation expressed in the act allowed no latitude for a utilitarian calculation of the relative value of the fish and the dam. On the final day of the 1978 Congress, the act survived with only one major change: a cabinet-level committee was authorized which, by five out of seven votes, could exempt specific projects in which conflicts had not been resolved. The committee proposal had been supported by some environmental groups who feared that without it the act might not be renewed. In 1979 the committee voted unanimously to save the fish and kill the dam. The economic justification for the dam appeared dubious, despite the $100 million already spent.[62] However, after voting against the dam three times, Congress finally approved it in an amendment tacked onto a popular public works bill that Carter was unwilling to veto. The snail darter lost, but the act at least survived.

To sum up, in the case of clean air and water and the protection of wilderness and endangered species, environmental values have been defended with considerable success through political processes, despite strong pressures from the economic interests affected. While environmental measures have not gone as far as environmentalists have hoped, the substantial gains of the early 1970s have not been lost in times of recession and inflation, despite public awareness of the economic costs of pollution abatement. Public opinion polls cited in the previous chapter show continuing support for government spending to reduce pollution.

To be sure, many deadlines have been postponed—especially after threats of plant closings—but air quality had definitely improved by the late 1970s, and water quality was only slightly worse (mainly because of municipal wastes and urban and agricultural runoff, not industrial wastes).[63] While there is great room for improvement, particularly in urban areas, pollution certainly would be far worse without the actions already taken. Significant advances also have been made in the protection of natural beauty, forest areas, wilderness, and wildlife, though we have barely begun to deal with solid wastes. Environmental groups have lacked the political power and the technical expertise of industry or federal agencies, but they often have made effective use of opportunities to be heard through legislative lobbying, congressional and agency hearings, and especially in the courts, which have played a major role in the interpretation of environmental legislation.

By the end of the decade, most legislation called for a balancing of environmental and economic impacts and at least some consideration of long-term as well as short-term consequences. The next two chapters examine some of the analytical techniques and assessment methods through which such balancing may be attempted. There is still great reluctance to face resource limits and escalating energy use, or to question historic assumptions about industrial growth and ever-increasing levels of consumption. In Part Three we will look at more fundamental changes in life-styles and patterns of consumption and at alternative technologies that might contribute to the reconciliation of environmental and human values in a sustainable society.

NOTES

1. Council on Environmental Quality, *Environmental Quality 1979* (Washington, D.C.: CEQ, 1979), chap. 1.

2. F. P. Grad et al., *The Automobile and the Regulation of Its Impact on the Environment* (Norman, Okla.: University of Oklahoma Press, 1975); and H. D. Jacoby et al., *Cleaning the Air: Federal Policy on Automotive Emissions Control* (Cambridge, Mass.: Ballinger, 1973).

3. Clarence and Barbara Davies, *The Politics of Pollution*, 2d ed. (Indianapolis: Pegasus, 1975), chap. 6.

4. "1977 Clean Air Act Amendments," *Congressional Quarterly*, August 13, 1977, pp. 1713-18.

5. Constance Holden, "Auto Emissions," *Science* 187 (1975): 818-22; and Eugene Seskin, "Automobile Air Pollution Policy," in *Current Issues in U.S. Environmental Policy*, ed. Paul Portney (Baltimore: Johns Hopkins University Press, 1978).

6. Janice Crossland, "Cars, Fuel and Pollution," *Environment* 16 (March 1974): 15-27. See also the series by Arnold W. Reitze, *Environment* 19 (April, May, and June, 1977).

7. National Academy of Sciences, *Air Quality and Automobile Emissions Control, Vol. 4: The Costs and Benefits of Automobile Emission Controls* (Washington, D.C.: NAS, 1974).

8. James Canon, *A Clear View: Guide to Industrial Pollution Control* (Emmaus, Pa.: Rodale Press, 1976); and A. Myrick Freeman, "Air and Water Pollution Policy," in Portney, op. cit.

9. Council on Environmental Quality, *Environmental Quality 1976* (Washington, D.C.: CEQ, 1976).

10. Walter Rosenbaum, *The Politics of Environmental Concern*, 2d ed. (New York: Praeger, 1977), chap. 3.

11. National Academy of Sciences, *Air Quality and Stationary Source Emission Control* (Washington, D.C.: Senate Public Works Committee, 1975); EPA, *Community Health and Environmental Surveillance System (CHESS)* (Washington, D.C.: GPO, 1976); and Lester Lave and Eugene Seskin, *Air Pollution and Human Health* (Baltimore: Johns Hopkins University Press, 1978); and Richard Tobin, *The Social Gamble: Determining Acceptable Levels of Air Quality* (Lexington, Mass.: Lexington Books, 1979).

12. James Sawyer, "The Sulfur We Breathe," *Environment* 20 (March 1978): 25-32.

13. Luther Carter, "Uncontrolled SO_2 Requirements Bring Acid Rain," *Science* 204 (1979): 1179-82. See also Chapter 10.

14. Freeman, op. cit.; and Davies, op. cit.

15. Rosenbaum, op. cit., chap. 5.

16. Bruce Ackerman et al., *The Uncertain Search for Environmental Quality* (New York: Free Press, 1974).

17. National Commission on Water Quality (NCWQ), *Report to the Congress* (Washington, D.C.: NCWQ, 1976).

18. "Clean Water Act of 1977," *Congressional Quarterly*, December 24, 1977, pp. 2666-75; and "Water Act Revision Complete," *Science* 198 (1977): 1130.

19. Kathy Barton, "The Other Water Pollution Problem," *Environment* 20 (June 1978): 12-20; and Walter Westman, "Problems in Implementing U.S. Water Quality Goals," *American Scientist* 65 (1977): 197-203.

20. Lave and Seskin, op. cit.

21. Frederick Anderson et al., *Environmental Improvement through Economic Incentives* (Baltimore: Johns Hopkins University Press, 1977).

22. National Commission on Water Quality, op. cit.; and H. M. Peskin and E. P. Seskin, *Cost-Benefit Analysis and Water Pollution* (Washington, D.C.: Urban Institute, 1975).

23. Anderson et al., op. cit.; A. M. Freeman, R. H. Haveman, and A. V. Kneese, *The Economics of Environmental Policy* (New York: John Wiley, 1973); and Allen Kneese and Charles Schultze, *Pollution Prices and Public Policy* (Washington, D.C.: Brookings Institution, 1975).

24. Susan Rose-Ackerman, "Market Models for Pollution Control: Their Strengths and Weaknesses," *Public Policy* 25 (1977): 383-406.

25. Freeman in Portney, op. cit.; and Anderson et al., op. cit.

26. Council on Environment Quality, *Environmental Quality 1978* (Washington, D.C.: CEQ, 1978), pp. 431-32.

27. Robert Haveman and V. Kerry Smith, "Investment, Inflation, Unemployment and the Environment," in Portney, op. cit.; and League of Women Voters (LWV) Education Fund, "Are Jobs Really the Price of a Clean Environment," (Washington, D.C.: LWV, 1977).

28. Wallace Johnson, "Social Impact of Pollution Control Legislation," *Science* 192 (1976): 628-31; and A. Myrick Freeman, "Income Distribution and Environmental Quality," in *Pollution, Resources and the Environment*, ed. A. Enthoven and A. M. Freeman (New York: W. W. Norton, 1973).

29. Gus Spaeth, "A Small Price to Pay," *Environment* 20 (October 1978): 25-29.

30. Henry Peskin, "Environmental Policy and the Distribution of Benefits and Costs" in Portney, op. cit.; see also David Harrison, *Who Pays for Clean Air?* (Cambridge, Mass.: Ballinger, 1975); and Leonard Gianessi et al., "The Distributional Implications of National Air Pollution Damage Estimates," in *Distribution of Economic Well-Being*, ed. F. T. Juster (Cambridge, Mass.: Ballinger, 1977).

31. C. J. Ciccetti et al., *The Demand and Supply for Outdoor Recreation* (New Brunswick, N.J.: Rutgers University Press, 1969).

32. A. Myrick Freeman, "The Incidence of the Cost of Controlling Automotive Air Pollution," in Juster, op. cit.

33. Peskin, op. cit.

34. Robert Dorfman, "Incidence of the Benefits and Costs of Environmental Programs," *American Economic Review* 67 (1977): 333-40.

35. David Braybrooke and Charles Lindblom, *A Strategy of Decision* (New York: Free Press, 1970).

36. Gerald O. Barney, ed., *The Unfinished Agenda* (New York: Thomas Crowell, 1977), chap. 9.

37. *Environmental Quality 1978*, op. cit., p. 159.

38. Thomas Maugh, "Toxic Waste Disposal, a Growing Problem," *Science* 204 (1979): 819-23.

39. Thomas Maugh, "Burial is Last Resort for Hazardous Wastes," *Science* 204 (1979): 1295-98; and R. B. Pojasek, ed., *Toxic and Hazardous Waste Disposal*, vols. 1 and 2 (Ann Arbor, Mich.: Science Publishers, 1979).

40. G. Tyler Miller, *Living in the Environment* (Belmont, Calif.: Wadsworth, 1975), p. E62.

41. Environmental Protection Agency, *Resource Recovery and Waste Reduction: 4th Report to Congress* (Washington, D.C.: EPA, 1977); and William Franklin et al., "Potential Energy Conservation from Recycling Metals in Urban Solid Wastes," in *The Energy Conservation Papers*, ed. Robert Williams (Cambridge, Mass.: Ballinger, 1975).

42. Office of Technology Assessment, *Materials and Energy from Municipal Waste* (Washington, D.C.: OTA, 1979).

43. Talbot Page, *Conservation and Economic Efficiency* (Baltimore: Johns Hopkins University Press, 1977), chap. 5.

44. Bruce Hannon, "System Energy and Recycling" (Urbana, Ill.: Center for Advanced Computation, University of Illinois, 1972).

45. Page, op. cit., p. 97. See also Rosenbaum, op. cit., chap. 9.

46. Marion Clawson, ed., *Modernizing Urban Land Policy* (Baltimore: Johns Hopkins University Press, 1973); and National Academy of Sciences, *Urban Growth and Land Development: The Land Conversion Process* (Washington, D.C.: NAS, 1972).

47. Rosenbaum, op. cit., chap. 7; and Natural Resources Defense Council, *Land Use Controls in the United States* (New York: Dial Press/James Wade, 1977), chap. 7.

48. Ibid., chap. 2; and Fred Bosselman et al., *The Taking Issue* (Washington, D.C.: GPO, 1973).

49. Natural Resources Defense Council, op. cit., p. 258.

50. Elizabeth Haskell, "Land Use and the Environment: Public Policy Issues," *Environmental Reporter*, monograph no. 20 (November 8, 1974).

51. Robert Healy, *Land Use and the States* (Baltimore: Johns Hopkins University Press, 1976), chap. 4.

52. Ibid., chaps. 3 and 5; and Luther Carter, *The Florida Experience: Land and Water Policy in a Growth State* (Baltimore: Johns Hopkins University Press, 1974).

53. Natural Resources Defense Council, op. cit., chap. 13; and Haskell, op. cit.

54. Ian McHarg, *Design With Nature* (New York: Natural History Press, 1969); Donald McAllister, ed., *Environment: A New Focus for Land-Use Planning* (National Science Foundation, 1973); Charles Haar, *Land Use Planning* (Bloomington, Indiana: Indiana University Press, 1968); and Arnold Reitze, *Environmental Planning: Law of Land and Resources* (Washington, D.C.: North American International, 1974).

55. Fred Bosselman and David Callies, *The Quiet Revolution in Land Use Control* (Washington, D.C.: Council on Environmental Quality, 1971).

56. Task Force on Land Use and Urban Growth (William Reilly, ed.), *The Use of Land: A Citizen's Policy Guide to Urban Growth* (New York: Thomas Crowell, 1973), p. 7.

57. Ibid., chap. 3; R. W. G. Bryant, *Land: Private Property, Public Control* (Montreal: Harvest House, 1972); Lynton Caldwell, "Rights of Ownership Or Rights of Use? The Need for a New Conceptual Basis for Land Use Policy," *Environmental Law Review* (1975).

58. Natural Resources Defense Council, op. cit., chap. 11; G. Tyler Miller, *Living in the Environment* (Belmont, Calif.: Wadsworth, 1975), chap. 10; and Herbert Kaufman, *The Forest Ranger* (Baltimore: Johns Hopkins University Press, 1960).

59. Roderick Nash, *Wilderness and the American Mind*, rev. ed. (New Haven: Yale University Press, 1973), chap. 12. For other references see Chapter 5.

60. Miron Heinselman, "Crisis in the Canoe Country," *Living Wilderness* 40 (January 1977): 12-24.

61. "Final BWCA Legislation," *Wilderness Report* 15 (November 1978): 1.

62. Constance Holden, "Endangered Species: Review of Law Triggered by Tellico Impasse," *Science* 196 (1977): 1426-28; and Luther Carter, "Lessons from the Snail Darter Saga," *Science* 203 (1979): 730.

63. National Wildlife Federation, "1979 Environmental Quality Index;" and *Environmental Quality 1979*, op. cit., chap. 1.

8
COSTS, BENEFITS, AND RISKS

We have seen that the setting of standards for air and water pollutants frequently involves the comparison of costs and benefits. A similar balancing of costs and benefits is a feature of many environmental and technological decisions today. What contribution can such quantitative techniques as cost-benefit analysis and risk-benefit analysis make to these decisions?

Can trade-offs between conflicting values be resolved rationally by formal analytic methods—or is it impossible to compare economic costs, loss of human life, and scenic beauty, for example, on a common scale? What roles do scientific judgments and value judgments play in decisions about health, safety, and the environment? Do quantitative techniques facilitate communication among elected representatives, agency administrators, technical experts, and citizens—or are they too complex for the nonspecialist to understand? Is risk-benefit analysis helpful in setting regulatory standards for such toxic substances as pesticides, food additives, and suspected cancer-causing chemicals (carcinogens)—or do such purportedly objective methods disguise institutional biases and rationalize decisions that were made primarily on political grounds?

I. COST-BENEFIT ANALYSIS

The goal of cost-benefit analysis (CBA) is to maximize the balance of benefits over costs, calculated in monetary units—which are familiar and easily aggregated. We will start with an example from water resource planning, and then consider the relation of economic costs and benefits to intangible environmental and human values. Lastly, the uses and problems of CBA as an instrument of policy choice in particular institutional contexts are discussed.

Cost-benefit analysis is widely used to evaluate *public projects*, such as dams and airports, which require large investments of public funds. The National Environmental Protection Act (NEPA) requires a "balanced weighing of costs and benefits" of proposed projects, using "a systematic interdisciplinary approach." Formal CBAs, while not required under NEPA, often are included in environmental impact statements. But CBAs are specifically mandated by Congress in water resource planning. The Flood Control Act of 1936 instructed agencies to measure the costs and benefits of water projects in dollars, and the mandate has been renewed and revised several times since. Debates over dams and irrigation schemes of the Army Corps of Engineers and the Bureau of Reclamation often have revolved around cost-benefit calculations.

In many cases the marketplace already assigns *monetary quantities* to the benefits and costs of a proposed project. Reliable economic estimates can be made of the increased production, wages, and local tax benefits that can be anticipated. The construction costs for labor and materials can be calculated from engineering estimates. But how are dollar equivalents to be assigned for benefits or costs that do not normally enter the market? One method is to make estimates based on the prices of the closest *market substitute.* In some cases, however, there simply are no comparable items on the market, especially such common-property resources as clean air and water, which are not bought or sold by individuals. Another method is to infer *willingness to pay* from indirect evidence concerning people's behavior, or to conduct surveys in which respondents are asked what they would be willing to pay for a particular benefit, or accept as compensation for a particular loss.[1]

These various methods of assigning "shadow prices" can be illustrated in the case of *recreational benefits* from a lake that would be created by a proposed dam. The closest market substitute would be the fees charged for private fishing, boating, and swimming rights or beach admission charges. People's willingness to pay for recreational benefits can be inferred indirectly from what they spend on travel and lodging to reach similar lakes and beaches. There also have been surveys and interviews in which respondents were asked what they would be willing to pay for such recreational opportunities.[2] In practice, however, most federal projects are subject to the guidelines laid down by the Water Resources Council, which in the 1960s stipulated the somewhat arbitrary figure of $1.35 per visitor-day for all recreation, but by the early 1970s allowed for some differentiation ($0.75 to $2.25 per day for general recreation, $3 to $9 for special kinds of hunting and fishing).

A Case Study: The Tocks Island Dam

In 1960 the Corps of Engineers drew up plans for a dam on the Delaware River, 75 miles above Philadelphia; in 1962 funds for it were authorized by Congress. Before the project finally was rejected in 1975 by the Delaware River Basin Commission, it had been the subject of 50 studies and protracted controversy, much of it centering on the cost-benefit analyses prepared by the Corps. The dam was to be a multipurpose project for which four main benefits were calculated.[3]

1. Recreation

44% of the projected benefits were attributed to mass recreation, assuming 10 million visitors annually once the beaches were fully developed. The specified figure of $1.35 per visitor-day was used. In the ensuing debate, environmentalists defended the more "natural" recreational possibilities of the river without the dam, such as canoeing, hiking, and moderate-density swimming in selected shoreline areas. The Corps replied that urban families from the Philadelphia and New York areas preferred lake recreation, and that more people could be accommodated by a lake than a river. Yet when it estimated costs the Corps made no mention of public transportation, which would be needed if low-income urban families were to benefit, or of access highways for middle-income families. (Access highways would have cost $680 million in New Jersey alone—more than the cost of the dam itself—but the Corps said transportation was not in its jurisdiction.) Not until 1975, in a six-volume study commissioned from two independent counsulting firms, was any comparison made with other alternatives for providing recreation for urban families (such as swimming pools and expanded state parks and beaches).

2. Flood Control

The protection of downstream property from damage, such as the damage that occurred in the 1955 floods, accounted for 11% of the benefits. The dam would provide complete protection from the moderately severe floods expected, on the average, once in ten years, but not from the extreme storm that could be expected once in 100 years. The possibility of a dam collapse—an event of very low probability but catastrophic consequences—was not even considered. The benefit was calculated as the difference between the average annual flood damage with and without the dam, assuming that the dam would encourage flood-plain use, new development, and increased land and property values. The Corps made no attempt to compare alternative ways of reducing flood damage, such as flood-plain zoning or insurance requirements that would retard growth in vulnerable areas. It claimed that it was evaluating a multipurpose project and was not obligated to study single-purpose alternatives. The independent study in 1975 recommended flood-plain zoning as the cheapest way of reducing flood damage.

3. Water Supply

34% of the total benefits were ascribed to the maintenance of adequate flow for the water intakes of Trenton and Philadelphia and the prevention of excessive salinity during low-flow periods. In this case the benefits were calculated from the least-cost alternative, a dam and reservior that could be built on another river; interest was calculated at prevailing rates for private financing, since Congress had not authorized single-purpose dams. Environmentalist critics proposed a number of alternatives, including a storage reservoir to be filled in high-flow periods, an alternate water source for Camden, and a program for reducing water consumption and waste.

4. Hydroelectric Power

12% of the benefits were derived from electric power generation. Here again the figures were obtained from the cost of the cheapest alternative, a new coal-fired plant. Critics claimed that the figures were inflated and made no allowance for improvements in generating efficiency. They urged once again that efforts be directed at curbing the growth in demand.

The critics also attacked the omission of *environmental costs* from the Corps' evaluation. Because of the slow turnover of lake water and the high level of nutrients from farms and cities upstream, rapid eutrophication of the lake could be expected. Algae blooms would limit recreational use unless expensive tertiary treatment were added in upstream sewage treatment plants, mainly in New York State. There also were no estimates of the effects of the dam on shad in the river, oysters in the lower estuary, or ecological diversity in the region around the dam. Underlying the debate, then, were the conflicting values of economic growth and environmental preservation. Opponents of the dam emphasized the scenic beauty and natural diversity of the valley and the preservation of one of the last free-flowing rivers in the East, whose loss was not included as a cost.

Nor did the Corps include *the human costs* of the dam. The rural character of the valley communities would be destroyed and their style of life would be completely changed. The sense of community stability would be lost; the local population already felt helpless facing a huge government bureaucracy. There also would be substantial social impact on the access regions, particularly in New Jersey. Little attempt was made to find out the views of the affected populations.[4]

The defeat of the dam can be attributed partly to the broadening of *the scope of evaluation* and partly to *the political processes* that replaced the Corps and its allies in Congress as the locus of decision making. The geographical boundaries of analysis were enlarged with the formation of the five-state Delaware River Basin Commission, and the indirect costs were extensively debated. The governor of New Jersey felt strongly about the cost of access highways, and the governor of New York objected to the cost of tertiary sewage treatment. In addition, new political constituencies were developing as people became more aware of environmental values. The Corps has had strong support from certain interest groups, including construction contractors, real estate developers, and unions in the building trades. As indicated earlier, it always has been particularly susceptible to pork barrel politics in which members of Congress seek generous federal funds for their constituents and approve similar requests from other districts without careful scrutiny. The Corps considers projects in isolation, with virtually no comprehensive planning.[5] Its opponents have been supported by the growing environmental movement and a new recognition of the far-reaching character of secondary impacts. In this case, the final decision occurred in 1978 when the middle stretch of the Delaware was designated by Congress for protection as a Scenic River.

A retrospective study of the Tocks Island controversy concludes that cost-benefit analysis is an inadequate means of resolving *fundamental value conflicts.* CBA, the study suggests, is inherently biased against the representation of such intangible values as ecological balance and natural beauty. Moreover, CBA is carried out by engineers and economists who, according to these authors, tend to be "insensitive to fragile, soft and unquantifiable environmental and human values." Finally, they suggest, CBA can be used only when there is an agreed framework within which plans are formulated and criteria applied. But in a society deeply divided over social priorities, CBA cannot resolve conflicts over such basic values as economic growth and environmental preservation.[6]

The *planning procedures* of federal agencies had been considerably broadened by the later years of the Tocks Island controversy. An environmental impact statement finally was prepared, but it was not a very thorough one. In 1973, the interagency Water Resources Council developed Principles and Standards for new projects. Planning is to be directed to two main objectives: national economic development (within which a formal CBA is to be included) and environmental quality (quantified where possible, but with qualitative descriptions also). Two additional objectives are mentioned but receive much less emphasis: regional development and social well-being.[7] Thus CBA no longer has the virtually exclusive role in justifying decisions that it once had, but it is still a central consideration. It was the main technique used in the Cross-Florida Barge Canal Review (1977), in which the loss of benefits of a free-flowing river was still omitted from the calculations. In the Florida case, the Tennessee-Tombigbee Canal, and the Lock and Dam No. 26 on the Mississippi at Alton, Ill., independent economists hold that the Corps inflated the benefits enormously by overestimating the barge traffic they will carry.[8]

Values, Costs, and Benefits

Evident in the Tocks Island case, as in most applications of CBA, are six issues that are important in relation to the values discussed in this volume.

1. Distributive Justice

CBA deals with aggregate costs and benefits. It does not ask who bears the costs and who reaps the benefits. As a leading exponent puts it, "Distributional questions are beyond the scope of cost-benefit analysis."[9] The analyst's concern for efficiency rather than equity has been defended on various grounds. Sometimes it is said that inequities will tend to average out; a person who is a gainer in one project may be a loser in the next one. However, this neglects the structures of economic and political power through which some people are consistent gainers and others are habitual losers. Again, it is said that if there is a net benefit, the gainers could compensate the losers and everyone would still benefit.

But such hypothetical reasoning is not convincing if there are no mechanisms for the actual compensation of the losers; the wealthy might get most of the benefits and the poor pay most of the costs.

It often is asserted that if individual projects are judged by criteria of net economic benefit alone, any inequities that ensue can be mitigated by the adoption of income transfer policies, such as progressive taxes and welfare payments. I have urged, by contrast, that equity considerations should be part of project evaluation itself, and that such decisions are basically political. A few companies with huge contracts may reap a large portion of the benefits. Moreover, in a society in which wealth already is unevenly distributed, willingness to pay is a dubious criterion of social benefit. The preferences of a small minority may be heavily weighted because of the economic resources at their disposal. For example, the preference of an affluent backpacker, figured at $6 a day, counts as much as four low-income swimmers at $1.50 a day. The Rawls difference principle (Chapter 4) would lead to particular concern for recreational opportunities for the urban poor, but often, as in the Tocks Island case, alternatives nearer to urban centers would provide such recreation more effectively and at lower cost.

2. Discounting the Future

Over how long a time period, and at what discount rate, should future costs and benefits be calculated? The annual interest for the construction costs of the Tocks Island Dam was calculated at 3 1/8%. If instead the more realistic rate of 7% were used, the benefit/cost ratio would drop from 1.6 to 0.9. Benefits from dams usually are calculated for 100 years, though 50 years probably would be a more realistic figure. What happens after that, when the lake fills in? In the CBAs submitted by electric power companies in applying for licenses for new nuclear plants, time frames of only 20 or 30 years are used; the companies thereby dismiss problems of radioactive waste storage that will continue for hundreds of thousands of years. The deferred costs are passed on to future generations.[10] This is an extreme case of what often occurs: long-term indirect costs are delayed and discounted or omitted.

We noted (Chapter 5) some of the reasons for discounting the future. People prefer present benefits to future ones. A dollar now is worth more to me than the promise of a dollar ten years hence because I could invest the first dollar during the interim. But the use of current interest rates to discount environmental and health costs is dubious. Questions of resource depletion and harm to future generations are effectively excluded; any effects more than 30 years off are treated as insignificant for decisions today. No distinction is made between renewable and nonrenewable resources.[11] The Rawlsian concept of justice and some versions of utilitarianism count present and future people equally. This would imply a zero discount rate except where there are actual mechanisms by which present investment will produce augmented future benefits. We will see that there are no such mechanisms for compensating our distant descendants for long-lived risks.

3. Environmental Values

A number of economists have maintained that the market process in general, and CBA in particular, systematically undervalues irreplaceable natural assets. In the past, technology could be counted on to reduce the cost of natural resources or to provide substitutes. But nonrenewable resources are becoming scarcer and often there are no substitutes. Wilderness and unspoiled scenic areas will be more and more scarce, and technology will only make them scarcer. While the decision to *preserve* such areas is reversible (since the option to develop remains open), the decision to *develop* may be irreversible; wilderness once destroyed seldom can be recovered. The extinction of an endangered species is similarly irreversible; the genetic information once lost is irretrievable. CBA makes no allowance for the importance of keeping future options open or for preserving unique or rare natural environments. Nor can ecological balance and diversity be converted into dollar equivalents.[12]

4. Social Values

It has been argued that CBA is oriented to material values. The requirement of quantification leads to the neglect of intangible human values.[13] What dollar estimate could one attach to the disruption of a neighborhood divided by a freeway? The Tocks Island CBA considered property losses but not the destruction of valley communities or the social impact on the access regions. In what units can community cohesion and stability be measured? There have been attempts to quantify aesthetic values, but they are of limited applicability.[14] One can ask people what they would pay to visit an area of natural beauty, or to prevent its destruction, but this may not always be a valid way of judging its importance to people. In response to the Roskill Commission's CBA of alternative sites for a third London airport, there was a strong public reaction, prompted in part by the small weight given to natural beauty in comparison to the travel time of the affluent. The loss of irreplaceable Norman churches and other historical sites did not even enter the calculations.[15]

CBA is carried out in the framework of utilitarian assumptions. It deals with consequences and with quantitative totals. It cannot directly include deontic considerations (Chapter 4) such as duties, rights, and obligations. It is anthropocentric and excludes duties to animals, consideration of the welfare of nonhuman beings, and the stewardship of creation—except as they result in human economic benefits. Moreover, individual rights play no role, except as they are embodied in legal constraints that limit the options analyzed. The analyst may be firmly committed to justice or freedom, but these commitments will have to be expressed at other points in the decision-making process and not among the costs and benefits calculated.[16]

5. Institutional Biases

The previous four problems are inherent limitations in CBA as a method; the fifth and sixth are problems arising from the institutional setting in which it usually is carried out. In practice, CBA is almost always a means by which an agency justifies and promotes its own programs. The formulation of problems and the preselection of alternatives, which are frequently the most important decisions, occur before the analysis is made. In the analysis itself, an agency typically overstates benefits and understates costs. The use of artificially low interest rates already has been mentioned. There are numerous examples of costs that have been understated because of subsidized capital, preferential tax treatment, and neglected environmental effects.[17] Secondary benefits are assiduously sought, while secondary costs are conveniently overlooked. Double counting occurs when the same benefit appears under two headings, such as water supply and water quality. One author speaks of "the self-serving assumptions" of agency analyses.[18]

6. The Role of the Expert

While the assignment of monetary values appears to be a technical question, it often reflects the biases of analysts or their judgments of what the public wants. The Department of Transportation's CBA of the supersonic transport overestimated the demand for high-speed travel and the dollar value of the time saved. Differing weights would be assigned by various social groups to the incommensurable benefits of such projects. Value conflicts that should be resolved politically are concealed in what look like rational, neutral, objective calculations. This may appeal to administrators, but it hinders public debate of the policy issues and lessens the accountability of bureaucratic officials. Numbers carry an unwarranted authority when used to legitimate decisions that are basically political in character.[19]

In the Tocks Island case, as in many others, the experts were initially all on one side—employees of a government agency. Only gradually were other experts, the public, and elected officials drawn into the controversy, and the underlying assumptions and value judgments subjected to scrutiny and public discussion. In theory, CBA yields a document with explicit assumptions, open to public discussion, forming one among other inputs to democratic decision making. In practice it is often an instrument of technocratic planning, accessible only to specialists, and difficult for citizens or legislators to challenge. Because it usually is expressed in the language of professionals, public participation is inhibited.

The Uses of Cost-Benefit Analysis

What can be concluded concerning the potentialities and limitations of CBA? Some of the criticisms above are attacks on inadequate or biased CBAs; they can be answered by requiring *better analyses* and modifications in their *institutional context.* Many of the neglected costs can be included if the geographical boundaries of analysis are enlarged and if indirect impacts on the environment and various population groups are considered. Federal guidelines now specify more realistic discount rates, namely the current interest rate for government borrowing.

The problem of institutional bias can be mitigated if CBAs are carried out in *a variety of institutional settings.* Some authors have urged that analyses be done by disinterested experts on behalf of the public, perhaps at regional centers.[20] Others question whether any organization is disinterested, and they recommend that alternative plans and evaluations be prepared by several organizations representing diverse interests, including government agencies, independent consulting firms, university teams, and public interest groups.[21] Another proposal outlines a two-stage process in which alternative CBAs are prepared by adversary groups and presented before a technical-analytic court.[22] On particularly important decisions involving new technological options, the Office of Technology Assessment in the U.S. Congress can do its own CBAs.

Citizen participation provides a partial answer to reliance on experts, but it is not easily achieved. Public hearings early in the planning process can allow objectives and alternative options to be discussed by diverse stakeholders. Planning documents should be understandable and accessible for use by citizens' organizations. Participation can be time-consuming, and is sometimes dominated by local and parochial interests, which usually are more vocal in opposing projects than in suggesting constructive alternatives. Today there is widespread disillusionment with technocratic planning, and if opponents of development are not heard early in the planning process, they are likely to cause even more costly delays later through court challenges.

Other criticisms above refer to *inherent limitations* of CBA as a method. These limitations are not serious when a project has narrow and clearly defined objectives, when the main impacts are physical and readily quantifiable, and when there is a small number of options for achieving the objectives. In the extreme case when there is only one objective and no major indirect costs, CBA can be replaced by *cost-effectiveness* studies, which seek the cheapest way of accomplishing a given task or attaining a specified level of a benefit. One can compare the cost-effectiveness of alternative technologies for achieving a specific reduction in the level of a pollutant in a factory effluent, for example, or the cost-effectiveness of several proposed programs for early detection of cancer. Here the benefit does not have to be calculated in dollar equivalents, though it does have to be clearly defined. The analysis determines only how a benefit can be obtained in the cheapest way, not whether it is worth the cost.

When benefits and costs are more diverse, CBA can be a useful input into decision processes, provided people are aware of its *limitations* and its *institutional context.* In general, it introduces a bias toward development because many of the indirect costs are difficult to quantify. It requires some technical expertise, which is more easily available to some groups than to others. In the past, it often has been used late in the planning process to justify agency decisions already made on other grounds. Usually only options within a relatively narrow range of assumptions have been considered. But when it is used cautiously, as one input into an open and pluralistic decision process, it can contribute to rational discussion, the accountability of agency officials, and accessibility to review by Congress, the judiciary, and the public.

My own conclusion is that CBA can be a useful component of project planning, provided 1) there is pluralistic participation, public discussion, and recognition of assumptions and limitations, and 2) it is viewed as only one input into wider decision-making processes that can take unquantifiable consequences and intangible values into account. An environmental impact statement, in particular, can combine a CBA with an assessment of indirect social and environmental costs. The NEPA legislation specifies "consideration of presently unquantified environmental values along with economic and technical considerations." Technology assessment also provides a wider framework for evaluation of unquantifiable impacts. Both these assessment methods are discussed in the next chapter.

II. RISK-BENEFIT ANALYSIS

Potential harm to human life and health constitutes a particular kind of cost that often must be weighed against potential benefits. The techniques of cost-benefit analysis have been extended to allow quantitative treatment of such risks. We look first at ways of quantifying the value of a human life. We then take up ways of dealing with uncertainty, especially in estimating the effect of low-probability events such as those in low-level exposures and major accidents.

Risks to Life and Health

A *risk* is an uncertain potential harm to human life or health (or to property or environment, though these are not our concern here). It usually is expressed as the probability or chance of a particular harmful effect (death, injury, illness, genetic damage, and so on). The uncertainty in prediction may arise from random phenomena, unforeseeable events, or scientific ignorance. In some cases, such as auto fatalities, a risk can be estimated with considerable accuracy for a population, even though no one can predict which individuals will be killed. In other cases, such as the possibility that changes in the carbon dioxide content of the atmosphere will affect global climate (the "greenhouse effect"), estimates of risks to populations are uncertain because scientific data are limited. Estimates of risks and their relation to benefits have been central in many policy debates, including those on DDT, saccharin, asbestos fibers, auto emissions, and nuclear power.[23]

The *estimation of risks* is basically a scientific judgment. Usually the scientist first must estimate the expected levels of human exposure to a harmful agent (such as radiation, chemical pollutant, or carcinogen) and then estimate the harm from such exposure. Data on the latter sometimes can be obtained from occupational exposure, or from statistical studies of diverse population groups subject to differing exposure levels. Alternatively, dose-response data can be obtained from laboratory experiments on animals and extrapolated to human beings, assuming that the biological processes are similar. The risks from technological accidents can be estimated from past history and from engineering tests on system components.

In traditional risk-benefit analysis (RBA), risks then must be converted into *monetary costs* for comparison with benefits. The cost of illness or injury is calculated from data on medical expenditures plus wages lost. There are several ways of calculating the dollar equivalent of a human life. 1) *Projected future earnings* can be estimated. $300,000 is a typical average figure for a U.S. worker. 2) Inferences can be made from *social decisions*, such as court awards for injury or death, expenditures for safety equipment, or wage premiums for high-risk occupations. A $300-per-year bonus for a job with a one in 1,000 annual fatality rate, for instance, implies a valuation of life at $300,000. Court awards and safety expenditures vary widely, however, and persons choosing high-risk jobs may have motives that are not representative of the general population. 3) In *interviews and surveys*, people can be asked what they would pay for a small reduction in risk. The limited data that have been obtained by this method show very wide variations among differing types and levels of risk.[24]

The calculation of *the dollar value of a human life* by the projected future earnings method, used in most RBAs, may be criticized on conceptual grounds. If applied consistently, the method would require that the lives of the elderly would be considered valueless. If future earnings are discounted, a child's life would be worth much less than an adult's.[25] Future generations may value life very differently from the present generation. I would maintain that there are distinctive characteristics of human life that should make us hesitant to treat it as if it were a commodity on the market. Life cannot be transferred and its loss to a person is irreversible and irreplaceable. To speak of human life as sacred does not mean that it can be assigned an infinite value in practice, or that it cannot enter trade-offs with other costs and benefits. However, it does suggest that the cost in human lives should not be aggregated with economic costs, but kept as a separate kind of cost concerning which accountable decision makers, rather than technical analysts, should make the inescapable value judgments.

There are some circumstances in which a benefit and a risk to life can be directly compared in the same *nonmonetary units.* The number of deaths that would be averted by a life-saving drug or by diagnostic X-rays can be compared with the number of deaths that would be caused by their use. In other cases in which the only benefit is the saving of lives (as with safety equipment or such medical technologies as kidney dialysis), cost-effectiveness can be calculated as cost per life saved—without requiring that the benefit be expressed in dollars. In general, however, costs and benefits are diverse, and *partial aggregation* seems the most suitable procedure: the risk in lives lost per year is kept as a separate figure, and other costs and benefits are aggregated in dollars.[26] If one option is less costly in both lives and dollars, the choice is clear. If lives and dollars must be weighed against each other, the value judgment can be made by the responsible decision maker rather than hidden in the analysis. Safety is always a relative matter; the public has to decide how much safety it wants.

The *public acceptability of risks* is strongly influenced by many factors in addition to expected economic benefits. *Voluntary risks* (in sports, automobiles, and smoking, for instance) are widely accepted at levels roughly 1,000 times higher than involuntary ones (such as public hazards over which a person has no control). *Occupational hazards* often have been accepted as "part of the job," though occupational safety and exposure standards are now closer to standards for the general public than they were a decade ago. We will ask later whether acceptance is really voluntary and whether compensation for extra risks is adequate. *Risks to identifiable individuals* (a trapped miner or a critically injured child) are taken more seriously than statistical risks in which specific individuals cannot be identified in advance (such as carcinogens and auto accidents). *Major catastrophes* (10,000 people killed in one event, for instance) are publicized and feared more than equivalent risks from many smaller incidents (one person killed in each of 10,000 separate accidents). *Risk aversion* varies considerably with the context and with individual temperament; some people will gamble if there is a chance of gain, while others will give up a possible gain to avoid a possible loss or a period of uncertainty.[27] In sum, the magnitude of a risk is a scientific question, but the acceptability of a risk is a value judgment which must be decided through political processes and not by technical experts.

Dealing with Uncertainty

Many of the problems of RBA are similar to those of CBA. *Distributive justice* is an issue if one group of people is exposed to risks and another stands to benefit. People living downwind of a coal-fired plant are exposed to health hazards out of proportion to the benefits they receive from it. In some cases, such as occupational injuries, there are provisions for the victims to be *compensated* by the employer, the consumer, or society under liability and workman's compensation laws. In other cases some pooling of risks occurs through medical insurance, life insurance, and social security. But such provisions are seldom adequate, so the distribution of risks should be scrutinized in any policy decision, especially if people might be unaware of the risks or have no opportunity to accept them voluntarily.

Problems of *discounting the future* and of intergenerational justice are also similar to those discussed earlier. Often the benefits of a project are immediate and obvious while the risks are delayed and uncertain. Persistent chemicals, changes in carbon dioxide in the atmosphere, and radioactive wastes are among the long-term and possibly irreversible hazards to future generations. The problem of irreversibility compounds that of uncertainty. Later information may show that an action with which we have had little previous experience has irreversible consequences that had not been anticipated. In particular, genetic changes may not be evident for several generations and, if widespread, might be impossible to reverse. It seems to me irresponsible to apply a time-discount which would essentially dismiss such long-term risks.

Scientific uncertainties are a crucial problem in many RBAs. Health hazards from chemicals in the environment often involve extraordinarily complex interactions about which information is very limited. The impact of two chemicals together may be much more serious than that of either one alone (synergistic effects). While some ecosystems are resilient, others are quite fragile, and a small perturbation may have unexpected indirect effects. In some cases there are causal chains with many links. There is evidence, for example, that fluorocarbons from aerosol cans enter the atmosphere and interact with and deplete the ozone layer, allowing more ultraviolent radiation to reach the earth, thereby increasing the incidence of skin cancer. There is considerable uncertainty in estimates of the magnitude of several of these effects. Better scientific data are the answer to such uncertainties, but often decisions cannot await the results of further research. With aerosols, distributive justice is grossly violated because the entire human population is potentially at risk, whereas only a small group that prefers spray cans is the beneficiary. With uncertain but possibly catastrophic consequences, and only minor benefits, policy should err on the side of safety. The banning of fluorocarbon propellants (effective in 1979) was clearly justified, despite the uncertainties.

Data are particularly difficult to obtain on *low-probability events* such as the effects of low levels of radiation or suspected carcinogens. It is extremely difficult to show causal or even statistical relationships in human populations, since people are exposed to many pollutants and move among jobs, and there may be latency periods of 20 to 40 years between exposure and illness. As noted earlier, increased lung cancer from asbestos inhaled by shipyard workers in the 1940s was not detected until the 1970s. Even animal experiments at low levels would require experiments on thousands, and in some cases millions, of animals. The common practice is to obtain data with large dosages using a smaller group of animals, and then extrapolate the results to lower dosages—assuming that the dose-response curve is linear and there is no threshold or "safe" level. A typical study might require 500 animals, two years, and cost $250,000. There are short-term tests on mutations in bacteria that have shown fairly high correlations with animal tests; batteries of three or four such quick tests are being used to screen some families of chemicals.[28]

Catastrophes are low-probability, high-consequence events. In decision theory, a risk is defined as the probability of an occurrence multiplied by the magnitude of the consequences; probability and magnitude are treated symmetrically. But in policy decisions there may be particular reasons for taking greater precautions than the theory would suggest to avoid potentially large-scale disasters because their social disruption would be greater than a succession of smaller accidents. Again, there usually is greater uncertainty in estimating low probabilities, and the consequences of error or unexpected sequences of events in large-scale systems can be enormous.[29] In the case of dam failures, there is a previous history on which estimates of risks can be based (though subject to uncertainties about geological conditions, earthquake predictions, and the behavior of dams under stress). In the case of nuclear reactor accidents, there is little previous history and no possibility of direct experimentation, though experiments and theoretical calculations can be carried out concerning the failure of separate components. However, these calculations have been criticized for neglecting possible interaction among failure modes, and unlikely combinations of unforeseen circumstances. In addition, the contribution of human error is difficult to estimate, quite apart from deliberate human acts of war, sabotage, or blackmail by terrorist groups (see Chapter 10).

The status of *uncertain risks* has been central in several significant legal cases involving judicial review of agency decisions. Traditionally, court action has required a high level of proof, verging on virtual certainty for criminal conviction, or at least evidence "beyond reasonable doubt." But now courts are the scene of disputes about uncertain future risks. In *Ethyl Corporation* v. *EPA*, the majority ruled that EPA regulation of lead in gasoline does not require proof of actual harm or imminent danger, since the action was a precautionary measure under legislation designed to protect public health. In another landmark case the courts had to weigh the uncertain health hazard from the asbestos-like fibers in the 67,000 tons of iron-ore wastes dumped each day into Lake Superior by the Reserve Mining Company. The Circuit Court accepted "suspected but not completely substantiated relationships" as grounds for "a reasonable medical concern for public health" and precautionary actions by state agencies. But it held that the probable severity of risks should be balanced against social and economic benefits—especially to the 3,200 persons whose jobs would be affected if the plant closed. The court therefore ruled that the company should be given "a reasonable period" to convert to on-land disposal. The on-land facility was scheduled for completion in 1980 after a decade of litigation.[30]

A National Academy of Sciences report on *the regulation of chemicals in the environment* discusses the problems of incommensurability, uncertainty, irreversibility, and distributional equity mentioned above. Loss of life is kept as a separate entry so that the decision makers can provide their own explicit value judgments of lives versus dollars. Environmental effects are also kept separate, and only market costs and benefits are aggregated in dollars. The panel concludes:

> There is no satisfactory way to summarize all the costs or benefits of regulatory options in dollars or other terms which can be mathematically added, subtracted, or compared. In short, there is no sub-

> stitute for an experienced decision maker exercising good judgment. However, the techniques developed by decision theory and benefit-cost analysis can provide the decision maker with a useful framework and language for describing and discussing trade-offs, noncommensurability, and uncertainty. They also can help to clarify the existence of alternatives, decision points, gaps in information, and value judgments concerning trade-offs. . . . Value judgments about noncommensurate factors in a decision such as life, health, aesthetics, and equity should be explicitly dealt with by the politically responsible decision makers and not hidden in purportedly objective data and analysis.[31]

It appears that when risks and benefits are very diverse, or when there are significant scientific uncertainties, RBA should be used only with great caution, lest its results acquire an unwarranted authority. One detailed study of RBA concludes that it has been successfully applied to well-defined decisions between specific options having similar risks and benefits, but has been of limited use in broad policy decisions.[32] There is perhaps an intermediate level of decision in which RBA is useful primarily as a stimulus to discussion—provided the analyst makes clear the assumptions and limitations of the analysis so that it is accessible for debate and criticism. Especially when there are great uncertainties, institutional and personal biases are likely to influence assumptions, and multiple analyses by independent groups could clarify the issues. Diverse types of risk should be listed separately so that accountable decision makers can supply their own weights—within the guidelines provided by legislative processes and interpreted by the courts. After all, the balancing of risks to life and health against economic costs and benefits is a value judgment and a political act, not a scientific judgment.[33]

III. TOXIC SUBSTANCES

How might these general principles be applied in the regulation of toxic substances, which have become an increasingly serious problem? Between 1953 and 1960, 52 people died and 150 suffered serious brain and nerve damage in Minamata, Japan, before the cause was found: a plant discharging mercury into the bay where the residents fished. In Virginia, kepone from industrial wastes in the James River killed millions of fish and threatened human health. Fishing in the Hudson River was forbidden because of contamination by polychlorinated biphenyls (PCBs) from a General Electric plant.[34] Studies of the carcinogenic effects of vinyl chloride on animals were known to the chemical industry for two years before precautions were taken to protect workers exposed to it. Recent controversy has centered on food additives and consumer products, such as saccharin, Red Dye #2, DES added to cattle feed, and TRIS, a flame retardant used on children's sleepwear. Asbestos, benzene, chlorine, chloroform, and coal tar are among the identified carcinogens.

Cancer is the number two killer in the United States. The National Cancer Institute estimates that one-fourth of living U.S. citizens will die of it. Environmental causes are now believed to be responsible for 60 to 90% of all cases, and cancer incidence is increasing (even after allowance for the greater average age of the population).[35] Cigarettes and alcohol also are important causes, but pesticides, food additives, and industrial chemicals must be assigned a large share of the blame. Many of these chemicals contribute substantially to human welfare. How should risks be balanced against benefits in setting standards for toxic substances? What ethical issues arise in the policies of the regulatory agencies responsible for the protection of human life and health?

The chief *benefit from regulation* of toxic substances is the reduction of risks to human health and the environment (and the corresponding reduction in the costs of illness and pollution). The chief *costs of regulation* are the direct and indirect economic costs of compliance, plus governmental administrative costs. Congressional legislation has usually stated the dual objectives of reducing risks and minimizing economic costs, but in most cases it has left to agency discretion the way these conflicting goals are to be balanced in setting regulatory standards. A House subcommittee report in 1976 opposed the use of cost-benefit analysis in regulatory decisions because of the prevalence of unquantifiable factors and institutional biases.[36] But there has been renewed interest in the use of CBA to prevent "costly and excessive regulation" during a period of inflation.

Regulatory Standards

Congressional mandates for standard setting have varied from specific legislative determination to broad agency discretion. In some cases Congress has said that safety alone is to be the criterion. In other cases regulatory agencies have been instructed to balance risks to life and health against such economic and social costs as pollution control equipment, the loss of productivity, and the impact of compliance on inflation. In the previous chapter, the debates over air and water pollution standards were discussed. There have been similar debates over almost every toxic substance.[37]

The Nuclear Regulatory Commission has broad discretionary powers and makes extensive use of RBA techniques in setting *radiation standards.* There is an upper limit of 5 millirems per year for the most exposed member of the public, but within this limit the NRC sets design criteria for effluents from nuclear power plants "as low as is reasonably achievable." The emission standards determined with the use of CBA are not fixed and absolute, but are responsive to changes in technology, economic costs, and medical knowledge. In the calculations, a figure of $1,000 per man-rem of exposure is used—an arbitrary number, but a cautious one that probably attributes a higher value to human life than the future earnings method. Assumptions about the relation of emission levels to population exposure are also thought to be conservative. The balancing is done on a case-by-case basis; emission standards are not as stringent for older plants, for which additional control equipment would be more costly than for new plants. Some flexibility is allowed by the use of quarterly averages, and temporary violations are sometimes permitted in the interest of continuity of service, on the assumption that the standards provide a substantial margin of safety.[38]

A study of such uses of analytic methods in *the setting of radiation standards* was carried out in 1977 by the NAS. The study underscores many of the issues that we have discussed. "Irreversible processes, quality of life, risk avoidance, distributional effects, incommensurability, and ethical considerations are not adequately addressed in conventional benefit-cost analysis."[39] The study concludes that calculations in monetary units are a useful starting point, but that additional weighting factors should be introduced for costs or benefits (especially in relation to life and health) that are undervalued by marketplace economics in comparison with societal value judgments. "The values of the weighting factors have to be established by society in general, whether through the political process, public survey, or other means."[40]

In the case of *pesticides*, EPA must weigh benefits and costs of regulation, but does not have to use a formal CBA. Costs to farmers and to consumers from increased crop damage are to be considered. There is some built-in bias because the benefits of reduced risks are more uncertain than the costs; food chains are only partially understood, and data on delayed long-term or chronic effects are inadequate. DDT, aldrin, and dieldrin have been banned and EPA has restricted use of endrin and kepone. New pesticides must be tested before marketing, and the burden of proof of both benefit and safety lies with the manufacturer. However, the burden of proof in canceling the registration of a pesticide, once approved, is on EPA.[41]

The *Toxic Substances Control Act* of 1976 covers substances not included under pesticide or food and drug legislation. For a chemical in use, the burden of proof of harm is on EPA, though the act specially requires regulation of PCBs. For new chemicals, of which a thousand come on the market each year, the burden of proof of safety is on the manufacturer. Premarket screening usually can be done with animal skin tests and bacterial tests, but full animal testing is required for some types of chemicals. The EPA is dependent on data from the regulated industries and has funds for only occasional tests of its own. In reaction to earlier laws specifying absolute bans based on safety alone, the 1976 act specified that EPA should consider costs and should "balance economic and environmental imperatives." The EPA is therefore under pressure from environmental and consumer groups to tighten standards, and from industry and other government agencies to avoid regulations that would add to costs or inflation.[42]

The *Occupational Safety and Health Administration* (OSHA) is responsible for safety in the workplace. Standards have been set for asbestos, vinyl chloride, coke oven emissions, and a group of 14 carcinogens. The agency is supposed to weigh the interests of industries against those of workers and the public, but it is not told how a balance is to be struck. On acrylonitrile (used in making plastics), for example, OSHA reduced standards from 20 to 2 parts per million (ppm) when it was linked to higher cancer rates. Dupont says compliance will cost $3.5 million, and will prevent an estimated seven deaths per year from cancer, but that a further reduction to .2 ppm, which has been proposed, would cost $126 million to save only one additional life. The Council on Wage and Price Stability opposes the lower standards as inflationary.[43] OSHA has resisted any direct comparison of costs and benefits because the data are so uncertain. But its proposed reduction of workplace standards for benzene (from 10 to 1 ppm, which would cost industry an estimated $500 million) was rejected by the Court of Appeals because it was not based on at least a rough estimate of benefits and costs.[44]

The *Food and Drug Administration* (FDA) must weigh benefits against risks in the case of drugs and natural or unintentional food contaminants. But in the case of deliberate food additives, safety alone is to be considered. According to the Delaney amendment to the Food, Drug, and Cosmetic Act of 1958, "no additive shall be deemed safe . . . if it is found, after tests which are appropriate for the evaluation of the safety of food additives, to induce cancer in man or animal." In principle an agency cannot weigh benefits of an additive, no matter how large, against a cancer risk, no matter how small. In practice there is some latitude in defining what tests are "appropriate" and what substances are "food additives." Additives are easily removed and there are usually substitutes. The benefits to the consumer usually are small (except in the case of some preservatives). The chief beneficiary is the food company, which sells a more attractive product. In the case of artificial sweeteners there may be more substantial benefits, especially to diabetics and weight watchers.[45]

In 1977, FDA announced its intention to *ban saccharin* because Canadian experiments had shown a high incidence of bladder tumors in rats exposed to massive doses. The public outcry led Congress to forbid action for 18 months to allow further study; the period was subsequently extended. An NAS technical panel supported the extrapolation from animals to humans, and from high to low doses; it said that saccharin is indeed a weak carcinogen and its consumption is rising, especially among children. But it found the evidence of benefits to weight watchers inconclusive and made no definitive recommendations. The majority of an NAS policy panel in 1979 advocated repeal of the Delaney amendment in order to give the FDA greater flexibility in weighing costs and benefits, and a variety of options short of an outright ban. Saccharin could be sold separately, or sold over-the-counter if it can be classified as a drug, or sold in products with a warning label (though such labels seem to have had little effect on cigarette use). A minority of the panel (seven of 37 members) supported the Delaney clause because it ensures a margin of safety and gives the FDA some protection from political pressures.[46]

There are strong arguments for *absolute bans* of carcinogenic food additives. The preponderance of scientific opinion is that there are no thresholds or safe levels. We are exposed to an increasing variety of chemicals, some of whose effects may be cumulative, or synergistic, or may harm children yet unborn. Moreover, the drawing of an absolute line strengthens the hand of the FDA in resisting pressures from the industries affected.[47] My own conclusion, however, is that benefits as well as risks from saccharin should be taken into account, though not by a formal CBA. Life and health should be given very great but not infinite weight in such balancing. Controlled use may be a better option than a total ban if risks can be greatly reduced while many of its benefits are retained. Congress has given regulatory agencies too little guidance for establishing standards in the case of radiation, and too much guidance in the case of food additives. To counteract industry pressures on the agencies there should be greater public access and participation in agency decisions and more extensive use of outside experts representing a spectrum of viewpoints, in addition to review through congressional oversight hearings and the courts. Political pressures have not been absent in the past; they should be brought into the open.

Ethical Issues in Regulating Risks

Institutional biases are likely to be influential when there are great uncertainties in the data, as there often are with toxic substances. Costs usually are more immediate and more easily identified and quantified than the benefits of reduced risks. In addition to the ethical issues discussed earlier in looking at RBA, there are additional issues that arise in the context of regulatory agencies.

1. Paternalism versus Individual Freedom

It has been argued that people should be free to accept risks voluntarily if they want to, as long as they are fully informed and other people are not harmed. The libertarian tradition since J. S. Mill has held that the state should intervene to prevent people from harming others, but not to prevent them from harming themselves.[48] In the case of cigarettes, individuals are free to decide for themselves whether the pleasure is worth the risk; a warning label is attached only to inform them of the risk. But in an interdependent society there are few risks that a person can accept without affecting other people. Cigarette smoking increases insurance rates and medical costs for everyone. The cigarette policy adopted was more the product of the power of the tobacco industry and tobacco-growing states than of cost benefit analysis or libertarian traditions.

"Free, informed consent" to risks from toxic substances in the environment is a totally unrealistic goal. Most exposure is indirect, involuntary, and difficult to observe. Even when people are aware that there are risks, they have great difficulty evaluating complex statistical data, and little opportunity to choose their level of exposure. How free is a worker, even if fully informed, to turn down an unsafe job when work is scarce and mobility is limited? Future generations cannot give their consent. The control of toxic substances in the environment should be a matter for public policy, not individual choice.

2. Access to Information

In theory, adjustments in wages, land prices, and product prices could provide compensation for workers, homeowners, and consumers who voluntarily accept toxic risks. In practice, trade secrets limit the information available to the public concerning the substances to which they are exposed. Even the regulatory agencies are heavily dependent on the regulated industries—hardly an impartial source—for data on product safety and control costs. Unions and environmental and consumer groups have supported citizens in securing more adequate data and have begun to acquire technical expertise of their own. EPA can obtain trade secrets if it protects their confidentiality. But better access to information is a precondition of intelligent public decisions on regulatory standards.[49]

3. Pluralistic Participation

Who decides? I have suggested that toxic risks are not a matter for individual choice nor for technical expertise alone, but for debate in open decision-making processes. Since there are scientific uncertainties, differences in perceptions of risks, and value judgments in comparing incommensurable risks and benefits, it is particularly important that the procedure be seen as fair and open to all interested parties. In the past, producers have been better represented than consumers. Those who bear the risks should have an active voice, rather than being treated paternalistically as passive recipients of government protection. Input of the parties affected can occur through legislative and judicial channels, but also directly in regulatory agencies. Accountable officials are responsive to mobilized constituencies and to informed participation in hearings. Unions or public interest groups were initiators in 22 of 26 actions by EPA, OSHA, and FDA to regulate carcinogens.[50] Advice by outside scientific experts and external review by people representing a range of interests also should be more extensively used.

4. Economic Impacts

Industry has emphasized the economic costs of compliance and has usually favored CBAs as a way of preventing costly regulations. Regulatory agencies have been under frequent pressure from mission-oriented agencies with strong constituencies (agriculture, transportation, and energy, among others). Executive offices have more recently attacked the inflationary impact of proposed regulations. President Ford issued an executive order requiring economic impact statements, and President Carter in 1978 issued an order requiring a regulatory analysis on any regulation expected to have an annual effect of more than $100 million on the economy.[51] The guidelines call for a tabulation of "burdens and gains," including the costs of compliance, thereby encouraging the use of CBAs. OSHA backed down on proposed cotton dust limits after pressure from the Council of Economic Advisors. EPA was urged by the Council on Wage and Price Stability (which reviews regulatory analyses) to defer implementation of clean air standards. The Office of Management and Budget scrutinizes the administrative costs of regulations. These executive agencies have often prepared their own CBAs based on quite arbitrary assumptions, and with no provisions for a record subject to public discussion and congressional or judicial review.[52]

In looking to the future, *better scientific data* on health effects are essential for credible regulatory standards. In the previous chapter we noted the difficulties in establishing air emission standards for sulfur dioxide when information about the effects of sulfates is so limited. Congress has been reluctant to provide funds for epidemiological studies of human populations. Such studies are expensive, but the benefits of reducing the uncertainties are likely to be far greater. More research funds for testing new chemicals should go to scientists who are independent of chemical industries, on the one hand, and government regulators on the other. Scientists at the National Institutes of Health, universities and independent research institutes are not value-free, but they are less subject to the particular institutional biases which have influenced much of the past work. Research on the costs of compliance should also be carried out by regulatory agencies and academic economists so that decisions will not be as dependent on industry estimates, which are often inflated. Outside peer review of both scientific and economic data should be more frequently sought.[53]

Even with better data, value judgments are inescapable in setting regulatory standards. *Open decision-making procedures* in agencies are therefore essential. The courts can foster administrative accountability by insisting that agencies articulate the assumptions and the methods used in regulatory decisions. There should be a detailed record of how each decision was reached, including the treatment of distributional issues, unquantifiable factors, and rates for discounting the future. The limitations of the data and the presence of value judgments should be clearly indicated rather than buried in supposedly objective calculations. The visibility of the process will enhance its credibility and the opportunity for review by Congress, the courts, and the public.[54]

Will such procedures lead to *more rational decisions?* Risks may sometimes be overemphasized in political processes, especially in cases where the public has an emotional fear of an unfamiliar hazard or a new technology. The public can also be very complacent about risks from a familiar product or process. Studies of "risk perception" show that most people overestimate dangers from large-scale catastrophic events and underestimate low-level chronic effects.[55] Other studies suggest that when there is scientific uncertainty, public access to information is likely to stir up additional controversy.[56] Yet public debate is crucial in a democracy. In the last analysis it is the citizens who must decide how much safety they want, knowing the costs. We should aim for a better-informed public, while recognizing that risk acceptability is never a purely rational question. I agree, therefore, with the NAS study cited earlier that "Congress should provide increased and consistent statutory guidance as to the relative importance that should be given to health, environmental and economic factors in regulating chemicals," while leaving the agencies "some flexibility to respond to changes in scientific knowledge, technological feasibility and economic cost."[57]

I have suggested that CBA can be one useful input into decisions by accountable officials, as long as human lives are considered separately and unquantifiable effects are not ignored. *Analytic techniques* can help to identify and classify the effects of regulation and establish priorities for research and action. They can provide a framework for discussion, a tool for organizing information and a component of a documented record of the reasoning of regulatory officials which can facilitate congressional review and public accountability. They can help in identifying the uncertainties that are most crucial for policy decisions and the research that would be most useful. But the final decision is inescapably political and should not be hidden in the assumptions of a formalism.

Throughout the chapter I have stressed *the institutional contexts* in which costs and benefits are balanced. Formal techniques are seldom as objective as they appear, and they have inherent limitations in dealing with unquantifiable environmental and human values. Their use is a political act. I have therefore urged greater reliance on outside experts, informed participation by a variety of interest groups, and continuing congressional and judicial review. My own conclusion is that formal techniques can be used to obtain definitive answers only in relatively technical lower-level administrative decisions when there are clearly defined and widely accepted objectives which are not too diverse. When there are multiple and conflicting objectives on action options that have many different kinds of impact, analytic methods can stimulate dialogue and discussion and provide one input to responsible decision makers. In the next chapter I will suggest that technology assessment and environmental impact assessment provide a broader framework within which intangible values and pluralistic participation as well as quantitative analysis by technical experts can be included.

NOTES

1. Edward J. Mishan, *Cost-Benefit Analysis: An Introduction* (New York: Praeger, 1976); Orris Herfindahl and Allen Kneese, *Economic Theory of Natural Resources* (Columbus, Ohio: Charles Merrill, 1974); and A. Myrick Freeman, *The Benefits of Environmental Improvement* (Baltimore: Johns Hopkins University Press, 1979).

2. M. Clawson and J. Knetsch, *Economics of Outdoor Recreation* (Baltimore: Johns Hopkins University Press, 1966); and J. Kneese and R. K. Davis, "Comparison of Methods for Recreation Evaluation," in *Economics of the Environment*, 2d ed., ed. Robert and Nancy Dorfman (New York: W. W. Norton, 1977).

3. Data are taken from Harold Feiveson, Frank Sinden, and Robert Socolow, eds., *Boundaries of Analysis: An Inquiry into the Tocks Island Dam Controversy* (Cambridge, Mass.: Ballinger, 1976).

4. David Bradford and Harold Feiveson, "Benefits and Costs, Winners and Losers," in Feiveson et al., op. cit. Similar criticisms of the use of CBA in setting pollution control policy on the Delaware River below Philadelphia are given in Bruce Ackerman et al., *The Uncertain Search for Environmental Quality* (New York: Free Press, 1974), chaps. 6 to 9.

5. Walter Rosenbaum, *The Politics of Environmental Concern*, 2d ed. (New York: Praeger, 1977), chap. 6.

6. Laurence Tribe, Corinne Schelling, and John Voss, eds., *When Values Conflict: Essays on Environmental Analysis, Discourse and Decision* (Cambridge, Mass.: Ballinger, 1976), especially the preface and the chapter by Robert Socolow.

7. Water Resources Council, "Principles and Standards for Planning Water and Related Land Resources," *Federal Register* 38 (1973): 24778-841.

8. Jacquelyn Luke, "Environmental Impact Assessment for Water Resource Projects: The Army Corps of Engineers," *George Washington Law Rev.* 45 (1977): 1095-1122; and "Cost-Benefit Trips Up the Corps," *Business Week*, February 19, 1979, 96-97.

9. Henry Peskin and Eugene Seskin, eds., *Cost Benefit Analysis and Water Pollution Policy* (Washington, D.C.: Urban Institute, 1975), introduction.

10. See Kenneth Goodpaster and Kenneth Sayre, "An Ethical Analysis of Power Company Decision-Making," and Alasdair MacIntyre, "Utilitarianism and Cost-Benefit Analysis," in *Values in the Electric Power Industry*, ed. Kenneth Sayre (Notre Dame, Ind.: University of Notre Dame Press, 1977).

11. Robert M. Solow, "Intergenerational Equity and Exhaustible Resources," *Rev. Economic Studies* (1974): 29-46; idem, "The Economics of Resources or the Resources of Economics," *Amer. Economic Rev.* 64 (May 1974): 1-14; and Shaul Ben David, Allen Kneese and William Schulze, "A Study of the Ethical Foundations of Benefit-Cost Analysis," (Albuquerque, N.M.: Dept. of Economics, University of New Mexico, 1979).

12. John Krutilla and Anthony Fisher, *The Economics of Natural Environments* (Baltimore: Johns Hopkins University Press, 1975); and idem, "Valuing Long-Run Ecological Consequences and Irreversibilities," *J. Envir. Economics and Management* 1 (1974): 96-108, reprinted in Peskin and Seskin, op. cit. See also Allen Kneese and Blair Bower, eds., *Environmental Quality Analysis* (Baltimore: Johns Hopkins University Press, 1972).

13. E. F. Schumacher, *Small is Beautiful* (New York: Harper & Row, 1973), chap. 3; and Hazel Henderson, "Ecologists versus Economists," *Harvard Business Rev.* 51 (July 1973): 28-36 and 152-57.

14. Luna Leopold, "Landscape Esthetics: How to Quantify the Scenics of a River Valley," *Natural History* 78 (October 1969): 36-44; R. Burton Litton, "Aesthetic Dimensions in the Landscape," in *Natural Environments: Studies in Theoretical and Applied Analysis*, ed. John Krutilla (Baltimore: Johns Hopkins University Press, 1972).

15. Christopher Freeman, "Technology Assessment and its Social Context," *Studium Generale* 24 (1971): 1038-50; and Denys Munby, "The Christian Ethics of Cost-Benefit Analysis," *Anticipation*, no. 7 (April 1971): 16-22.

16. Laurence Tribe, "Policy Science: Analysis or Ideology," *Philosophy and Public Affairs* 2 (1972): 66-110; idem, "Technology Assessment and the Fourth Discontinuity," *Southern Cal. Law Review* 46 (1973): 617; and idem, "Ways not to Think about Plastic Trees," in Tribe et al., op. cit.

17. Krutilla and Fisher, op, cit., chap. 11.

18. Peskin and Seskin, op. cit., p. 30.

19. Alan Coddington, "Cost-Benefit as the New Utilitarianism," *Political Quarterly* 42 (1971): 320-25; John Benington and Paul Skelton, "Public Participation in Decision-Making by Governments," in *Benefit-Cost and Policy Analysis 1974*, ed. Richard Zeckhauser (Chicago: Aldine, 1975); and Thomas Means, "The Concorde Calculus," *George Washington Law Rev.* 45 (1977): 1037-65.

20. Krutilla and Fisher, *Economics of Natural Environments*, op. cit., pp. 281ff.

21. Bradford and Feiveson, op. cit., p. 157.

22. Harvey Brooks in Tribe et al., op. cit., pp. 133-34.

23. Excellent surveys are William Lowrance, *Of Acceptable Risk* (Los Altos, Calif.: William Kaufman, 1976); and Andrew Van Horn and Richard Wilson, *The Status of Risk-Benefit Analysis* (Cambridge, Mass.: Harvard Energy and Environmental Policy Center, 1976). See also National Academy of Engineering (NAE), *Perspectives on Benefit-Risk Decision Making* (Washington, D.C.: NAE, 1972).

24. Richard Zeckhauser, "Procedures for Valuing Lives," *Public Policy* 23 (1975): 419-64.

25. Ida Hoos, *Systems Analysis in Public Policy: A Critique* (Berkeley: University of California Press, 1972) chap. 6.

26. A procedure for partial aggregation is given in National Academy of Sciences (NAS), *Decision Making for Regulating Chemicals in the Environment* (Washington, D.C.: NAS, 1975), chap. 6 and appendix H.

27. Lowrance, op. cit., chap. 3; Chauncey Starr, "Social Benefit versus Technological Risk," *Science* 165 (1969): 1232-38; Paul Slovic, Baruch Fischoff, and Sarah Lichtenstein, "Cognitive Processes and Societal Risk Taking," in *Risk-Benefit Methodology and Application*, ed. David Okrent (Los Angeles, Calif.: University of California School of Engineering and Applied Science, 1975).

28. Thomas Maugh, "Chemical Carcinogens: The Scientific Basis," *Science* 201 (1978): 1200-05.

29. Joel Yellin, "Judicial Review and Nuclear Power: Assessing the Risks of Environmental Catastrophe," *George Washington Law Rev.* 45 (1977): 969-93. Other references are in Chapter 10.

30. Bowden Brown, "Projected Environmental Harm: Judicial Acceptance of a Concept of Uncertain Risk," *J. Urban Law* 53 (1976): 497-531; Marcia Gelpe and A. Dan Tarlock, "The Use of Scientific Information in Environmental Decision-Making," *Southern Cal. Law Rev.* 48 (1974): 371-427; and Norman Vig, "Environmental Decision-Making in the Lower Courts: The Reserve Mining Case" in *Energy and Environmental Issues: The Making and Implementation of Public Policy*, ed. Michael Steinman (Lexington, Mass.: Lexington, 1979).

31. NAS, *Decision Making for Regulating Chemicals*, op. cit., pp. 2 and 7.

32. Van Horn and Wilson, op. cit., pp. 24-25.

33. See articles by Harold Green in National Academy of Engineering, op. cit., and in Okrent, op. cit.; also "The Risk-Benefit Calculus in Safety Determinations," *George Washington Law Rev.* 43 (1975): 791-807. See also British Council for Science and Society, *The Acceptability of Risk* (Chichester, England: Barry Rose, 1977).

34. G. Tyler Miller, *Living in the Environment* (Belmont, Calif.: Wadsworth, 1975), pp. E40 and E98.

35. Council on Environmental Quality (CEQ), *Environmental Quality 1978* (Washington, D.C.: CEQ, 1978), p. 209; and World Health Organization, *Prevention of Cancer*, technical report, series 276 (Geneva: 1974). See also Samuel Epstein, *The Politics of Cancer* (San Francisco: Sierra Club, 1978); and Larry Agran, *The Cancer Connection* (Boston: Houghton Mifflin, 1977).

36. *Federal Regulation and Regulatory Reform* (Subcommittee on Oversight and Investigations, Committee on Interstate and Foreign Commerce, U. S. House of Representatives, 1976), p. 515.

37. Paul Hendrickson, Craig McDonald, and A. Henry Schilling, *Review of Decision Methodologies for Evaluating Regulatory Actions Affecting Public Health and Safety* (Richland, Wash.: Batelle Northwest, 1976).

38. Nuclear Regulatory Commission, *Rulemaking Hearing on Numerical Guides for Design Objectives and Limiting Conditions for Operation to Meet the Criterion "As Low as Practicable" for Radioactive Material in Light-Water Cooled Nuclear Power Reactor Effluents*, docket no. RM-50-2 (April 30, 1975).

39. National Academy of Sciences, *Considerations of Health Benefit-Cost Analysis for Activities Involving Ionizing Radiation Exposure and Alternatives* (Washington, D.C.: NAS, 1977), p. 8.

40. Ibid., p. 70. See also Michael Baram, "Regulation of Environmental Carcinogens: Why Cost-Benefit Analysis May Be Harmful to Your Health," *Technology Review* 8 (July-August 1976): 40-42.

41. Wilma McCarey, "Pesticide Regulation: Risk Assessment and Burden of Proof," *George Washington Law Rev.* 45 (1977): 1066-94.

42. *Environmental Quality 1978*, op. cit., chap. 4; and Paul Portney, "Toxic Substances and the Protection of Human Health," in *Current Issues in U.S. Environmental Policy*, ed. idem. (Baltimore: Johns Hopkins University Press, 1978).

43. Peter Behr, "Controlling Chemical Hazards," *Environment* 20 (July 1978): 25-29.

44. Luther Carter, "Dispute over Cancer Risk Quantification," *Science* 203 (1979): 1324-25.

45. National Academy of Sciences, *How Safe is Safe Enough: The Design of Policy on Drugs and Food Additives* (Washington, D.C.: NAS, 1974); President's Science Advisory Committee, *Chemicals and Health* (Washington, D.C.: National Science Foundation, 1973); and Barbara Culliton, "Saccharin: A Chemical in Search of an Identity," *Science* 196 (1977): 1179-83.

46. National Academy of Sciences, *Saccharin: Technical Assessment of Risks and Benefits* (Washington, D.C.: NAS, 1978); and idem, *Food Safety Policy: Scientific and Societal Considerations* (Washington, D.C.: NAS, 1979); summaries in *Science* 202 (1978): 852 and 203 (1979): 1221-24.

47. See, for example, the symposia on the Delaney amendment in *Preventive Medicine* 2 (1973): 125-70; and *Chem. and Engineering News* 55 (June 27, 1977): 24-46.

48. Deborah Johnson, "Individual Consent and Toxic Substances," in *Toxic Substances: Decisions and Values, Vol. I* (Washington, D.C.: Technical Information Project, 1979).

49. *Toxic Substances and Trade Secrecy* (Washington, D.C.: Technical Information Project, 1977).

50. Jeffrey Smith, "Toxic Substances: EPA and OSHA are Reluctant Regulators," *Science* 203 (1979): 28.

51. Executive Order 12044, "Regulatory Analysis," *Federal Register* 43 (1978): 12661.

52. Michael Baram, "Regulation of Health, Safety, and Environmental Quality and the Use of Cost-Benefit Analysis" (Washington, D.C.: Administrative Conference of the U.S., 1979).

53. *Toxic Substances: Decisions and Values, Vol. II* (Washington, D.C.: Technical Information Project, 1979); and Richard Tobin, *The Social Gamble* (Lexington, Mass.: Lexington, 1979), chap. 10.

54. David L. Bazelon, "Risk and Responsibility," *Science* 205 (1979): 277-80; Michael Baram, op. cit.

55. Baruch Fischoff et al., "Weighing the Risks," *Environment* 21, no. 4 (1979): 17-20, 32-38; Paul Slovic et al., "Rating the Risks," *Environment* 21, no. 3 (1979): 14-20, 36-39.

56. Dorothy Nelkin and Susan Fallows, "The Evolution of the Nuclear Debate: The Role of Public Participation," *Annual Review of Energy* 3 (1978): 275-312.

57. NAS, *Decision Making for Regulatory Chemicals*, op. cit., p. 20.

9
ASSESSMENT METHODS

We have seen some of the limitations of cost-benefit analysis, and similar analytic techniques, in dealing with unquantifiable effects, value judgments, and public participation. Environmental impact assessment and technology assessment provide procedures within which environmental and human values can be more adequately considered. We will look particularly at environmental impact statements and technology assessments and their use in decision making. The last section of the chapter explores assessment from the standpoint of two recurrent issues of this volume: 1) the relation of value judgments and scientific judgments in policy choice, and 2) mechanisms for integrating the roles of scientists, agency officials, elected representatives, and citizens in decision making about technology and the environment.

I. ENVIRONMENTAL IMPACT ASSESSMENT

The requirement of an assessment of the environmental impact of major federal projects was established by the National Environmental Policy Act (NEPA) of 1969. How has the act been interpreted by the courts and applied by federal agencies? Have environmental impact statements been an effective procedure for the inclusion of environmental values in project planning? Has NEPA facilitated public participation in agency decisions?

The National Environmental Policy Act

NEPA has been called the Magna Carta of environmental protection. It hardly warrants such an appraisal, but it did initiate some significant new procedures. A *new national policy* concerning comprehensive federal responsibility for the environment is proclaimed in section 101 of the act. The policy not only contrasts with the historic U.S. dedication to economic growth and development, but also goes beyond the resource-management ideals of Theodore Roosevelt and the conservation agencies of the 1930s. The new goals include "the widest range of beneficial uses of the environment without degradation," "a balance between population and resource use," and "the maximum attainable recycling of depletable resources." Each generation, the act states, has responsibilities "as trustee of the environment for succeeding generations." Section 102 contains the crucial *action-forcing provision* that "every recommendation or report on proposals for legislation and other major federal actions significantly affecting the quality of the human environment" must include a detailed statement of environmental impacts.[1]

NEPA also established a three-member *Council on Environmental Quality* (CEQ), with responsibilities to review the environmental activities of all federal agencies, develop guidelines for the preparation of impact statements, issue annual reports, and advise the president on national environmental policies. President Nixon had opposed the legislation leading up to NEPA, but once it was passed he recognized that environmental protection was a popular issue, and for a while gave it considerable support. He signed the bill on January 1, 1970 and appointed capable members to CEQ. But neither the Nixon nor the Ford administration provided sustained or aggressive environmental leadership, and the CEQ had a very limited staff and statutory authority. Meanwhile the attention of Congress was turning to energy and economic issues. Consequently much of the burden of interpreting the requirements for environmental impact statements fell on the judicial system.

NEPA provided unprecedented opportunities for *public participation* in environmental decision making. Along with the Freedom of Information Act, it gave citizens access to a vast amount of information that would otherwise have remained hidden in agency files. An agency's draft environmental impact statement (EIS) is circulated to other federal and state agencies and citizens' organizations for comments, which are to be included in the final EIS. The initiating agency is supposed to discuss and answer these comments in detail, though these responses often have been rather cursory. The EIS requirement, coupled with the possibility of later court action, gives to environmental groups new points of access to agency decision making. Each EIS must consider alternative action options, including the option of "no action"—though few impact statements to date have given the detailed comparisons of alternatives that would make them genuine instruments of rational planning. Nevertheless, the EIS process represents a significant vehicle for public disclosure, agency accountability, and citizen participation.

NEPA at first had little influence on *agency decisions.* Bureaucracies are notoriously resistant to change. Each agency has its traditional constituency which in general is development oriented. It wants to promote its own dominant mission and interests. Most agencies thus initially followed their established procedures, and tacked on an EIS to justify a decision already made—rather than using it as a real input in their planning. They resented the uncertainties of litigation and the delays created by the backlog of projects to be evaluated. But a survey by the CEQ after six years' experience claims that the delays were no longer substantial if work on an EIS was initiated at an early stage of planning. Agencies at first prepared brief and superficial statements, which often were deemed unacceptable by the courts. They then went to the opposite extreme and presented huge multivolume reports as impressive court exhibits–though much of the material was of dubious relevance and comprehensibility.[2] In 1978 CEQ introduced new regulations that call for a shorter EIS (normally under 150 pages) focused on the central issues. There is a new "scoping" procedure, in which all interested parties participate, to identify the most significant issues early in the planning process. Environmental groups, industry, and agency officials have welcomed the changes, which should reduce delays and produce better decisions without a loss in public involvement.[3]

The most effective mechanism in *the EIS process* has been the opportunity for public review and comment. Citizen activity, access to information, and judicial receptiveness have combined to bring agency decisions under scrutiny. The EIS often has served as a rallying point in mobilizing local opposition to a project. The second mechanism, review by other agencies, has been less effective, since agencies usually try to minimize conflict by accommodation. The third mechanism, the use of the EIS in planning within agencies, has only begun to be significant. As noted earlier, every agency has its traditional clients who benefit from and support its primary mission. But the effectiveness of environmental groups in court action brought new constituencies into the orbit of agency deliberation. Administrative officials became more aware of environmental consequences, and a wider range of goals and criteria are entering the earlier stages of planning.[4]

The EIS requirement at least forces an agency to look systematically at the environmental consequences of proposed projects, and in a number of cases has resulted in *the modification of decisions.* The decision of the Department of Transportation to discontinue the construction of I-66 in Washington, D.C. is attributed mainly to the EIS. The U.S. Corps of Engineers has altered substantially several major projects, and abandoned others, because of NEPA. There also have been more subtle modifications in agency attitudes, and efforts have been made to obtain more diverse staff with greater environmental expertise.[5] In addition, the courts have begun to require "program" or "generic" impact statements, covering a type or group of projects, even earlier in the planning process before specific sites are considered. This was required for the breeder reactor, for example, and for coal development on public lands. Such generic impact statements resemble technology assessments, which are discussed later.

The future effectiveness of the EIS will depend mainly on the extent to which it is *integrated within project planning.* If environmental factors are considered early in planning, when alternative options are really open, the EIS can be a significant policy and decision tool, rather than an after-the-fact justification or a burden of paperwork endured to satisfy the courts. But NEPA does not prevent an agency from ignoring an EIS once it is prepared. The 1978 guidelines at least require a record of how an EIS is used. It would be unrealistic to require an agency to adopt the option with the smallest environmental impact; environmental values do not always outweigh other national goals or the primary mission of the agency. NEPA could be strengthened, however, if an agency were required to adopt the mitigation measures identified as practicable in the EIS (such as changes in design and location, or techniques for reclamation and pollution control). Courts could then review the implementation of the EIS as well as its preparation. Subsequent scientific observation and monitoring should also be required to determine whether estimates of predicted impacts were valid. The comparison of predicted and observed effects of mitigation measures should lead to EIS studies of greater scientific credibility.[6]

The future of the EIS also depends on the extent to which it receives *support from Congress*. The Alaska Pipeline Bill (1973) legislated that reports already submitted had satisfied NEPA requirements—which in effect exempted the pipeline from further scrutiny under those requirements. Senator Henry Jackson favored the pipeline but opposed the exemption, which he realized would weaken NEPA (of which he had been a leading advocate). The Senate vote was exactly divided, and Vice-President Spiro Agnew cast the deciding vote, favoring exemption because of the energy crisis. NEPA would be seriously undermined if licensing and EIS requirements are waived for future energy projects. In the long run, then, the effectiveness of the EIS depends on the attitudes of Congress, and ultimately of the U.S. public—attitudes not only toward the environment, but also toward energy growth, consumption, and nonrenewable resources.

Environmental Impact Statements

Impact statements are required not only from federal agencies, but also from agencies in states that have passed legislation similar to NEPA, and from industries that need federal or state permits for construction or development activities. There is no prescribed format, but an EIS must include: a description of the current environmental situation; a description of the proposed project; an analysis of probable environmental consequences of all phases of the project, including short- and long-run effects; a comparison with feasible alternative options; and suggestions for minimizing adverse effects.

One format frequently used is a matrix chart described in a U.S. Geological Survey circular by Luna Leopold et al.[7] The chart has 90 *horizontal rows*, each representing a physical, chemical, biological, or cultural characteristic of the environment. For example, under biological characteristics are nine rows for fauna (birds, fish, insects, endangered species, and so on). Under cultural characteristics are four rows for cultural status (cultural patterns, health and safety, employment, and population density). The form also is divided into 100 *vertical columns*, each representing a component of the proposed project activity, grouped by categories (including construction, resource extraction, processing, transportation, and waste production) and broken down into more specific subcategories. The resulting form has 9,000 boxes or cells, each representing the impact of one project activity component on one environmental characteristic. Usually only a small fraction of these cells will be relevant to any given project, but the two-dimensional checklist of impact categories helps to ensure that no significant effects are overlooked.

The assessor is instructed to record two numbers (on a scale from one to ten) in each relevant cell, the first number representing the *magnitude* and the second the *importance* of that particular impact. According to the circular, the magnitude of the impact is a factual, objective, scientific question, while the assignment of importance is a value judgment on the part of the assessor. It is recommended that several members of an assessment team, with differing interests and disciplinary backgrounds, should make independent ratings of importance and examine the reasons for their disagreements before preparing a composite matrix. The text of the EIS then will give details on all of the major entries, and suggest ways in which impacts could be reduced. The matrix serves as an initial search list and as a summary, but the text carries the burden of analysis. No attempt is made to aggregate the numbers to obtain overall ratings. The problems of incommensurability and aggregation, which arise in cost-benefit analysis, are thereby avoided. However, the numbers in the corresponding cells for alternative project options can be directly compared, allowing a point-by-point comparison.

A recent textbook manual on EIS preparation emphasizes that there is no one correct method; *the procedure should be adapted to the situation.* With or without the use of a matrix, the EIS should include a thorough ecological inventory and a very careful analysis of potential forms of pollution (such as air, surface water, ground water, noise, and chemical wastes) and their consequences. This manual offers an optional numerical evaluation scheme, simpler than a matrix, in which each member of a panel of experts is asked to assign an overall rating to the impacts of each aspect of the project under four classifications: biotics, natural resources, socioeconomics, and aesthetics. It is claimed that each expert will have biases, but that these will tend to average out. The manual stresses the preparation of a broad, well-documented, interdisciplinary report; the technical information should be presented in ways that are understandable by citizens as well as by experts.[8]

Many *criticisms of impact statements* are indictments of the way they have been carried out in practice. Prior to the 1978 guidelines calling for briefer reports, many statements included voluminous details on minor issues, while important impacts were overlooked or minimized. The typical EIS compiled existing scientific data, which often were inadequate and out of date; little original investigation was carried out. Particular value judgments given by assessors also have been attacked. For example, the widely used USGS circular gives, as a model example, an impact statement for the leasing of an area of federal lands in California for phosphate mining. The example discusses the project's impact on the Sespe Condor Sanctuary, one of the few remaining nesting areas of the California condor. Critics have said that the impact on an endangered species should have been given much heavier weight in this assessment.[9]

One study of a group of impact statements prepared by industries and government agencies suggests that the *implicit value system* was "that of the developer." Negative impacts were dismissed as minimal or easily mitigated, while the value of industrial expansion and economic growth was assumed. Interviews with the university scientists who carried out the scientific work on which these impact statements were based showed that they saw themselves as data-gathering technicians fulfilling contractual agreements; few acknowledged any responsibility to the public for the way their data was used. They usually were not aware of how their reports were rewritten, ignored, or selectively quoted in the final EIS. There was some evidence that EIS preparation involved perfunctory or superficial scientific work in which normal professional standards were compromised.[10]

A common deficiency in the past has been *the neglect of social impacts.* NEPA itself mandates "a systematic, interdisciplinary approach which will insure the integrated use of the natural and social sciences," and the CEQ has called for a broad consideration of the human costs and benefits of proposed projects. Yet, with the exception of economics, which always has been strongly represented (whether or not cost-benefit analyses were included), the social sciences have played a small part in most impact statements. Suggestions for ameliorating adverse social impacts have been rare compared to those for physical and biological phenomena. There have been cases where the EIS of a project likely to lead to the doubling of a county's population within a few years has made only passing reference to schools, social services, and community patterns of life. One study has suggested that another common shortcoming, the failure to analyze the distribution of impacts among diverse social groups, may be partly a product of the myth nourished by many agencies that they are servants of an undifferentiated public interest.[11]

There is, however, a growing recognition of the importance of *social impact assessment.* The EIS manual mentioned above devotes two chapters to such questions as political institutions, social services, the character of neighborhoods, community cohesion, the social distribution of costs, and land use and development patterns. Studies of highway projects today sometimes go beyond such obvious effects as noise and air pollution to consider the more intangible psychological and sociological results of dividing neighborhoods or isolating and connecting living spaces in new ways.[12] A study of assessment procedures for projects of the Corps of Engineers dwells on the importance of social science research, identification of the perceptions of the groups affected, and the use of social indicators and quality of life indexes as measures of a project's consequences. As mentioned earlier, the Water Resources Council advocates the evaluation of projects under four separate sets of accounts: national economic development, regional development, environmental quality, and social well-being.[13]

In a study of the *methodology of social impact assessment*, many of the authors urge careful analysis of public perceptions, subjective responses, and the distribution of impacts. They outline the contribution of psychologists and sociologists to multidisciplinary teams studying the human repercussions of proposed projects. Several of these authors hold that interaction between experts and citizens should occur at many points throughout the planning process—both as a source of information about local interests and probable responses, and as a way of broadening the base of the value judgments that occur in any assessment process. Planning should be pluralistic, they suggest, because there are ideological components in the assessment of the consequences of alternative options.[14] These broader forms of EIS, with extensive provisions for public participation, seem to provide a framework within which questions of social justice and intangible human and environmental values can be appropriately raised.

II. TECHNOLOGY ASSESSMENT

Technology assessment (TA) is in many ways similar to environmental impact assessment, but it is broader in scope. It examines the potential consequences of a new technology, or a new application of an existing technology, rather than a particular project. In this respect it resembles a generic or program EIS that analyzes the environmental impact of a whole class of projects. It therefore occurs earlier in the history of a technological development, before particular sites are considered. A TA deals comprehensively with the full range of social and environmental impacts. It is oriented toward policy decisions by a legislature, a government agency, or, occasionally, an industrial corporation. It provides decision makers (especially legislators) with analyses of various consequences that are likely to follow from alternative policy options for developing or controlling a technology.

The Goals of Technology Assessment

Four general characteristics of technology assessment can be identified:[15]

1. Early Anticipation

TA is an attempt to anticipate consequences beforehand rather than waiting for them to become evident. It is a future-oriented inquiry, an "early warning system." The crucial decisions in the social management of a new technology should be made at an early point in its deployment, before heavy financial investments and employment patterns have built up pressures for its perpetuation. The benefits of a technology usually are immediate and obvious; the costs and risks often are delayed, remote, and cumulative. Careful study and foresight can identify some of these effects before the new technology has acquired a momentum that is difficult to control.

2. Diverse Impacts

In TA studies, a wide range of impacts are considered, beneficial as well as adverse, social and political as well as environmental and economic. Particular scrutiny is given to unintended effects, second-order consequences, indirect costs, and long-term repercussions. Because these impacts are diverse, the assessment should be carried out by an interdisciplinary team including sociologists and economists, along with biological and physical scientists and engineers. The analysis should deal with both quantifiable and unquantifiable variables.

3. Diverse Stakeholders

In the past, representatives of government and industry have presented the benefits of a new technology, whereas people who may have to face the indirect costs often have had no effective voice. TA requires the identification of the main parties who may be affected and an analysis of the consequences of the technology for each party. The impacts on different social and economic groups may vary widely. The assessment team must examine and take into account the interests of all impacted parties, which is a broader task than that of an EIS. Provision can be made for public participation at certain stages of the assessment process to ensure that diverse interests and viewpoints are represented.

4. Alternative Policies

Assessments not only trace current trends but also analyze the effects of alternative policies. As a policy-making tool, TA is designed to present decision makers with information about the probable consequences of various options for legislative or agency decision (or for industrial management, though industry rarely has sponsored wide-ranging assessments). Just as an EIS must be prepared early enough so that project options are really open, a TA must be prepared early enough so that policy options are indeed open. Further, if the studies are to be useful to decision makers they must be understandable to the nonspecialist, and some thought must be given to modes of implementation and regulation.

The relationship of assessment to the decision-making process takes various forms, depending on *the sponsoring agency.* The National Academy of Sciences has carried out many studies that are essentially TAs; these reports, prepared by panels of experts, are transmitted to the congressional committees or government agencies that requested and funded them. Many assessments are carried out within federal or state agencies, though these usually are narrower in scope. A series of comprehensive TAs has been commissioned by the National Science Foundation on such subjects as solar energy, advanced auto engine design, and electronic funds transfer. The Office of Technology Assessment (OTA), established by Congress in 1972 has at the request of particular congressional committees prepared interdisciplinary assessments of technologies under their jurisdiction.[16]

The actual work of assessment is carried out in a variety of *institutional contexts*, each of which has characteristic strengths and weaknesses. 1) In universities there are experts from many disciplines available, but there are barriers to interdisciplinary research. The career patterns and reward structures of academia operate along departmental lines. Problems of scheduling and management in universities also hinder the rapid preparation of reports. 2) Research institutes and consulting firms usually have a narrower spectrum of disciplines on their staff, but have developed managerial techniques to produce studies for clients with short deadlines. 3) Assessment teams within government agencies are closer to the officials with power to act, and the data they assemble can be used subsequently for other purposes. But their reports often have reflected institutional biases and inadequate staffing; in a few cases, unwelcome conclusions have been suppressed. 4) OTA has been gradually appointing its own in-house interdisciplinary staff, whose reports have been in general more satisfactory than those done under contract by outside consulting firms. However, there have been political pressures on the OTA staff, which at times may have compromised its independence.

Another issue is *the timing of the assessment* relative to the stage of technological development. An assessment early in the development of an emerging technology can investigate a broad range of options, including not deploying the technology at all. But early assessments are more speculative, and both their projections and the delineation of impacts are more uncertain. Moreover, it may be difficult to generate interest among legislators or the public in applications that are too far in the future. Assessments late in the history of an existing technology, on the other hand, may be ineffective in controlling the momentum already acquired or in influencing the financial and political forces already committed. If positions have polarized, a TA is more likely to generate controversy and be criticized as biased.

Finally, there are differences of opinion as to how far a TA should go in making *policy recommendations.* According to one expert, a TA should only "identify societal options and clarify trade-offs to provide an input to public decision-making and policy formulation."[17] The Deputy Director of OTA insists that its task is to inform Congress of the predicted consequences of various policy options, but not to recommend any one policy.[18] A leading proponent of TA seeks a middle position between neutral studies without recommendations, at the one extreme, and partisan studies that strongly advocate predetermined views, at the other. He sees a place for assessments that go "a step beyond the evenhanded analysis of consequences" to "highlight various desirable policy options," giving clear reasons for the preferred choices.[19] At a conference of assessors, some speakers held that in the interests of objectivity and neutrality TAs should avoid value judgments and recommendations, presenting only policy menus from which decision makers can make informed choices. Others, with whom I am more sympathetic, maintained that claims of neutrality only conceal covert value judgments, and that busy decision makers appreciate recommendations, as long as the alternatives, criteria, tradeoffs, and areas of uncertainty are clearly indicated.[20]

The Office of Technology Assessment

In addition to the general need for technology assessment described above, there were specific needs of Congress that led to *the establishment of OTA.* In the past, a Congress with few technical resources of its own faced federal agencies and a presidential office with extensive technical expertise. OTA was proposed partly to redress this balance by establishing a technical capability independent of the executive branch. The mandate of OTA as established in 1972 was "to provide early indications of the probable beneficial and adverse impacts of the applications of technology and to develop other coordinate information which may assist the Congress."[21]

The Board of OTA consists of 12 members—six from the Senate and six from the House, evenly divided between the two political parties to ensure bipartisanship. A separate Advisory Council of scientists and public representatives was created, but its role was not clearly defined and its advice has not been frequently sought. Requests for assessments can originate with any congressional committee or the OTA Board or Director. At first, assessments were contracted out mainly to research institutes or consulting firms (the deadlines usually have been too short for university teams), but a larger fraction of recent work has been done by the OTA staff itself. In-house assessments seem to have been more thorough, more balanced, and more oriented toward policy options than contracted ones. After analyzing impacts and examining alternative policy options, most reports have explored possible legislative actions, such as research support, tax incentives and penalties, or standards and regulations. OTA has no decision-making or legislative powers itself, and is clearly the servant of Congress.

A good example of an OTA assessment is a study of the effects of *offshore energy systems* (oil wells, deepwater ports, and floating nuclear power plants) along the coast of New Jersey and Delaware. After outlining the probable social and environmental impacts of these systems, the report proposes a number of congressional options for planning, monitoring, and regulating these developments, including possible modifications of agency jurisdictions to ensure better enforcement of standards. Among specific options discussed are stricter regulations and design standards for oil tankers, legislation on liability for oil spills, and mechanisms (such as the zoning of ocean areas) for resolving the increasingly serious conflicts among diverse users of the ocean. The report also concludes that the Nuclear Regulatory Commission has not adequately studied the special risks of floating nuclear plants. This assessment involved some public participation, including workshops, questionnaires, and interviews with citizens and officials in the affected regions to obtain a variety of views about various kinds of impacts.[22]

A number of questions can be raised about the relationship of OTA to the political process.[23] First, OTA has concentrated on *immediate short-run problems* at the expense of early warning assessments and long-run issues. Congressional committees usually have sought quick analyses relevant to pending legislation. As a consequence, OTA has not been able to devote as much effort to long-term anticipatory assessments, or to delayed social impacts, as had been originally contemplated. But in 1978 OTA drew up a list of priorities for assessment topics, representing a balance of short- and long-term issues, and studies were initiated in several emerging technologies. However, with continuing pressure from Congress for reports useful in immediate decisions, and with limited funds, it is not clear how much attention will be given to long-run issues.[24]

Second, there have been *organizational problems* in the operation of OTA. After oversight hearings and a comprehensive review in 1978, a House subcommittee concluded that both the Board and the Advisory Committee had been too closely involved in OTA administrative decisions, and had not given the director sufficient authority and indepencence.[25] Most of the topics chosen for assessment, for instance, reflected the legislative interests and committee assignments of the board members. As a result, OTA had excellent relationships with some committees but not with others. Both the board and its first director avoided such controversial topics as weapons systems and nuclear reactor safety—though it was understandable that a new agency, dependent on Congress for funds, should seek to minimize controversy until its credibility had been established. Moreover, some congressional committees—especially those that are allied with an agency and an industry in the promotion of a technology—have seen OTA as a potential threat to their own power.[26]

Third, *the actual use of assessments* by Congress has a mixed record. OTA has been credited with playing a significant role in several congressional decisions. Its reports were cited repeatedly during Senate floor debates on offshore oil exploration, for instance, and during committee deliberations on priorities for funding new mass transit technologies. The congressional review in 1978 gave OTA a qualified vote of confidence. In a survey of the staff of committees that had requested TAs, half said the assessments had had a significant impact on hearings and legislation, and two-thirds said the reports were evenhanded and unbiased. However, the review recommended better liason between OTA and Congress, and some efforts in this direction were made while Russell Peterson was director (1978–79). Of course, Congress responds mainly to pressures from constituents, and TAs alone will have little influence unless there is considerable public concern on the same issues. Legislative decisions are made primarily on political grounds; the outcome is more the result of compromise between conflicting groups and of coalitions among multiple power bases than of rational choice. Nevertheless, TAs can help to clarify the environmental consequences, stakeholder interests, and new policy options that will be considered within this political context.

Despite these problems, OTA has a strong *positive potential* in the social control of technology. Its publications are open and available to the public as well as to Congress. It offers a source of expertise independent of industry and administrative agencies, and a perspective different from that of other government offices. As it gains acceptance and staff capability, it seems to be more willing to undertake controversial topics, social impact analyses, and long-term early-warning studies. The Advisory Council and the separate advisory panels for each assessment could be more diverse and could be given a larger role in project design and the review of drafts of reports. Public participation activities, which to date have been included in only a few assessments, could be expanded as a corrective to the professional or institutional biases of the OTA staff and outside contractors. It has been proposed that assessments could give more attention to the reasons for conflicting views, which would help the staff of congressional committees in planning hearings. We will return to these questions after examining the assessment procedure itself.

Assessment Methods

In the early 1970s there were several attempts to formulate a series of steps that should be followed in technology assessment.[27] But no formal procedure has received universal acceptance. A study of 24 TAs (mainly done by teams at universities or consulting firms under the sponsorship of the National Science Foundation) finds considerable diversity among the methodologies actually employed.[28] But the study indicates that the assessment process in all cases could be described functionally under three headings:

1. Technology Description and Projection

Each TA team first assembled data on the current state of the technology and the patterns that its future development might take. Typically this was done by interviews with a series of technological experts and an extrapolation of current trends, with some allowance for forces that might alter these trends. Since both the technological development and its impacts would be affected by the future social context, some social forecasting had to be included, implicitly if not explicitly. Usually a basic continuity was assumed and past social trends were projected; for example, it was assumed that life-styles and social institutions would not change significantly. In some cases, more varied alternative futures were visualized by the presentation of scenarios (imagined, plausible sequences of events) reflecting a wider range of assumptions.

2. Impact Assessment

Environmental, economic, political, and social impacts on critically affected population segments were listed. Secondary and higher order impacts then were traced from each of these primary impacts. Typically this was done by a series of checklists and by reliance on experts in several disciplinary fields (though social scientists were not strongly represented, except for economists). A preference for the use of mathematical models to predict quantifiable impacts was evident in these TAs. There were difficult judgments in bounding each study—establishing its temporal horizon, geographical scope, and the impacts selected for detailed analysis. Uncertainties in estimating impacts created other problems; sometimes at least upper and lower limits could be given, or a range of estimates or probabilities could be agreed on.

3. Policy Analysis

The relative importance of the various favorable and unfavorable impacts was indicated. In a few of the TAs quantitative magnitudes were assigned to each impact on each social group, with weights according to estimated importance, and an aggregate sum obtained, a "net social benefit." The aggregation of diverse impacts is problematic, as we have seen, and it is not surprising that most of the teams avoided such techniques. Most assessors recognized that value judgments were present, and in some cases made provisions for representatives of affected groups to offer their own judgments on the anticipated impacts. The final stage was the formulation of policy options that might modify these impacts, and a comparison of the changes that would result. This survey concludes that "policy analysis is typically the weakest aspect"—partly, it appeared, because of the technical orientation of the teams, and partly because of the tendency to use up most of the time and funds on earlier phases of assessment. This deficiency in TAs sponsored by NSF is less evident in assessments carried out by OTA, which usually gives considerable thought to policy options that Congress might consider.

There have been some TAs that have given very careful consideration to *social impact assessment.* An interesting example is an assessment of energy conservation strategies prepared for NSF by a consulting firm.[29] The impact of alternative technologies on individuals is examined in terms of a wide range of quality-of-life indicators (including standard of living, job satisfaction, psychological well-being, recreational opportunity, and political participation). Similarly, the impact of energy consumption policies on households, organizations, communities, and institutions is analyzed from the perspective of each affected group. This assessment combined expert opinion with small surveys and interviews to predict the responses of various population segments to the policies proposed. Evaluation criteria discussed included freedom (absence of overt social control or coercive measures) and justice (perceived equity and fairness). Social justice enters primarily as a pragmatic consideration; actions felt to be unfair might evoke opposition and disruptive protest. The experts were asked to "estimate the strength and effectiveness of opposition from each group affected." The study did recognize the importance of looking at social values, but mainly as an aid in predicting the responses of impacted parties. I will argue that more extensive provision for public participation is desirable.

III. VALUES AND ASSESSMENT

In this final section the role of value judgments in assessment is explored—primarily in the work of technology assessment, though the same issues arise in environmental impact assessment. The possibilities of public participation in the assessment process also are examined. The section concludes that TA can provide significant opportunities for consideration of environmental and human values, provided there are provisions for diverse assessment teams, public participation, and interaction with decision makers in legislatures or agencies.

Value Judgments and Technical Judgments

The study of 24 TAs mentioned above found considerable disagreement among team members interviewed about *the role of value judgments* in the assessment process. Some insisted that professionals make great efforts to produce objective, value-free reports and to avoid normative judgments or biased recommendations. Others said that objectivity is impossible and that the only way to obtain evenhanded reports is to be sure that a variety of different biases is represented. The study concludes that value-laden decisions are inescapable at a number of points in assessment procedures. It suggests that there is some safeguard against bias if team members "really listen to different stakeholders" and take seriously their perceptions and attitudes.[30]

I would want to go even further in emphasizing *the inescapability of value judgments*—and therefore the importance of pluralistic participation. Any long-range assessment requires some portrayal of the social context within which the impact of a technology is to be evaluated. Such forecasts are based partly on beliefs about trends and projected events, and partly on assumptions about possible and desirable changes in institutions and attitudes. The selection of impacts considered significant enough to investigate and the bounding of the study are exercises in both scientific and political judgment. Next, the comparison of impacts entails trade-offs between incommensurable economic, social, and environmental values, as we have seen. The distribution of impacts raises questions of social justice in the case of new technologies, just as it does in the case of particular projects subjected to cost-benefit analysis or environmental impact assessment. Finally, only "feasible policy options" are explored in detail, and criteria of feasibility are always value-laden.[31]

There are thus many opportunities for *the biases of assessors* to be expressed. The assessors are almost exclusively white, male, middle class, and college educated, representing only a narrow portion of the cultural and socioeconomic spectrum. The teams are interdisciplinary, but engineers and economists have been dominant, especially in the leadership of assessments made by consulting firms and universities. Every discipline has its preferred conceptual models, types of explanation, and habits of mind. Engineers, for example, are understandably intolerant of ambiguity and enthusiastic about mathematical models and computer simulations. They may be reluctant to see the team deal with social factors that are difficult to quantify, such as job satisfaction, worker alienation, or mental health. In another study of NSF-sponsored TAs, the authors conclude that a lower status usually was accorded to social scientists than to natural scientists on assessment teams. They note the frequency with which physical scientists made statements about social phenomena (often only on the basis of "common sense" or "group brainstorming")—whereas social scientists rarely made such forays outside their fields of expertise.[32]

Moreover, technology assessment operates within a *political environment.* If it is to be relevant to policy decisions it cannot be totally apolitical. TA is a component in the control of technology and therefore a source of power. There are subtle pressures on assessors not to rock the boat; there are rewards, including chances for further contracts, if the contracting agency is not offended. But a careful TA is likely to generate bad news for someone—which runs against the desire of most administrators to minimize conflict. Even when there are no external pressures to avoid politically sensitive issues, assessment teams tend to censor themselves against potentially controversial conclusions, and to tone down recommendations that might alienate their sponsors.[33]

The framework of technology assessment is almost exclusively that of *existing attitudes and institutions.* In proposing an Office of Technology Assessment, Congressman Emilio Daddario (who was subsequently the first director of OTA) wrote:

> Possible changes in values, attitudes or institutions are important but not easily predicted. These changes are usually long term and fall beyond the primary focus of Technology Assessment. Therefore, because of their slow evolution, present human values and political institutions will serve as the frame of reference for purposes of measurement and appraisal.[34]

While a few long-range TAs of emerging technologies do look at basic social goals, most TAs implicitly accept the dominant values of U.S. industrial society, such as technical efficiency and the pursuit of economic growth. People who reject the consumer society and its life-styles as incompatible with environmental preservation and human fulfillment are unlikely to have a voice in the preparation of TAs. Unconventional or alternative technologies, novel institutions, and radical policy options are seldom evaluated.[35]

Some critics have attacked the whole technology assessment movement for *failing to raise fundamental questions* about the nature of technology and the technological society. One interpreter sees TA as a one-sided apology for contemporary technology by people with a stake in its continuation. The real expertise needed for a critical assessment, he says, is social and moral, not technical.[36] Another thinks that failure to examine basic issues in the relation of humanity and technology has led to the neglect of alternative forms of production and more meaningful forms of work, which would require major institutional changes.[37] I would agree that these fundamental questions about the character of a technological society should be raised (as in Chapters 3 and 13 of the present volume). But such broad-gauge discussion of cultural attitudes and social institutions need not exclude the detailed assessment of particular technologies. The analyses at the two different levels should, in fact, be mutually illuminating.

There have been some attempts to develop a specifically *value-oriented approach* to technology assessment. One version starts from diverse portrayals of desirable futures. Alternative scenarios are developed that correspond to a variety of human aspirations. The team then asks how a particular new technology might fit into each scenario and what its impact might be on human values. The proponents of this approach insist that no TA is objective and that values should be made explicit. They urge us to "reopen the ethical debate on the meaning of the public good," which is neglected when assessors are preoccupied with tracing impacts. Most TAs, they point out, assume that material growth and increased consumption are high-priority goals. Instead, emphasis might be given to other values, such as work satisfaction, equality, freedom, and the revitalization of neighborhoods. Technological developments then would be considered in the context of diverse value assumptions and alternative futures.[38]

I myself would want to seek *the inclusion of a wider range of values* within the more commonly used TA procedures described earlier. Though some TAs may be accused of protechnology bias, there are others that have advocated stopping or redirecting a technological development (for example, the NAS study that opposed the expansion of New York's Kennedy Airport). In recent NSF announcements soliciting applications for the funding of TAs, several unconventional categories are listed, including innovative, resource-conserving technologies and alternative or intermediate technologies. Some TAs have raised far-reaching questions about social goals, resource consumption, and the kind of society we want. For example, a Stanford study of solar energy looks at the social and political impacts of small-scale decentralized systems amenable to individual or community control. It acknowledges that a small but growing group of citizens is critical of large-scale centralized technologies.[39] OTA hopes to give some attention to appropriate technology and the question of centralization versus decentralization.[40]

Public Participation

Public participation in decisions about technology occurs mainly through *elected representatives* and *accountable officials* who are the primary decision makers in a democratic society. Through elections, legislative and agency hearings, lobbying efforts, and other political channels, the citizen can have an influence on policy decisions. Some interpreters see TA only as an input of technical information transmitted to these decision makers by experts. However, I would defend the inclusion of public participation *in the TA process itself* for several reasons. On purely pragmatic grounds, citizen participation is a source of information about public attitudes and perceptions of which assessors might otherwise be unaware. Later protests and opposition might be reduced if potential objections were anticipated and alternatives were developed to take them into account. Further, a diversity of perspectives would help to broaden the range of impacts that are identified for study.[41]

I have argued, moreover, that technology assessment is in many respects *a political activity*. It deals not only with the estimation of impacts but also with judgments of their importance and their distribution among social groups. It therefore should include provisions for the direct involvement of the parties affected. In a pluralistic society, a wide range of alternative options should be explored—not just for the benefit of decision makers but for the sake of intelligent public debate—and the trade-offs should be examined from the standpoint of a diversity of human and environmental values. The demand for public participation here is one more form of the reaction against technocratic and elitist modes of policy analysis. I see dangers whenever essentially political choices are disguised as technical decisions. TA provides another vehicle for the citizen participation that I have defended as an expression of social justice and individual freedom.

In the EIS process, judicial challenges are a major opportunity for citizen involvement, but technology assessment is not part of an agency action subject to court review. However, there are other forms of public participation in *the preparation and review of TAs*. In the OTA study of offshore energy technologies along the New Jersey and Delaware coasts, local meetings, opinion surveys, and workshops were held to obtain the views of a cross section of citizens and officials in the areas that would be affected. Draft versions of the report were reviewed by a panel of citizens and government officials. Some other OTA studies have similar provisions for public involvement.[42] NSF-sponsored TAs have included opinion surveys, interviews with impacted citizens, and workshops with representative stakeholders and interest groups.[43]

More diverse *oversight committees* would help to broaden the range of values acknowledged in TAs. Hazel Henderson, who is on the Advisory Council of OTA, has urged a stronger representation of labor, environmental, consumer, and other citizens' groups on the oversight committees of particular assessments. They could review the study design and the team composition with an eye to diversity in disciplines and perspectives. They could critique work in progress, point out unexamined assumptions, encourage the search for imaginative new options, and make sure that issues of justice in the distribution of costs and benefits among social groups are not neglected. By more actively monitoring the TA process, such a diverse advisory panel would promote "the vigorous articulation of values, ethical norms and societal goals, so that the broadest range of technical and societal options can be presented to the public for debate."[44]

There also should be provisions for the circulation of draft versions for comment by agencies and outside groups, as occurs in the EIS process.

Another proposal is the preparation of *adversary assessments* by two teams with strongly contrasting assumptions and viewpoints. For example, an industrial and an environmental organization could prepare counterassessments. This would "highlight diverse value-based perspectives rather than trying vainly to produce value-free assessments."[45] However, it is unlikely that busy policymakers would take time to study and sort out the conflicting values in divergent assessments. A single integrated TA can better summarize diverse viewpoints and policy options; comments by various interest groups could be appended, as in the EIS procedure. In particularly controversial cases, a full-scale TA could be followed by two adversary mini-assessments prepared by groups with differing viewpoints (if they had access to all the data of the longer study and adequate funds to enlist competent scientific expertise).[46]

In sum, I believe there are significant opportunities for *environmental and human values* to be considered in the context of EIS and TA procedures, provided there are diverse disciplines and viewpoints on assessment teams, mechanisms for public participation, and consideration of a wide range of options. In contrast to cost-benefit analysis, the assessment framework readily can incorporate a long time-perspective, a broad spectrum of impacts, and a diversity of criteria. Significant input from various stakeholders and interest groups can be obtained. Assessments always will reflect the biases of the institutions in which they are carried out, but public scrutiny can encourage greater care in documentation and even-handedness in dealing with conflicting views.

If it is effectively integrated into political processes, assessment can be a vehicle of *democratic decision making* rather than of technocratic planning by experts. The final responsibility for policy choice lies not with the assessors but with elected representatives and accountable administrators. Throughout Part Two I have maintained that the concentration of economic and political power in industrial corporations and government bureaucracies, as well as the technical complexities of decisions about technology and the environment, present serious obstacles to the operation of democratic institutions. But I also have maintained that political democracy does work, through a variety of judicial, administrative, and legislative channels, when there is a concerned and informed citizenry.

NOTES

1. *The National Environmental Policy Act of 1969*, USCA 42 sec. 4321 ff.; Bill Shaw, *Environmental Law* (St. Paul, Minn.: West Publishing, 1976), part II; and Lynton Caldwell, *Man and His Environment: Policy and Administration* (New York: Harper & Row, 1974), chap. 4.

2. Council on Environmental Quality, *Environmental Impact Statements: An Analysis of Six Years' Experience by 70 Federal Agencies* (Washington, D.C.: CEQ, 1976). See "Symposium on Environmental Impact Statements" in *Natural Resources J.* 16 (1976): 243-356, especially articles by Daniel Dreyfus and Helen Ingram and by Hanna Cortner.

3. Council on Environmental Quality, *Federal Register* 43 (June 9, 1978): 25230; *Environmental Quality 1978* (Washington, D.C.: CEQ, 1978) chap. 9; and Luther Carter, "New CEQ Regulations Reform NEPA Process," *Science* 203 (1979): 345.

4. Richard Andrews, *Environmental Policy and Administrative Change* (Lexington, Mass.: Lexington, 1976); and Richard Liroff, *A National Policy for the Environment: NEPA and its Aftermath* (Bloomington: Indiana University Press, 1976). For a more negative appraisal, see Sally Fairfax, "A Disaster in the Environmental Movement," *Science* 199 (1978): 743-48.

5. Articles by Richard Andrews, Allan Miller, Robert Cahn, and Lynton Caldwell in *Workshop on the National Environmental Policy Act* (Subcommittee on Fisheries, Wildlife Conservation, and the Environment; Committee on Merchant Marine and Fisheries, U.S. House of Representatives, 1976).

6. H. Paul Friesma and Paul Culhane, "Social Impacts, Politics, and the EIS Process," *Natural Resources J.* 16 (1976): 339-56; and "Implementation of the EIS," *Yale Law Journal* 88 (1979): 596-611.

7. Luna Leopold et al., *A Procedure for Evaluating Environmental Impact*, U.S. Geological Survey Circular no. 645 (Washington, D.C.: USGS, 1971).

8. Sherman Rosen, *Manual for Environmental Impact Evaluation* (Englewood Cliffs, N.J.: Prentice-Hall, 1976). See also Ruthann Corwin and Patrick Hefferman, eds., *Environmental Impact Assessment* (San Francisco: Freeman, Cooper, 1975); and Marlan Blissett, ed., *Environmental Impact Assessment* (Austin: University of Texas Press, 1976).

9. Stahrl Edmunds and John Letey, *Environmental Administration* (New York: McGraw-Hill, 1973); and Jens Sorenson, "Some Procedures and Programs for Environmental Impact Assessment," in *Environmental Impact Analysis: Philosophy and Methods*, ed. Robert Ditton and Thomas Goodale (Madison: University of Wisconsin Sea Grant Program, 1972).

10. Gordon Matzke et al., *An Examination of the Moral Dilemmas of University Scientists Participating in the Preparation of Environmental Impact Statements* (Stillwater, Okla.: Oklahoma State University Press, 1978).

11. Friesma and Culhane, op. cit.

12. Per Johnson, "Social Aspects of Environmental Impact," in Ditton and Goodale, op. cit.

13. *Social Impact Assessment: An Overview* (Ft. Belvoir, Va.: U.S. Army Engineers Institute for Water Resources, 1975); and U.S. Water Resources Council, "Principles and Standards for Planning Water and Related Land Resources," *Federal Register* 38 (1973): 24778-841.

14. K. Finsterbusch and C. P. Wolf, eds., *Methodology of Social Impact Assessment* (Stroudsburg, Pa.: Dowden, Hutchinson & Ross, 1977), especially chapters by Peter Sassone and by George Peterson and Robert Gemmell.

15. National Academy of Sciences, *Technology: Processes of Assessment and Choice* (1969), and National Academy of Engineering, *A Study of Technology Assessment* (1969) (both prepared for Committee on Science and Astronautics, U.S. House of Representatives); and François Hetman, *Society and the Assessment of Technology* (Paris: Organisation for Economic Cooperation and Development, 1973).

16. Raphael Kasper, ed., *Technology Assessment* (New York: Praeger, 1972); and Joseph Coates, "Technology Assessment," in *1974 McGraw-Hill Yearbook of Science and Technology* (New York: McGraw-Hill, 1974).

17. Vary Coates, *Technology and Public Policy: The Process of Technology Assessment in the Federal Government* (Washington, D.C.: George Washington University Press, 1972).

18. Daniel DeSimone, December 2, 1976, quoted in Joe Armstrong and Willis Harman, *Strategies for Conducting Technology Assessment* (Stanford, Calif.: Stanford University, Department of Engineering-Economic Systems, 1977), p. 125.

19. Joseph Coates, "Technology Assessment: The Benefits, the Costs, the Consequences," *Futurist* 5 (1971): 225-31; and idem, "The Role of Formal Models in Technology Assessment," *Technological Forecasting and Social Change* 9 (1976): 140.

20. Sherry Arnstein and Alexander Christakis, *Perspectives on Technology Assessment* (Jerusalem, Israel: Science and Technology Publishers, 1975), chap. 10.

21. Anne Cahn and Joel Primack, "Technological Foresight for Congress," *Technology Review* (March-April 1973): 39-48; and Vary Coates, "Technology Assessment in the Political Context," *J. Int. Soc. for Tech. Assessment* 1, no. 2 (1975): 45-49.

22. Office of Technology Assessment, *Coastal Effects of Offshore Energy Systems* (Washington, D.C.: OTA, 1976).

23. Craig Decker, "A Preliminary Assessment of the Congressional Office of Technology Assessment," *J. Int. Soc. for Tech. Assessment* 1, no. 2 (1975): 5-26; and *The Office of Technology Assessment: A Study of Its Organizational Effectiveness* (Washington, D.C.: Commission on Information and Facilities, U.S. House of Representatives, 1976).

24. Office of Technology Assessment, *Annual Report to the Congress for 1978* (Washington, D.C.: OTA, 1978); and *OTA Priorities, 1979* (Washington, D.C.: OTA, 1979).

25. *Review of the Office of Technology Assessment and its Organic Act* (Washington, D.C.: Subcommittee on Science, Research, and Technology; Committee on Science and Technology, U.S. House of Representatives, 1978).

26. Barry Casper, "The Rhetoric and Reality of Congressional Technology Assessment," *Bull. Atomic Scientists* 34 (February 1978): 20-31.

27. Martin Jones, *A Technology Assessment Methodology* (Washington, D.C.: MITRE Corp., 1971), summarized in *Futurist* 6 (1972): 19-26; Marvin Cetron and Bodo Bartocha, eds., *The Methodology of Technology Assessment* (New York: Gordon and Breach, 1972); Hetman, op. cit.; and *Methodological Guidelines for the Social Assessment of Technology* (Paris: Organisation for Economic Cooperation and Development, 1975).

28. Armstrong and Harman, op. cit.

29. Kurt Finsterbusch and Patricia Weitzel-O'Neill, *A Methodology for the Analysis of Social Impacts* (Vienna, Va.: Braddock, Dunn, and McDonald, 1974); and Kurt Finsterbusch, "A Policy Analysis Methodology for Social Impacts," *J. Int. Soc. for Tech. Assessment* 1, no. 1 (1975): 5-15.

30. Armstrong and Harman, op. cit., p. 100.

31. Louis H. Mayo, "Evaluation of Social Justice of Anticipatory Project Assessments" (Washington, D.C.: George Washington University, Program of Policy Studies in Science and Technology, 1973).

32. Stanley Carpenter and Frederick Rossini, "Value Dimensions of Technology Assessment," in *The General Systems Paradigm: Science of Change and Change of Science*, (Washington, D.C.: Society for General Systems Research, 1977), pp. 463-69.

33. Marc Berg, "The Politics of Technology Assessment," *J. Int. Soc. for Tech. Assessment* 1, no. 4 (1975): 21-32; and Arnstein and Christakis, op. cit., chap. 10.

34. *Report of Subcommittee on Science, Research and Development* (Committee on Science and Aeronautics, U.S. House of Representatives, 1968), p. 12.

35. Carpenter and Rossini, op. cit.

36. Henry Skolimowski, "Technology Assessment in a Sharp Social Focus," *Tech. Forecasting and Social Change* 8 (1976): 421-25.

37. Langdon Winner, "On Criticizing Technology," *Public Policy* 20 (1972): 35-59.

38. Arnstein and Christakis, op. cit., chaps. 15 and 16. See also Marc Berg, Kan Chen, and George Zissis, *Technology Assessment Methodologies in Perspective* (Ann Arbor: University of Michigan, Program in Decision Analysis and Processes, 1974); and idem, "A Value-Oriented Policy Generation Methodology for Technology Assessment," *Technology Forecasting and Social Change* 8 (1976): 401-20.

39. Stanford Research Institute, *Solar Energy in America's Future*, 2d ed. (Washington, D.C.: U.S. Energy Research and Development Administration, 1977).

40. Office of Technology Assessment, *OTA Priorities, 1979* (Washington, D.C.: OTA, 1979).

41. H. Paschen et al., "Some Problems of Evaluation in TA Studies," and E. K. Jochem, "Technology Assessment and Participation," in *Methodological Guidelines*, op. cit.; and Michael Baram, "Technology Assessment and Social Control," *Science* 180 (1973): 456-73.

42. Office of Technology Assessment, *Annual Report to the Congress* (Washington, D.C.: OTA, 1977), p. 74.

43. Armstrong and Harman, op. cit.

44. Hazel Henderson, "Technology Assessment and the Citizen," *Trends in Engineering* (April 1974): 9-16. See articles by Joseph Coates, Sherry Arnstein, Hazel Henderson, and Willis Goldbeck in "Symposium on Public Participation in Technology Assessment," *Public Admin. Rev.* 35 (1975): 67ff. See also Byron Kennard, "Some Models and Criteria for Public Participation in Technology Assessment," *J. Int. Soc. for Tech. Assessment* 1, no. 3 (1975): 43-46.

45. Arnstein and Christakis, op. cit., pp. 172 and 173.

46. Frederick Rossini, Alan Porter, and Eugene Zucker, "Multiple Technology Assessments," *J. Int. Soc. for Tech. Assessment* 2, no. 1 (1976): 22-28; and Hugh Folk, "The Role of Technology in Public Policy," in *Technology and Man's Future*, 2d ed., ed. A. Teich (New York: St. Martin's Press, 1977).

PART THREE: SCARCE RESOURCES

10
ENERGY OPTIONS

Part Three deals with conflicting values in decisions about scarce resources. We will continue to focus on U.S. policy, but will give greater attention to the global context of national choices. Most environmental impacts are limited geographically to an airshed, a watershed, or an ecosystem (though we will note several global environmental issues, such as ocean pollution and climate changes). But resource problems are inherently international; one country's decisions on oil, grain, or minerals will affect other countries. Justice among nations is not an issue in pollution control or land use but it is in policies for energy, food, and materials.

The United States has made considerable progress in formulating policies to preserve the environment, but it has hardly begun to respond to resource scarcities. It has remained confident that technology can overcome resource constraints. Only a minority has questioned the American dream of ever-growing production and consumption. In Part Three it will be suggested that the transition to a sustainable society requires some fundamental changes in attitudes, policies, and social institutions.

Energy is treated first because it is basic to agricultural production, mineral extraction, and other resource problems. We are not literally running out of energy, but the era of cheap energy is over. The rising economic, environmental, and human costs of energy will affect almost every phase of life.

Value judgments play a major role in energy decisions. There are inescapable trade-offs among diverse values: health and safety, environmental preservation, employment, justice, participation, and sustainability. All of the energy options entail risks; in estimating many of these risks there are scientific uncertainties about which experts disagree. For each option, we will look mainly at the social and ethical issues, but we cannot ignore technical and economic questions. We must ask both what is technically feasible and what is ethically desirable. Who benefits and who bears the costs and risks of various energy systems?

Which technologies accelerate the concentration of economic and political power? How can the public participate in energy decisions when the technical issues are so complex? Which sources are most promising for developing countries? Which policies take into account the needs of future generations?

Until 1700 all world civilizations were based on wood, wind, water, and animal power. The Industrial Revolution was the product of technologies powered by coal. The direction of twentieth-century growth in industrial nations has been largely determined by the availability of cheap oil. The society of the twenty-first century will be shaped to a significant degree by the energy technologies that are chosen in the next two decades to replace oil. We are choosing alternative futures, not just alternative technologies. Behind the debates between nuclear and solar advocates, for example, there are divergent visions of human fulfillment and conflicting ideals of the good life.

Three time-scales for energy decisions can be distinguished. In the short term (the next ten years), oil and natural gas still will be dominant. In the middle term (ten to 25 years), as oil and gas taper off, increased use of coal or nuclear reactors is likely, though solar energy can begin to play an important part. In the long term (beyond 25 years), breeder reactors and solar energy (and possibly fusion, if it proves practical) seem to be the main options.[1] Since it takes a decade or two for new equipment to be developed and widely installed, the transition to middle- and long-term technologies has to begin now, and it will be our main concern here. I will argue that conservation is crucial in all three time-spans: to reduce our disproportionate consumption of world oil, to minimize the expansion of coal and nuclear technologies, and to facilitate the shift to renewable sources.

I. FOSSIL FUELS

Fossil fuels represent the energy of the sun captured in the chemical energy of plant life over periods of millions of years and preserved in the geological strata of the earth. They are therefore essentially nonrenewable; once they are used they cannot be replaced within the span of human history. Reserves of oil and natural gas are measured in decades at current depletion rates. Reserves of coal are considerably larger, but their use has high costs in human health and environmental degradation.

The Politics and Ethics of Oil

U.S. growth in the twentieth century was fueled by inexpensive oil. High-energy agriculture has made intensive use of fertilizer and machines—both based on oil. Transportation accounts for one-third of energy end-use, mainly in the form of oil. Our patterns of housing, work, and daily life are built around the automobile. Oil, and the natural gas associated with it, provides 75% of our total energy supply. Out of a total U.S. energy budget of 78 quads (quadrillion BTU) in 1978, the sources in quads were: oil 38, natural gas 20, coal 14, hydropower 3, and nuclear, 3.[2]

Estimates of *recoverable oil reserves* vary widely, depending on assumptions about prices, the state of technology, and future oil discoveries. Deep wells and offshore wells are very expensive, but are being extended as oil prices rise. U.S. production peaked in 1970 and will not be a major source after 2000. World oil is expected to peak around 2000, and to be prohibitively expensive for most uses after 2020. Shale oil may extend these dates, but at a high cost in capital, extraction energy, water use, and residual wastes.[3]

The politics of oil are extremely complicated because oil plays such a large part in national life. In the past, we have heavily subsidized oil use through depletion allowances, import quotas, highway construction funds, and other means. As mentioned earlier, a powerful lobby was formed by a coalition of diverse interests: oil companies and oil-producing states; auto, insurance, and highway construction companies; and labor unions. A small number of oil companies have immense assets and have effectively limited competition by "vertical integration" (control of oil production, refining, pipelines, and service stations) and "horizontal integration" (control of coal and uranium as well as oil resources).[4] These companies have made extensive use of television and magazine advertising to present their interpretation of the energy crisis and the policies they favor. Congress has been virtually paralyzed by pressures from conflicting constituencies: Sunbelt oil-producing states and Snowbelt oil-consuming states; industrial, environmental and consumer lobbies, and so forth. The politics of oil extends heavily into international relations. Almost half of our oil comes from overseas, creating huge balance-of-payment deficits that have led to falling exchange rates for the dollar. United States dependence on foreign oil makes us highly vulnerable to price rises, embargoes, and political instability in the Middle East (such as in Iran). Our competition with other countries for a vital resource is a growing source of international conflict.

There is a major issue of *distributive justice* in the disproportionate U.S. consumption of global energy resources. With 5% of the world's population, we use one-third of the world's commercial energy and half of all gasoline. On the average, a U.S. citizen uses twice the energy of a European, 20 times that of a Third World citizen, and 50 times that of a person in one of the poorer developing countries. United States oil imports drive up the price of oil, which is devastating for Third World nations that need it for fertilizer and irrigation pumps. In the next chapter, the close connections between food and energy are traced; our oil use contributes to hunger and starvation in other countries. Oil that affluent nations burn is forever unavailable for industrial development elsewhere. In a few decades, a small portion of humanity will have used up a resource that took hundreds of millions of years to accumulate. Future generations also will need oil for nonfuel uses: petrochemicals (including fertilizer), lubricants, drugs, and so on. At current rates, 80% of all oil and gas on earth will be consumed during the lifetime of one generation, and most of it by a few nations.[5]

A whole chapter could be devoted to the politics and ethics of oil, which loom large in the short run. There are important environmental impacts, especially ocean pollution from tanker spills and accidents, and from oil well blowouts such as the 1979 accident off the coast of Mexico. I wish, however, to deal mainly with policy for the immediate and long term, which should be central in current decisions. I will return later to some policies for reducing oil consumption, such as deregulation, taxes, auto efficiency standards, and other conservation measures. Reducing oil use also would reduce air pollution, particularly from auto emissions, though this can be controlled more easily than pollution from other energy sources.

Human and Environmental Costs of Coal

Coal reserves are much more abundant than oil, and coal is cheaper per unit of energy. United States coal could fill all our energy demands for several centuries at current levels, or for perhaps a century at historic growth rates. Several recent proposals call for a threefold increase in U.S. coal production by 2020. However, coal entails much higher human and environmental costs than oil:

1. Deep-Mining Fatalities

The long history of mine accidents and "black lung" disease is a classic instance of high risks falling on one group while benefits accrue to other people. The Coal Mine Health and Safety Act of 1969 finally legislated major improvements in mine safety. Fatalities per million working hours fell from 1.2 in 1970 to 0.3 in 1977. This still represents 139 fatalities per year, which could be considerably reduced by stricter enforcement of safety and ventilation standards; one-third of the working mine sections still did not meet the dust standards set in 1969.[6] It is expected that when all mines meet the standards, the incidence of "black lung" among new miners will be less than 1% of past rates. But disabling injuries have been high (15,000 in 1977) and have been increasing.[7]

2. Strip-Mining Damage

Coal from Western surface mines is safer, cheaper, and lower in sulfur than Eastern underground coal, but it has higher environmental costs. The Surface Control and Reclamation Act of 1977 will reduce surface damage greatly; environmentalists were in general pleased with the final bill, despite its limitations. Highwalls bordering stripped areas must be backfilled and the original contours restored "except where impractical." Reclamation efforts must be sustained for five years, or ten years in arid areas. Reclamation costs, estimated at 35 cents per ton of coal, will add only a very small amount to the cost of delivered energy.

But reclaimed land often can be used only for pasture, and in semi-arid regions it will take decades for natural plant life to be restored. In the Southwest, ecosystems are fragile and reclamation will compete with other demands for scarce water.[8]

3. Social Impacts of Western Coal Development

Projected increases in coal production would create a large population influx in several areas of the West, especially if coal gasification and liquifaction or oil shale extraction are heavily developed. "Boom town" growth will exceed the capacity of most communities to supply housing, schools, and health services, unless there is careful planning and extensive federal assistance. High turnover rates are disruptive of local communities, which feel that they are the victims of decisions in which they have little voice.[9] Native American life will be particularly vulnerable to disruption by the influx of thousands of workers from different cultural backgrounds. In southeast Montana, for instance, coal development threatens the extinction of the North Cheyenne and Crow reservations as distinctive tribal groups, and coal leases offer little protection to tribal lands.[10]

4. Air Pollution

The burning of coal inflicts even greater social costs than mining it. Sulfur dioxide (SO_2) in stack emissions produces sulfates at rates that vary greatly with moisture and the level of fine particulates. As indicated in Chapter 7, the added mortality from chronic respiratory disease aggravated by sulfates rises rapidly above 25 micrograms per cubic meter; but the evidence is uncertain at lower levels. Current sulfate levels in many urban areas are 15 to 20 micrograms. Respiratory illness is evident at lower prolonged average levels down to 10 micrograms.[11]

The 1977 Clean Air Amendments require "the best available control technology," which at the moment means *limestone scrubbers.* Scrubbers add 14% to capital costs, decrease energy output by 6%, and require daily disposal of 80,000 cubic feet of sludge from a typical plant.[12] The utilities have opposed the added expense and have advocated use of tall stacks, low-sulfur coal, and relaxation of air standards until better technologies are developed.[13] In 1979, EPA announced regulations that represent a compromise: strict requirements for new plants (half the previous allowable emissions), but lax standards on old plants (many of which still have no emission controls).[14] Environmentalists have urged the requirement of scrubbers on all plants, because SO_2 not only creates sulfates but also interacts synergistically with other pollutants (including nitrogen oxides and heavy metals); in addition, it creates "acid rain," which harms fish, forests, and agricultural yields. Fluidized-bed boilers appear promising to improve the efficiency of coal burning and to reduce nitrogen oxides; they are also expected to reduce sulfur emissions by 90%. Magnetophydrodynamic plants under construction in California and Russia may offer a clean and efficient method of producing electricity directly from hot gases.

5. Climate Changes

The greatest potential risk, but also the most uncertain, is the effect of carbon dioxide (CO_2) from burning coal and oil. If present trends continue, the concentration of CO_2 in the earth's atmosphere is expected to double by the middle of the next century.[15] The trend may be accelerated by the global loss of forests, which absorb CO_2. But scientists are divided about the effect on climates. CO_2 cuts down on heat radiation (the greenhouse effect) and temperatures might rise as much as 6° C in a century. If polar ice sheets melted, ocean waters would rise 200 feet, submerging many coastal cities and low-lying areas around the world, though the melting would be very slow. More immediate would be the alteration of air currents and rainfall due to the greater warming at the poles; world agricultural patterns might be drastically affected. While there may be a long-term cooling trend that would partly compensate for the heating from CO_2, any major alterations of climates would have incalculable consequences for humanity.[16]

All of these human and environmental costs would become more serious if the United States launches a major program to produce *synthetic oil and gas* from coal. In 1979, President Carter proposed a crash program for synthetic fuels, to be financed from a windfall tax on profits from the deregulation of domestic oil. The technologies available are expensive and inefficient, require large amounts of water, and generate huge quantities of waste. Environmentalists have said that such a massive program would absorb funds and technical talent that could better be directed to conservation and solar energy. They also have urged that new energy projects should not be exempted from environmental standards or procedural requirements.[17]

In short, coal reserves are adequate to sustain an expansion of coal production, which I will argue is preferable to additional nuclear facilities, but the expansion should be minimized by strong conservation measures. Current standards for mine safety and strip-mining reclamation should be strictly enforced, and scrubber requirements extended, despite the economic costs to the consumer. However, coal should be used only for a few decades *as a transition fuel*, because of the possible effects on the global climate. In the long term coal reserves will run out if they are heavily used. Global coal deposits also are very unevenly distributed: the United States and the Soviet Union have 74% of known world reserves, Europe and Japan have little, and all Latin America and Africa together have less than 1%.[18] Coal can help reduce our dependence on oil, but it cannot be a long-term source for the United States, much less for other countries.

II. THE NUCLEAR DILEMMA

In 1979, 72 nuclear reactors supplied 13% of the nation's electricity, or about 4% of its total energy. Another 91 reactors under construction or awaiting licenses would double these figures by the late 1980s. A few years ago it was predicted that 40 reactors a year would be built, adding a thousand reactors by the year 2000. But few orders for reactors have been placed in recent years because electric utilities face lower demand estimates, mounting capital costs, licensing delays, uncertainties about future regulations, and rising public opposition. On balance, how do the benefits and costs of nuclear energy compare with those of coal?

Nuclear Benefits and Costs

Nuclear plants produce no combustion products. In two decades of use, there has not been a single case of public injury or private property damage from a reactor accident. For equivalent energy production, uranium mining involves fewer fatalities than deep coal mining and less environmental damage than strip-mining. Uranium can be used only to produce electricity; its use would allow coal to be saved for other fuel and petrochemical uses. Low-grade uranium ore, which is becoming economical as prices rise, appears adequate for several decades—and could be extended a hundredfold by the breeder reactor.

Compared with coal plants, nuclear capital costs are higher but fuel costs are lower. At the moment, the total direct *economic costs* of nuclear electricity are slightly lower than those of coal-burning plants with scrubbers; the balance is close enough that it could be changed by future environmental and safety regulations or technological improvements.[19] But nuclear power involves far greater subsidies and hidden costs, including insurance liability for major accidents (under the Price-Anderson Act) and the unknown costs of plant dismantling and radioactive waste disposal, much of which the public is likely to bear. The wasted heat from nuclear plants seldom can be used in cogeneration systems because the plants should not be built near population centers. In addition, there are several unique risks:

1. Low-Level Radiation

The amount of radiation that the general population receives from routine plant operations is very small compared to radiation from other sources. Public exposure averages .003 millirems from all nuclear plants, which is negligible compared to the 70 millirems received from medical X-rays and 120 millirems from natural radiation, on the average.[20] The exposure limit for a person near a plant boundary is 5 millirems per year, which is probably less than people living downwind from a coal plant receive from radioactivity in the fly ash. An NAS study estimates that 2,000 cancer deaths will occur by the year 2000 from low-level radiation from nuclear plants; in the same period, at least 15 million people will die of cancer from other causes.[21] There is recent though still controversial evidence of increased incidence of cancer among nuclear shipyard workers and employees at the Hanford atomic facility, which suggests the need for stricter standards for occupational exposure (currently 5 rems per year).[22]

2. Reactor Accidents

The AEC-sponsored Rasmussen report concluded that with 100 plants operating, a person's chance of being killed in a year by a reactor accident would be one in 5 billion (compared to a one in 2 million chance of being struck by lightning).[23] However, this report has been criticized by a succession of review panels. One panel objected that immediate deaths were underestimated tenfold and delayed cancer deaths completely omitted.[24] A review panel appointed by the Nuclear Regulatory Commission concluded that in calculating accident probabilities the Rasmussen report had greatly underestimated the uncertainties in its figures and the role of human errors.[25] In 1979 the NRC told its staff that the numbers in the report were too uncertain to be used in licensing or policy decisions.[26] The report also neglected acts of deliberate sabotage, terrorism, or blackmail (hijacking), which are likely to be more common in an era of international conflict and politically alienated minorities. Such human actions are more difficult to quantify than technical failures, but may well constitute a larger risk.[27]

The reactor accident in 1979 at Three Mile Island was a product of maintenance negligence, mechanical failure, poor instrument and control room design and multiple human errors. Unknown to the operators, the reactor core was partially exposed for 13 hours and heated to 4000°F or more. 90% of the fuel rods burst and the core became partially molten. Safety experts had not even considered the possibility that a hydrogen bubble would form and would hinder reactor shut-down. There were two days of confusion in the control room during which the public was not alerted of the danger; communications were poor and evacuation plans inadequate.[28] The Kemeny Commission blamed human as much as technical failures and severely criticized the complacency of both the industry and the regulators. It recommended more rigorous training of reactor operators, closer supervision by the government, and approval of emergency evacuation plans before any more licenses are issued. It urged that new plants should be remote from population centers. The accident may lead to such safety improvements, but it has also increased public skepticism about reassurances from the industry—and about safety studies which did not even consider the accident sequences which actually occurred.[29]

3. Radioactive Wastes

A large backlog of highly radioactive military wastes and spent fuel rods already has accumulated. Early plans called for burial in salt mines, but one site was abandoned when old drill holes were found, and at another site pockets of brine were discovered. The wastes can be solidified into glass or cement blocks to limit dispersal if water does reach them.[30] Yet there still are many unanswered questions about geological stability and the interaction of wastes with water and rocks at high pressures and temperatures. A 1978 NAS study calls for more extensive research on alternative solidification methods and burial sites (including granite formations). The study holds that there are promising options, but that we do not know enough yet to make such an important decision and safe disposal is likely to be very costly.[31] An interagency review in 1979 concluded that there are still too many uncertainties to start building a permanent waste disposal facility.[32] None of the proposed methods has been rigorously tested.

The long-term security of burial sites also is problematic. Plutonium is highly toxic and remains dangerous for incredibly long periods, on the order of 100,000 years. Some types of disposal sites would have to be guarded to prevent flooding, drilling, theft, or sabotage. Alvin Weinberg has described the "Faustian bargain" of nuclear power: enormous quantities of energy at the price of "a vigilance and a longevity of our social institutions that we are quite unaccustomed to." He favors the bargain but says it will require a "nuclear priesthood" of technicians with high discipline to maintain essentially perpetual surveillance over buried wastes.[33] It seems to me utterly unrealistic to count on the stability of social institutions on such a time-scale. No social order in the past has lasted more than a few centuries. The occurrence of two world wars and several revolutionary upheavals in this century suggests that we should use methods that do not depend on disciplined human vigilance for thousands of years. Intergenerational justice requires that we minimize the risks exported to future generations to secure short-lived benefits for this generation.[34]

4. Nuclear Weapons Proliferation

Nuclear war would have consequences that are many orders of magnitude greater than any of the impacts above. The diversion of materials to make nuclear weapons is therefore a risk that must be taken seriously, even if the probabilities appear low. Weapons can be made from current reactor fuels, but only by complex processes of uranium enrichment, or by the extraction of plutonium from spent fuel. Plans to build U.S. reprocessing plants to extract plutonium (to use again in conventional reactors) have been postponed—largely because of the proliferation issue. The proposed breeder reactor also would produce large quantities of plutonum that would be transported as fuel for conventional reactors. Even with an international inspection system, it would be difficult to prevent a country with such facilities from diverting the small quantity of material needed to make nuclear weapons. If plutonium were in wide use in many countries, a terrorist organization could steal enough to make a bomb.[35]

Several major studies have recommended that U.S. research on reprocessing and the breeder be continued, but that actual deployment be deferred indefinitely to reduce the risk of proliferation.[36] The Carter administration adopted this policy, but was unsuccessful in persuading European countries to follow suit. Proliferation risks can be somewhat reduced by tighter security measures, and there is some possibility of new reprocessing techniques less susceptible to diversion (for example, some uranium or fission products can be left with the plutonium).[37] As in the case of reactor safety and waste disposal, the risks from proliferation are as much the product of social institutions as of technical systems. Proliferation risks from plutonium might be acceptable in a stable world, but I believe that they are unacceptable in the situations of intense social conflict and political instability that are likely to occur in many countries in the years ahead.

The Comparison of Risks

There have been a number of attempts to apply *risk-benefit analysis* to the choice between coal and nuclear plants.[38] The dollar value placed on injuries and on the loss of lives is based on medical costs and loss of earnings (see Chapter 8). For coal, the largest social cost is illness and death from air pollution. One study estimates the social costs of SO_2 without stack scrubbers at .42 cents per kilowatt-hour (kwh), or 17% of the direct economic costs of electricity (2.5 cents per kwh); with scrubbers the figure drops to 5%. For coal, occupational hazards (chiefly mining accidents) are 0.4% of economic costs, compared to .04% for nuclear power.[39] This study uses the Rasmussen report, together with highly subjective estimates of risk from sabotage, to calculate the social cost of accidents. The total nuclear social cost is 0.3% of economic costs—less than one-tenth of the figure for coal plants with scrubbers. But these figures are very uncertain, and data on the two most serious potential consequences—climate changes due to CO_2 from coal, and nuclear war from weapons proliferation—were considered too uncertain even to include in the calculation.

A more direct comparison can be made by asking *how many deaths* can be attributed to the operation of a 1000-megawatt plant for a year. For one such coal plant there are roughly 2 occupational fatalities per year (from mining, transportation, and plant operation), compared to 0.5 with a nuclear plant. Estimates of public fatalities from SO_2 vary from 0 to 23, depending on assumptions in extrapolating health data and on the extent of pollution control. EPA calculates that if the 1975 SO_2 standards are met there will be less than one fatality per plant-year. Calculations on major reactor accidents also vary widely. The Rasmussen report estimates .0004 immediate deaths per reactor-year; if this were multiplied by 25 to allow for unfavorable winds and delayed deaths, reactor accidents would contribute only .01 fatalities.[40] Even if this were off by a factor of ten, it would be much less than coal fatalities, though sabotage and proliferation introduce additional unknowns. The Ford-Mitre study concludes that neither the coal nor the nuclear option has a clear advantage on risks; it recommends that both be kept open as a hedge against uncertainty and as an opportunity to improve the safety of both. "The range of uncertainty in social costs is so great that the balance between coal and nuclear power could be tipped in either direction with resolution of the uncertainties."[41]

Some of the limitations of risk-benefit analysis discussed in Chapter 8 are evident in the following comparisons of coal and nuclear power:

1. Scientific Uncertainty

It is very difficult to obtain direct data on low-probability events, and the extrapolations of dose-response curves to low levels of SO_2 and radiation remain uncertain. The effects of CO_2 on the climate are not well understood. We have had little operating experience with the emergency reactor cooling system. Human errors and such unscheduled events as sabotage or diversion are subject only to speculative guesses. Under such circumstances, assumptions can vary widely and are likely to be influenced by institutional and personal biases as well as economic and professional interests.

2. Catastrophic Consequences

Two very uncertain possibilities should weigh heavily in any deliberation because they are global and conceivably irreversible: climate change from CO_2 and nuclear war from plutonium diversion. The aversion of the public to large-scale disasters suggests that reactor accidents also should be given greater weight than probability calculations would assign.[42] The media report an event in which 1,000 people are killed, while they ignore 1,000 events in which one person is killed. People fear large disasters, such as fires, hurricanes, and earthquakes, more than equivalent dangers from small, familiar sources or voluntarily accepted risks. One large-scale nuclear accident could lead to a public outcry and demands to shut down all reactors.

3. Distributive Justice

We should look at the distribution of risks as well as their totals. Coal miners and the urban poor bear most of the risks from coal. Western coal states should be compensated for the environmental destruction from which Midwestern cities benefit. Seepage of liquid radioactive wastes create local risks for the sake of national benefits. There is a growing conflict between state and federal jurisdiction; federal preemption of state authority would provoke strong public reaction, but no one wants such hazards in his or her back yard. By 1979, 14 states had bans or moratoria on nuclear waste depositories. There are also equity issues in the exposure of nuclear workers to higher radiation levels than the public. The justification often given is that occupational risks are voluntarily accepted and are usually accompanied by additional compensation. Nuclear workers probably have more options for alternative employment than coal miners have, but "informed consent" for job-related risks is seldom fully informed or fully voluntary.

4. Future Generations

Intertemporal equity is an issue in CO_2 production; future generations will bear the brunt of climate changes caused by our use of coal. Radioactive wastes might be a potential risk to hundreds of future generations for the sake of electricity for our generation. Reprocessing plants, which remove plutonium from spent fuel before waste disposal, would reduce the risk to future generations, but would increase considerably the short-term risk from theft and proliferation. Retrievable storage might allow some benefits to future generations if new processes are discovered for recovering useful materials, but the greater accessibility would entail greater risks. In this case, I suggest, it is more important to minimize possibly catastrophic risks than to seek possible additional future benefits. This would favor permanent deep burial, even at considerable expense to those who benefit from the energy today.

In addition to the problems of used fuel, there are hazards from carbon-14 and radon gas from *uranium mine and milling wastes.* It is estimated that if all U.S. electricity (at 1975 levels) were nuclear, the radon from a year's wastes would produce two fatalities per year. But the isotope which produces radon has a half-life of 80,000 years. Since it decays so slowly, fatalities would add up year after year, totalling 200,000 from one year's waste alone—plus another 200,000 from wastes produced each subsequent year. It is often assumed that a time-discount should be applied to future deaths. William Ramsey, for example, calculates that at a discount rate of 5% (the current return on utility assets) 200,000 future radon deaths would be equivalent to 40 deaths today, which appears small compared to the 200 deaths that would occur from coal-mining accidents if the same amount of electricity were produced from coal.[43] In Chapter 8 I argued that a time discount should be used only when there are specific mechanisms for transferring cumulative benefits from the present to the future. In theory we could invest funds to buy future radon safety or to compensate future radon victims, but such procedures appear implausible on this time-scale. If we counted present and future deaths equally, we would make much greater efforts to bury uranium mining wastes so that radon gas could not escape.

5. Freedom and Participation

Coal and nuclear systems both require massive capital investment that only federal agencies and giant corporations can make. Both options will increase further the concentration of economic and political power. But nuclear systems are exceptionally complex and have unusual risks that require a unique degree of government regulation and a large bureaucracy to protect the public. In a stable world the security measures that would reduce the risks of sabotage and theft would be no threat to civil liberties. In a violent and fast-changing world, however, the surveillance of dissident groups would require police powers that could be easily abused. Participation in decision making also is unusually difficult in the nuclear case because there are so many complex technical issues. In the past the public was willing to let the experts decide; decisions were made primarily on technical and economic grounds. Today many people feel they were not fully informed of the risks; they distrust government and industry experts, and end by deciding on purely emotional grounds—which is likewise a denial of informed participation. The credibility of future decisions requires that government agencies must stop treating nuclear policy decisions as if they were merely technical issues. The nuclear opposition must be represented at all levels of decision making, even though this will entail delay and compromise in policies adopted.

In the light of these difficulties, I would recommend the methodology adopted by William Ramsey in *Unpaid Costs of Electrical Energy.*[44] Ramsey presents detailed estimates of the social and environmental costs of coal and nuclear energy, but he concludes that there are too many uncertainties and value judgments to aggregate the costs in a single numerical measure. Instead, he makes *comparisons relative to four values.* 1) *Health*: nuclear energy produces fewer fatalities and illnesses from routine operations, even if one uses low estimates for the effects of air pollution from coal. 2) *Environment*: nuclear energy entails less extensive land damage than coal mining, though uranium wastes will increase as low-grade ores are used. 3) *Catastrophe avoidance*: coal comes out ahead, at least for the next few decades during which climate changes are unlikely to be serious; nuclear proliferation and reactor accidents are more immediate catastrophic risks. 4) *Equity*: coal involves less intractible problems. Inequitable risks to coal miners could be alleviated by stronger safety measures and by higher taxes on coal to pay "black lung" and accident compensation, whereas little can be done about radiation risks to future generations.

Which of the four values above is most important? Ramsey maintains that no objective judgment is possible and that citizens have *diverse value orientations.* Since no clear case can be made for either coal or nuclear energy, he advocates a mix in which both options are kept open. This would hedge our bets; future changes in our knowledge, our technology or our values could alter the balance. He advocates increased research on health uncertainties and technological improvements. In the meantime, says Ramsey, we should require full mitigation and abatement technologies (such as scrubbers), even though they are expensive, as a protection against uncertainties and as a premium necessary to muster a social consensus among citizens with divergent value structures—in order to reduce protracted controversy and delay. I differ from Ramsey mainly in assigning greater weight to catastrophe avoidance and equity—and in stressing conservation and solar sources.

A Policy for Nuclear Energy

There are no easy answers in the choice between coal and nuclear energy for the intermediate term. Direct economic costs are about equal, though indirect public subsidy of nuclear energy is greater and its costs are rising faster. In normal operation, nuclear energy has distinct advantages over coal in environmental impact, occupational safety, and public health, though there still are unresolved questions about radioactive waste disposal. The unknown effects of

CO_2 on climate probably rule out the long-term use of coal, but not its use for a few decades. Clearly we should use solar energy and conservation to minimize expansion of *both* coal and nuclear sources. But, on balance, I believe we should use coal as the main transition fuel because of the special nuclear risks that are not part of normal operation.

Nuclear war is the greatest single threat to humanity, and efforts to halt the arms race and achieve disarmament should have highest priority. There are many ways in which a nation can acquire nuclear weapons, but widespread circulation of plutonium would make acquisition much easier. I therefore am opposed to the deployment of reprocessing plants or breeders. In a world where political instability and social conflict are rampant, even conventional nuclear plants are highly vulnerable to sabotage and terrorism. Despite the efforts being made to improve reactor safety, there always can be unanticipated combinations of human errors and mechanical failures, and their consequences could be enormous. There always will be a tension between the industry's interest in profits and the public's interest in safety, which is only partially resolved by an ever-longer list of federal regulations. In the presence of human fallibility, institutional self-interest, and international conflict, I suggest we should be cautious about systems in which unexpected events can have such disastrous results. Except for CO_2, the hazards from coal are better known and are not potentially catastrophic.

I am not proposing that we should shut down the nuclear plants now operating or stop those under construction. The main economic cost of nuclear energy is the plant rather than the fuel, and it would be a social waste to throw away the large investment that these plants represent. We also should continue nuclear energy research and keep the nuclear option open as a long-term possibility. The British gas-cooled reactors, for example, are in some respects safer than U.S. high-pressure, water-cooled designs. Proliferation-resistant reprocessing and plutionium-free breeders (such as those using the thorium cycle) should be explored. Perhaps improved breeders and reactors in nuclear parks, under international management and located near disposal sites, will be an acceptable long-term option, despite its potential hazards, if solar electricity proves too costly—especially if nuclear disarmament or effective control of nuclear weapons has been achieved. We should keep these long-term options open but not deploy them in the intermediate term.

Another long-range possibility is *nuclear fusion*, whose main fuel, heavy hydrogen, can be obtained from ordinary water. It would produce no environmental damage or air pollution, relatively little radioactivity, and no weapons-grade material. Yet fusion faces formidable scientific and engineering obstacles in maintaining temperatures of millions of degrees. The magnetic confinement method places great demands on vacuum systems and structural materials; if it does work, it can be used only in large-scale installations. The laser fusion method might be adapted to smaller scales, but in some versions it could be used to produce nuclear weapons. Fusion may have a role in central electric production in the next century; research on it should be vigorously pursued, without attempting a rapid and perhaps premature commercialization.[45]

What about *other industrial nations*? The Soviet Union has large coal reserves, but Western Europe and Japan do not and so are likely to rely more heavily on nuclear energy. I still am hopeful that they will not embark on a major deployment of breeders or reprocessing. If reprocessing is expanded, diversion risks would be lower with partial separation techniques that do not produce weapons-grade material. European countries have very limited energy resources and have fewer options than the United States; several of them are in the midst of intense public debates, referenda, or legislative decisions about future nuclear policy.[46]

What about *Third World countries*, most of which have no coal? They resent being told by nations that are frantically stockpiling nuclear weapons that they cannot be trusted with peaceful reactors unless they sign the Nonproliferation Treaty. Argentina, Brazil, India, and Pakistan are eager to expand their nuclear technologies to produce electricity for industrialization. But many Third World representatives, while defending the right of equal access to nuclear energy, recognize that it will be of very limited use in their countries. Only a few urban areas have networks to distribute such huge blocks of electricity, and nuclear energy does little for agriculture, transportation, or rural development.

At a major conference of scientists and theologians sponsored by the World Council of Churches at MIT in 1979, most of the 130 delegates from the Third World supported a resolution calling for a five-year moratorium on nuclear plant construction to allow more informed public debate on the risks and benefits. They were less concerned about safety or health than about the *social and political impacts* of nuclear energy, especially the continued dependence of developing countries on foreign equipment and experts. Importing reactors and nuclear fuel from industrial countries, they said, would perpetuate patterns of economic and technical domination and dependency and accelerate the split between modern and traditional cultures within their countries. Many of them argued that solar energy is more adaptable to local cultures and offers better prospects for national self-reliance.[47]

III. THE SOLAR PROSPECT

Solar energy is renewable and, on the whole, safe and environmentally benign. But in some forms it is intermittent and many solar technologies are still expensive. We will examine the solar alternatives, first in terms of technical and economic possibilities, and then in terms of environmental and human values.

Solar Technology and Economics

There is a wide variety of types of solar technology at various stages of development. Solar collectors for *hot water* already are competitive with electric and gas heaters. The initial costs, currently in the $1,000 range, will fall with mass production, and can be repaid by fuel savings within a few years. An insulated water tank provides inexpensive thermal storage for cloudy days. Solar *home heating* systems have substantial initial costs but no fuel costs. Tax credits, low-interest loans, and mass-produced parts already are bringing down the average cost. Since the price of oil and gas is almost certain to rise, solar heating already appears competitive over the life of the system in most parts of the country. Good building design (insulation, glass-covered south walls, and so on) can cut energy needs drastically, and thermal storage at modest expense is provided by rocks or water tanks; however, some back-up (such as wood stoves or natural gas) is needed for prolonged periods of cold, cloudy weather.[48]

There are two ways in which *solar electricity* can be produced. In the "power tower" method, which has received stronger research funding, movable mirrors focus the sun's rays on a boiler that produces steam to run a generator. The mirror systems are expensive, however, and there is little prospect of major cost reductions or improvements in efficiency. Land use is substantial, perhaps 7 to 10 square miles for a 1,000 megawatt plant. In the photoelectric method, photovoltaic cells produce electricity directly from sunlight without any moving parts, and have proved reliable over long periods on space vehicles. Small panels can be installed on rooftops or the southern walls of homes or apartment buildings. Larger arrays can be built over parking lots, railway yards, and factory roofs. Solar farms with acres of cells could supply industries that need larger blocks of electricity; the largest units could be located only in desert or rocky areas where damage to vegetation would be minimal. Smaller systems have the advantage of shorter construction times, lower transmission costs, and the possibility of integration of electric and heating systems.

The chief problem with *photovoltaic cells* has been the cost, but this has been falling and is likely to drop dramatically with research and mass production. Cost reductions by factors of 10 to 100 have been common in solid-state R & D and are anticipated when silicon cells or thin films are produced in automated processes.[49] OTA estimates that by the late 1980s electricity from home or neighborhood solar units will cost 4 to 10 cents per kilowatt-hour (compared to 3 to 5 cents today from coal or nuclear plants). Expanded research funding and financial incentives to stimulate production could accelerate commercialization.[50]

In addition to these direct uses of the sun's rays, there are several ways in which energy from the sun can be obtained *indirectly* through natural processes. Water power is now the largest U.S. solar source, providing the equivalent of 3 quads in our total energy budget of 78 quads. In theory this could be tripled, but acceptable dam sites probably can be found for only another quad or two. Wind power might provide as much as 5 quads by 2020, but it is intermittent and would have to be used in conjunction with other sources or storage systems. Geothermal power from underground heat is available in only a few geographical areas. Ocean thermal gradients might be used to power floating generators, but the ecological impacts of circulating large volumes of cold water from the ocean bottom are uncertain.[51]

Biological sources, however, might make a considerable contribution—up to 5 quads from wastes alone by the year 2000, and possibly several quads more from deliberately cultivated crops. Agricultural and forest residues and mill and municipal wastes can be burned or digested to form methane gas. Manure from feedlots yields methane, and nutrients can be returned to the soil as fertilizer. Some crops can be fermented or distilled to produce alcohol, which can be burned directly or mixed with gasoline (gasohol); sugar cane and cassava are now used to make alcohol for autos in Brazil. "Energy plantations" of fast-growing trees have been proposed, though there are few regions where land and water could be spared for this purpose. None of these indirect solar sources is large and each is limited geographically, but together they could reduce dependence on nonrenewable fuels. Several of them are particularly promising for the Third World, such as hydropower in Africa and South America, energy plantations in some parts of the tropics, and methane generators in Asia.[52]

Estimates of total *U.S. solar potential* vary according to assumptions about the price of competing fuels and policies for conservation and solar energy. An NAS report says that 20% of the nation's energy could be solar by 2010.[53] A Stanford study says that with heavy subsidy 45% could be solar by 2020.[54] The CEQ estimates 25% by 2000 and at least 50% by 2020.[55] In 1979, President Carter set a goal of 20% solar by 2000, and proposed a solar budget of $1 billion. While there is some variation among these estimates, they are in agreement that solar energy has a major long-term potential, but that other sources will be needed in the intermediate term to replace oil and natural gas.

Environmental and Social Issues

The impacts of solar energy on nature and society are in general less destructive than other options:

1. Sustainability

Solar energy is sustainable indefinitely since it comes from the sun. In the long run we must learn to live off such "energy income" rather than the "energy capital" of depletable resources. Whatever funds are put into solar development will help us and other nations in both the intermediate and the long run. Solar energy used today does not lessen the solar energy available to future generations.

2. Low Environmental Impact

Solar heating and electric systems are nonpolluting. There is no way they can be used for weapons, no risk to pass on to future generations, and no opportunity for catastrophic accidents. There are, however, occupational and public health hazards from the manufacture of the solar components (especially air pollution from the coal used to make steel). While some authors claim that these hazards are comparable to those from coal-produced electricity, others say they are smaller than those from either coal or nuclear electricity.[56] The main environmental impact would be land use for "power tower" mirrors or large-scale photoelectric facilities, or for plantations of crops or forest to produce alcohol; but each of these methods could be limited to suitable areas. The very diversity of solar technologies and their geographical distribution will reduce the scale of damage from unforeseen consequences.

3. End-Use Matching

The adaptability and diversity of solar sources provide opportunity to match them to particular tasks according to function, scale, temperature, and location. For home heating, only low-quality energy (at low temperatures) is needed. Conversion and distribution costs are avoided if solar energy is obtained directly in the form of heat. Electricity has unique uses in communications and for stop-and-start motors, and it offers unmatched convenience and flexibility in home lighting and appliances; many other tasks for which electricity is currently used could be accomplished with less wasteful forms of energy. Where electricity is needed, it should be produced locally insofar as possible to save the large transmission and distribution costs. There are no economies of scale in direct solar technology (except in some storage systems); large-scale plants would be needed only where large blocks of electricity are required.[57]

4. Employment Opportunities

Power plants and oil refineries produce fewer jobs per dollar than any other major industry; solar technology is much more labor intensive. The manufacturing and installation of solar components would be well dispersed geographically. Solar heating and home insulation draw on local skills and shops and would integrate well with existing building trades. Working conditions are comparatively safe and there is no community disruption from temporary workers, as occurs with Western coal.

5. Decentralization

With solar energy there are opportunities for small businesses in production, installation, and repair. Community-level systems, including cogeneration of electricity and heat, could be run as cooperatives, small companies, or municipal utilities. Solar energy would facilitate individual involvement and citizen participation rather than dependence on experts. It would counteract the increasing concentration of economic and political power in a few giant energy corporations. As Amory Lovins points out, individuals and communities could be producers as well as consumers of energy and would have greater choice and control. Personal ownership is possible with solar equipment as it never could be with nuclear equipment.[58] Large companies will still be needed to mass-produce solar components, but the monopoly powers of energy corporations and regional utilities could be avoided. Moreover, decentralized systems are less vulnerable to deliberate disruption than large-scale ones; elaborate security provisions are unnecessary. The stakes are smaller when errors or accidents occur, and less extensive government regulation is required.

6. Developing Countries

Sunlight is rather evenly distributed among nations and regions, whereas oil, coal, and uranium are all very unevenly distributed. Cheap solar technology would be a great boon to the sun-rich Third World, most of which lacks the grids to distribute centrally-produced electricity. It would reduce dependence on other nations for expertise and fuel. Solar cookers, solar water pumps, methane digestors, small hydroelectric dams, windmills, and solar cells would be especially useful. They would encourage national self-reliance and fit in with local cultures. Solar energy would reduce international conflict from competition for oil without the dangers of nuclear proliferation. It thus can be a major instrument of global justice and peace.[59]

Critics of solar energy dwell on three issues: intermittency, decentralization, and cost. The *intermittency* of sunlight is a problem mainly in solar electric systems. In some regions solar electricity could be used to pump water up into reservoirs, from which the water could flow down to generate electricity during cloudy periods or at night. For the intermediate term, solar electricity can be coordinated with other sources and used mainly as a fuel saver while the sun is shining. It would be desirable to tie local solar systems into the grid so that electricity could be sold back to the utilities to help carry peak loads, which occur during hot summer days. In the long term, however, cheaper and more efficient storage systems will be needed. Promising lines of research include advanced batteries, hydrogen electrolysis, fuel cells, and cryogenic storage.[60]

Decentralization is attacked by critics as a romantic ideal. Solar enthusiasts are accused of nostalgia for the small community and local control that is no longer practical. Energy growth, say the critics, contributes to employment and economic growth, and electricity is convenient and reliable; only central electric systems can meet expanding demand. Most people do not really want to get involved in producing energy. If solar energy does develop, it will be controlled by large companies and utilities.[61] This centralization/decentralization debate involves fundamental differences in values and in visions of the future to which we will return shortly.

The other main objection to solar energy is *economic*.[62] But comparison of current costs is misleading for several reasons. First, the cost of nonrenewable fuels will go up as they are depleted, while *solar costs will fall* with further research, mass production, and new government policies. Less than 1% of U.S. energy R & D since World War II has gone to solar research; it is not surprising that solar technology is less well developed than other options. In particular, intense research on photovoltaic cells and storage systems is likely to lower their cost greatly.[63] Moreover, solar costs should really be compared not with the *average* cost of electricity from other sources, but with the *marginal* or replacement cost of electricity from new plants, which is considerably higher.

Second, market prices still omit many *environmental and human costs* borne by society, even when some of the externalities have been internalized through environmental and safety regulations. The market takes a short-term view and discounts costs to future generations. As irreplaceable resources are used, benefits are transferred from the future to the present, while some risks are transferred from present to future. By the Rawls criterion (Chapter 4), this would be justifiable only if future generations will have enough resources that after the transfers none of them would be the worst-off generation. Such a condition cannot be met if we continue to rely on depletable fuels.

Third, *depletable fuels have been heavily subsidized* in the past—adding up to at least $100 billion in the last 20 years, according to the CEQ.[64] Oil was subsidized by depletion allowances, tax credits, import quotas, and highway construction funds. Regulated prices gave oil and gas a competitive advantage and encouraged the technologies that use these fuels. Various nuclear subsidies have been mentioned: past R & D funds, present insurance coverage, future costs of waste disposal and the decommissioning of worn-out plants. For years nuclear energy received preferential treatment, being the only form of energy to have a separate congressional committee and federal agency to promote it. Because energy corporations have huge investments in fossil fuel and nuclear technologies, most of their research—and their political influence—supports continued use of these fuels. As a nation we have been very shortsighted in heavily subsidizing depletable fuels, which has placed renewable sources at a competitive disadvantage.[65]

Various national policies could be deliberately designed to encourage *the transition to solar energy*. Solar research, which by 1978 was finally up to 10% of the energy research budget, should be increased further; photoelectricity and storage systems, in particular, have been underfunded relative to their potential contribution. Solar purchases for federal installations would help to stimulate production. California has legislated a 55% tax credit on solar equipment. A Harvard Business School study advocates a 60% federal subsidy of solar purchases to correct for the market biases against solar energy and to represent its real value to society.[66] Because low-income families and renters seldom can afford heavy initial costs, low-cost loans repaid from fuel savings are essential. Building codes that discriminate against solar devices need to be revised, and a federal program of performance standards and testing should be instituted to protect consumers against inferior equipment. Assistance to other countries for solar development also would have far-reaching consequences. The CEQ has proposed a vast solar aid project on the scale of the Marshall Plan, and OTA has suggested joint research and training programs with other countries.[67]

How far should national policy go in tipping the balance from *depletable* to *renewable* resources and in ensuring that market prices are closer to true costs?

After long debate, Congress voted in 1978 to allow the price of natural gas to rise, and in 1979 did the same for domestic oil. The purpose was mainly the stimulation of oil and gas production, but one effect will be to make solar sources more attractive. A tax on all fossil fuels would slow depletion, encourage conservation, reflect indirect social and environmental costs, and speed the shift to renewable resources. But because higher energy prices will be a heavy burden for the poor (who spend a higher fraction of their income on energy than the rich do), there should be tax rebates to low-income families who keep their energy use low. These proposals would integrate well with conservation programs.[68]

IV. THE CASE FOR CONSERVATION

For the nation, it is cheaper to save a barrel of oil than to buy one, and it lessens dependence on foreign oil. Conservation seldom harms the environment. It extends the time to resolve problems with current energy technologies and to develop better alternatives. Conservation also is a response to important ethical issues in the waste of irreplaceable resources, the disproportionate U.S. use of world energy supplies, and the neglect of the needs of future generations.

Energy-Saving Measures

Let us ask first how energy can be saved by technical changes that improve efficiency, or by relatively minor behavioral changes, without significant changes in life-style.[69]

1. Heating and Cooling

Putting insulation and weather-stripping in existing houses is the most rapid way to reduce energy waste, and it pays for itself in fuel savings within a few years. Greater savings can be achieved by higher efficiency standards in building construction, though new houses only gradually replace older ones. Improved ventilation can reduce the use of air conditioning, which is extremely energy intensive. District heating of all the buildings in a community would effect large savings, especially if the heat were a by-product of electricity produçtion. Trenton, N.J. is building a 37-megawatt cogenerator with 65% efficiency (compared to 32% for conventional plants).[70]

2. Industry

Industry has begun to cut down on waste; from 1973 to 1978 energy used per unit product fell by 14%. More substantial savings can occur as old processes and plants are replaced. Process steam, which accounts for 45% of industrial fuel use, can be cogenerated with electricity, and the heat can be used more than once by cascading or heat recovery systems. New steel plants use half the energy of older ones. New processes, such as chloride extraction of aluminum, can replace more energy-intensive techniques. Recycling can cut down on raw material extraction, which uses 25% of industrial energy. Reclaiming a ton of steel from urban wastes (including collection and separation) uses only 14% as much energy as extracting a ton from ore. Recycling aluminum requires only 5% of the original extraction energy.[71]

3. Food

U.S. agriculture has become highly energy intensive through increased use of fertilizer and farm machinery. Additional use of chemical fertilizer produces greatly diminishing returns and growing runoff problems. According to several studies, organic farms produce comparable yields and income per acre with less than half the energy input.[72] More than half of the energy that goes into the U.S. food system is used in transporting, processing, packaging, and preparing food. It takes energy to refine flour—yet white flour is less nutritious than whole wheat flour. U.S. foods are overprocessed and overpackaged. Frozen foods are convenient but waste a lot of energy. Local food production saves on both transportation and processing. Many U.S. families now grow some of their own vegetables, but more extensive gardening would improve health while it saved energy.[73]

4. Electricity

Two-thirds of the energy used in generating electricity is now lost as waste heat, and there are additional losses in transmission. Electricity thus should be saved for essential uses (lights, motors, communications, and so on), which are estimated to be only 8% of all energy end-use.[74] In the past, low electricity rates were offered to large users—partly to promote sales, partly because of economies in distribution and billing. Today "inverted rate structures" are proposed to discourage heavy use. "Lifeline rates" would start with a low price for a basic block of electricity for minimal needs per person, with rising rates thereafter. This would offer some help to low-income families and would encourage conservation. Lower off-peak rates would produce some shift in demand away from peak-load periods and reduce the generating capacity needed. For appliances, the efficiency (work output divided by energy output) can be greatly increased.

5. Transportation

Gasoline for cars and trucks accounts for 25% of all U.S. energy use. Settlement patterns have been built around the auto, and one in four jobs is related to the auto. A car provides job mobility, social status, and personal freedom. U.S. models have been oversized and overpowered, mainly because profits on large cars are higher. Fuel economy can be greatly improved. Smaller cars save energy and reduce air pollution; radial tires, better ignition systems, and streamlining also help. Congress has mandated fuel economy standards averaging 27.5 miles per gallon by 1985, with further improvement expected thereafter. Major fuel economies can be affected by alternative designs: diesel motors, gas turbines, Rankine or Sterling engines, and stratified-charge engines.[75] For cars, as for buildings and electrical appliances, a 50% saving is a reasonable expectation from technical improvements alone within a couple of decades.[76] In the meantime, car pooling and stricter enforcement of the speed limit would effect immediate savings.

The shift to *more efficient modes of transportation* will be more difficult. We have subsidized autos and planes more heavily than buses and railroads, which are much more energy-efficient. Urban buses are rapid and convenient if they are assigned express routes and bus lanes. The electric trolley once was common in U.S. cities, and some trolley routes could be restored. Subways are expensive and save little energy unless they are very heavily used. But mass transit offers other benefits in reduced congestion, pollution, and fatalities. Federal funding of mass transit systems, long blocked by a coalition of auto-related interests, has started on a modest scale and should be expanded.[77]

But habits change slowly. Better, faster service attracts *mass transit* ridership more than lower fares do—though in some surveys only half the new riders were former car drivers. Public transportation does help the poor and elderly, however, who constitute most of the 20% of U.S. citizens who have no car. In the long run, better *town planning* is the most effective way to reduce transportation energy. The present urban sprawl creates long commuting distances. New urban designs with local work places and neighborhood stores would reduce travel distances. Better telecommunication systems can substitute for transportation for many purposes. According to one estimate, a combination of these measures could reduce the energy used in transportation to one-fourth of its present value.[78]

How much *reduction in demand* would all of these conservation measures produce? In the early 1970s it was assumed that demand would continue to rise at 3.5% per year, doubling in 20 years. But a 1978 NAS report calculates that, with rising energy prices and aggressive conservation policies, demand would drop by 20% by the year 2010, even though real GNP had doubled. There would have to be major shifts away from energy-intensive products and processes, but no sacrifice in living standards, employment, or economic growth.[79] A 1979 Harvard Business School study concludes that the United States could consume 30 to 40% less energy by 2000 with only minor life-style adjustments and no decline in economic growth. Its authors argue that solar energy and conservation could replace oil and make the expansion of coal and nuclear fuels unnecessary.[80] A study sponsored by Resources for the Future agrees that energy growth can be limited without endangering economic growth, but holds that for the transition period, some coal expansion will be needed.[81] The two studies differ mainly in their assumptions concerning the rate of solar technology innovation and the adoption of strong conservation measures.

Conservation, Justice, and Freedom

Reduction in the disproportionate U.S. use of world energy resources would clearly contribute to global justice. It would reduce the competition for oil, which drives up the prices that other countries must pay. It would allow more energy to go to the fulfillment of basic needs for all, such as food, rather than such luxuries for the few as air conditioning. By conserving for the future it would further intergenerational justice. But is conservation consistent with justice within the nation? Or would it result in unemployment and declining living standards that would especially hurt low-income families?

1. Energy and Economic Growth

In the past, energy consumption and GNP have grown together. But this historical correlation occurred while energy prices were falling, and it will not necessarily hold amid rising prices that will motivate greater efficiency and shifts to less energy-intensive goods and services. There is a correlation between energy use and GNP if one compares developing nations with industrial nations, but the correlation breaks down if one makes comparisons among industrial nations. Nations with a GNP per capita close to ours use half as much energy per person. In the short term, sudden drops in energy are harmful to the economy; the 1973-74 oil embargo led to job layoffs and economic recession. Over longer time-spans, however, energy and economic growth can be uncoupled by reducing waste (increasing efficiency), and by new patterns of production, transportation, and consumption.

There is a further question as to whether energy use correlates with quality of life. One study of 55 industrial nations found no correlation of energy consumption with a varied set of social indicators (including life expectancy, educational levels, cultural activities, and suicide and crime rates).[82] During the period of greatest U.S. energy growth (1955 to 1975), the quality of life as measured by most social indicators was falling. The per capita energy consumptions in New Zealand and Ireland are equal, but by the measures of human well-being they differ greatly.[83]

Sweden provides an illuminating comparison. It has a GNP per capita close to that of the United States and outranks us on almost all social indicators, but uses 40% less energy per capita. Heavy gasoline taxes, smaller cars, and excellent public transportation (aided by geographical compactness) have produced high transportation efficiency. In the United States, 68% of the energy used to run electric generators is wasted; in Sweden, only 47% is wasted, while 24% goes to cogenerated heat. Swedish commercial use of energy is 2.5 times as efficient as ours. Yet even Sweden can make further improvements in efficiency (at least 30%, by one estimate) with no sacrifice of living standards.[84]

2. Conservation and Employment

It has been widely assumed that energy growth promotes employment. But one cause of rising energy use has been the replacement of labor by machines, which has harmed employment. Again, energy production has been one of the least labor intensive of all industries (in jobs per dollar). Today a given investment would produce four times as many jobs in conservation or solar energy as in nuclear plants. Loan guarantees of $1.6 billion for building insulation would generate 400,000 jobs. Many conservation measures provide local jobs that integrate well with existing building trades and small businesses. Other energy-conserving projects can be specifically designed to create new jobs for the unemployed.[85]

The long-term shift to less energy-intensive products, new modes of transportation, and services rather than goods will decrease jobs in some industries and increase them in others. Such shifts will require programs of retraining and relocation assistance. Zero energy growth is not in itself incompatible with economic growth and high employment levels. Some labor-intensive processes do involve great drudgery and low productivity, and these should be avoided. Others offer creative and humanly fulfilling work roles and relatively high productivity, along with lower energy use (see Chapter 13).

3. Impact on Low-Income Families

Upper- and middle-class families should be the first target of conservation policies, since they use more energy per capita and could more readily afford energy-saving changes; low-income families enjoy few luxury uses. But a lot of energy is wasted by low-income houses, which usually are poorly insulated. Low-interest loans for insulation are an effective and equitable federal investment in conservation. Price decontrol and fuel taxes are strong conservation incentives but fall most heavily on the poor. One study found that the poorest tenth of U.S. families spent 27% of their income directly on energy, whereas the richest tenth spent less than 4%.[86] Elderly persons and large low-income families were hardest hit by the rise in oil and coal prices.[87] Energy taxes and rising prices have a very regressive impact unless accompanied by rebates or direct assistance to low-income families.

It has been proposed that the welfare system should include fuel stamps to provide a minimum amount of energy for heat. We treat minimum levels of food, housing, medical care, and education as basic components of human well-being to which everyone is entitled; surely energy is equally fundamental to a decent human life. Such a scheme would help to allocate an essential and increasingly scarce resource more equitably. However, it would lessen the incentives to conservation and add to an overburdened administrative bureaucracy. There certainly should be emergency funds available to alleviate special hardship situations and direct assistance to low-income families with large home heating bills. The underlying problem, of course, is income maintenance, which can be attacked only through overall income tax policies or a guaranteed annual income.[88]

4. Freedom and Coercion

With greater public awareness, voluntary conservation will be more widely practiced. But people are reluctant to make sacrifices that are not shared by others, and many forms of conservation can be introduced only through public policies. There is a minimum of overt coercion in public subsidies through tax credits, research funds, and federal grants. Mandatory efficiency standards apply coercion to industry rather than to individual citizens; this approach relies on technical changes, which often are easier to bring about than behavioral changes that people resist. The decontrol of energy prices provides incentives for conservation and preserves the freedom of the marketplace, but puts a very heavy burden on the poor.

Are stronger measures needed to effect conservation by individuals? An energy tax would provide additional economic incentives to save energy and some of its regressive impact could be mitigated by rebates to low-income families. However, prices are rising so fast that additional taxes would create great hardship without much reduction in consumption. If other methods fail to reduce oil use, I believe the rationing of gasoline and heating oil would be more effective than the alternatives, and on the whole more just in equalizing sacrifices. Judging from our own wartime experience, however, rationing would involve considerable inconvenience, an additional bureaucracy, and at least some black-market abuses. The establishment of social priorities among conflicting uses of fuel always will be controversial. For these reasons, rationing should be avoided unless shortages are severe.

5. Conservation and Politics

Along with solar energy, conservation lacks the strong constituency in government, industry, and labor that oil, coal, and nuclear energy enjoy. Powerful institutional forces support energy growth—the increase of production rather than the reduction of consumption. The market provides little incentive to seek alternatives until shortages are imminent. The political process focuses on immediate benefits that can pay off at the next election. In 1978 Congress adopted only a few of the conservation measures that Carter had proposed. It provided tax credits and some low-cost loans for home insulation and efficiency standards for autos and home appliances. It urged state commissions to reform utility rates, but took no federal action. It rejected all taxes on oil (wellhead, gasoline, gas-guzzler, and oil-burning plant taxes).[89] In short, only a few first steps in conservation were taken. By 1979 Congress appeared more receptive to energy-saving legislation, but had not yet acted.

The main obstacle to progress in conservation has been *public complacency*. Many people still think oil shortages are a temporary aberration attributable to

Arab diplomacy or manipulation by oil companies. Although citizens protest against local siting of refineries, power plants, power lines, or pipelines, they usually are quite willing for these facilities to be built somewhere else. In the early 1970s environmentalists showed that constituencies can be created around new values going beyond short-run economic benefits, though such support has been more difficult to achieve during times of recession and inflation. By 1979 rising energy prices, another round of lines at gasoline pumps, plus continued evidence of U.S. economic and political vulnerability because of dependence on foreign oil, were beginning to arouse greater public interest in conservation.

6. Energy and Life-Styles

New life-styles can make a major difference in energy consumption. Such changes start among individuals but spread to influence cultural attitudes and political decisions. Energy scarcity may bring about life-style changes whether we like it or not; we need to work out satisfying patterns of life compatible with lower energy levels. A study in Minneapolis found that communal living groups used 36% less gasoline, 40% less natural gas, and an astonishing 82% less electricity per capita than the average home. The savings arose mainly from sharing meals, appliances, transportation, and heating, and from forgoing air conditioners and freezers.[90] Rural groups using organic farming and solar equipment have even lower energy demands. Some of the same savings can be achieved by groups of families, communities with common facilities, and settlements designed as total energy systems. Ivan Illich maintains that lower energy use actually would help restore human interaction and the sense of community that are lost in a high-speed, energy-intensive civilization.[91]

The greatest energy savings would come from the adoption of *simpler life-styles* in which material goods are less central as sources of personal fulfillment. Changes in consumption patterns are the product of new values and attitudes as well as economic forces. In place of the American dream of ever-growing consumption, some people envision alternative images of the good life that do not require profligate use of energy. Their understanding of personal fulfillment is more compatible with global justice and harmony with nature than are prevailing American goals. While they are a small minority in the United States today, their numbers are growing and could rise with external pressures, changing values, and greater awareness of resource limits, as I will suggest in Chapters 12 and 13.

Conclusions

I have suggested that for the intermediate term coal should be used to bridge the gap as oil use declines, and that nuclear expansion can be avoided by aggressive programs of solar energy and conservation. For the long term we should keep the nuclear option open until we are in a better position to know whether solar energy could supply most of the needs of an energy-efficient society, and whether there are practical and safer forms of the breeder (or of fusion) that could be deployed—preferably under international management—during the twenty-first century. What are the distinctive risks and values in such an overall program?

There would be substantial human and environmental costs from *increased coal use*, especially health effects from SO_2, but also mine accidents and strip-mining damage. All of these can be reduced by strict enforcement of environmental and safety standards, though at an economic cost to the consumer. Climate effects from CO_2 are uncertain but potentially very serious, and are likely to limit the use of coal long before reserves are exhausted. Coal use therefore should be kept to a minimum that is determined by the rate at which solar and conservation programs can be deployed and by the risks of social and economic disruption from energy shortages.

Energy shortages, especially if they occur rapidly, can lead to economic destabilization and social unrest. Severe shortages could lead to recession, greater unemployment, human suffering, and perhaps authoritarian responses. The United States might then try to fill the gap by even greater oil imports, adding further to international tension and conflict. Some conservation measures can be introduced only slowly (for example, improved design of houses and industrial equipment, on which the turnover is slow), though other measures are more rapid (such as car pooling and home insulation). Some changes in consumption habits and behavior patterns are slower than technical fixes designed to improve efficiency. Here the risks of energy shortages in the United States in the short run have to be weighed against the harm to other countries—and to ourselves in the long run—if we are too slow to change. Past history suggests we are more likely to move too slowly than too fast.

There also are major *obstacles to implementation* of the policies proposed. There are powerful constituencies and strong institutional momentum favoring oil and nuclear energy. The rewards from solar energy and conservation are delayed and diffused through a wide public, whereas the rewards from conventional energy sources are rapid and concentrated in well-organized groups that have profited from public subsidies in the past. Additional economic incentives favoring the new options through subsidies and taxes must be provided if policies from which the public will benefit in the future are to be adopted.

What are the main values that such a program would help to realize? First, *sustainability*. Conservation would reduce the waste of irreplaceable resources. Solar energy in all forms is a renewable resource. We can start with solar heating, photoelectricity, and energy from biological wastes (agricultural and forest residues, urban sewage, and solid wastes). The long-term nuclear options (breeders and perhaps fusion) should be kept open in case the economic or environmental costs of solar electricity turn out to be too high. These policies would take seriously the needs of future generations. They are the sort of policies one would recommend if one did not know to which generation one would belong (see Chapter 5).

Second, *justice.* Conservation will help to correct the extreme disparities in energy use between the United States and other countries. It will reduce the competition for oil which drives up prices and creates international conflicts that threaten world peace. The solar energy that one nation uses does not lessen the supply to other nations or future generations; the materials needed for solar equipment seem to be relatively abundant. Justice requires putting energy for survival and basic human needs ahead of additional energy for the affluent, at home or abroad. Solar energy is particularly promising for developing nations. It would be adaptable to local conditions and needs, and would encourage self-reliance rather than continued dependence on foreign fuels and machines. While some U.S. solar research will help other countries indirectly, there is a significant opportunity for international cooperation in the rapid development of inexpensive technologies using renewable resources, from which all nations would benefit.

Third, *participation.* Energy decisions involve trade-offs and value judgments among very diverse (and often uncertain) costs, benefits, and risks. Public participation in political processes related to energy is thus essential, in place of reliance on technical experts or market forces. Nuclear decisions are particularly complex, and access to information has been difficult. The vulnerability of nuclear plants will demand increasing regulation and security measures. They are the epitome of large-scale, capital-intensive systems that contribute to the concentration of economic and political power. Solar energy, by contrast, offers significant opportunities for individual and community ownership and control. These values—sustainability, justice, and participation—were central in the series of reports and debates in the World Council of Churches and the National Council of Churches, which culminated in resolutions adopted by both bodies in 1979 supporting energy policies similar to those advocated above.[92]

In Chapter 13 I will suggest that there is no simple answer to the choice between *centralized* and *decentralized* systems. I will advocate a diversity of scales, since in some cases (such as steel production) economies of scale are great enough to outweigh other considerations. In the case of electricity, the matching of source to end-use is important; for many uses, decentralized systems have technical advantages (greater flexibility, lower distribution costs, the cogeneration of heat and electricity), in addition to social advantages in participation and local control. If we want to have a diversity of systems, we must deliberately reverse the trend toward the concentration of economic and political power. We will have to offset the momentum of hard systems by a major attempt to develop the largely untapped potentialities of soft systems.

NOTES

1. For an introduction to energy options, see Lon Ruedisili and Morris Firebaugh, eds., *Perspectives on Energy*, 2d ed. (New York: Oxford University Press, 1978); and Philip Abelson and Allen Hammond, eds., *Energy II: Use, Conservation and Supply* (Washington, D.C.: Amer. Assoc. for the Advancement of Science, 1978).

2. Monthly Energy Review (Department of Energy), Oct. 1979, p. 6.

3. Earl Cook, "Fossil Fuels," in *The Sustainable Society*, ed. Dennis Pirages (New York: Praeger, 1977).

4. See Raymond Vernon, ed., *The Oil Crisis* (New York: W. W. Norton, 1976); Carl Solberg, *Oil Power* (New York: New American Library, 1976); Robert Engler, *The Brotherhood of Oil* (Chicago: University of Chicago Press, 1977); and John Bair, *The Control of Oil* (New York: Pantheon, 1976).

5. Denis Hayes, *Rays of Hope: The Transition to a Post-Petroleum World* (New York: W. W. Norton, 1977), p. 48.

6. Carl E. Bagge, "Coal: Meeting the Energy Challenge," in Ruedisili and Firebaugh, op. cit. For overviews of coal, see Richard Schmidt and George Hill, "Coal: Energy Keystone," *Annual Review of Energy* 1 (1976): 37-63; and *U.S. Coal Development: Promises and Uncertainties* (Washington, D.C.: General Accounting Office, 1977).

7. Office of Technology Assessment, *The Direct Use of Coal: Prospects and Problems of Production and Combustion* (Washington, D.C.: OTA, 1979).

8. National Academy of Sciences, *Rehabilitation Potential of Western Coal Lands* (Cambridge, Mass.: Ballinger, 1974); Tom Alexander, "A Promising Try at Environmental Detente for Coal," *Fortune*, February 13, 1978, pp. 94-102; and "Strip Mining," *Congressional Quarterly*, July 23, 1977, 1495-1500.

9. John Gilmore, "Boom Towns May Hinder Energy Resources Development," *Science* 191 (1976): 535-40.

10. Joseph Jorgensen et al., *Native Americans and Energy Development* (Cambridge, Mass.: Anthropology Resource Center, 1978); and Rayna Green et al., *Report of a Conference on Energy Resource Development and Indian Lands* (Washington, D.C.: Amer. Assoc. for the Advancement of Science, 1978).

11. National Academy of Sciences, *Air Quality and Stationary Source Emission Control* (Washington, D.C.: NAS, 1975); and idem, *Implications of Environmental Regulation for Energy Production and Consumption* (Washington, D.C.: NAS, 1977).

12. Stephen Barrager et al., *The Economic and Social Costs of Coal and Nuclear Electric Generation* (Washington, D.C.: National Science Foundation, 1976), p. 11.

13. See Richard Gordon, "The Hobbling of Coal," *Science* 200 (1978): 153-58.

14. Luther Carter, "Uncontrolled SO_2 Emissions Bring Acid Rain," *Science* 204 (1979): 1179-82.

15. William Kellogg, "Is Mankind Warming the Earth?" *Bull. Atomic Scientists* 34 (February 1978): 10-19; and George Woodwell, "The Carbon Dioxide Question," *Scientific American* 238 (January 1978): 34-43.

16. National Academy of Sciences, *Energy and Climate* (NAS, 1977); and Stephen Schneider, *The Genesis Strategy: Climate and Global Survival* (New York: Plenum, 1976).

17. Nicholas Wade, "Synfuels in Haste, Repent at Leisure," *Science* 205 (1979): 167-68.

18. Nazli Choucri, "International Exchanges of Alternative Energy Sources," in Pirages, op. cit.

19. A. D. Rossin and T. A. Rieck, "Economics of Nuclear Power," *Science* 201 (1978): 582-89. For other pronuclear statements, see Arthur Murray, ed., *The Nuclear Power Controversy* (Englewood Cliffs, N.J.: Prentice-Hall, 1976); and H. A. Bethe, "The Necessity of Fission Power," *Scientific American* 234, no. 1 (1976): 21-31. For a critical view, see Irvin Bupp and Jean-Claude Derian, *Light Water: How the Nuclear Dream Dissolved* (New York: Basic Books, 1978).

20. Bernard Cohen, "The Impacts of the Nuclear Energy Industry on Human Health and Safety," *American Scientist* 64 (September 1976): 550-59; and Karl Morgan, "Cancer and Low-level Ionizing Radiation," *Bull. Atomic Scientists* 34 (September 1978): 30-41.

21. National Academy of Sciences, *Risks Associated with Nuclear Power: A Critical Review of the Literature* (Washington, D.C.: NAS, 1979); and idem *Biological Effects of Ionizing Radiation* (Washington, D.C.: NAS, 1979).

22. Jean Marx, "Low-Level Radiation: Just How Bad Is It?" *Science* 204 (1979): 160-64; and Constance Holden, "Low-Level Radiation: A High-Level Concern," *Science* 204 (1979): 155-58.

23. *Reactor Safety Study* (WASH-1400) (Washington, D.C.: Nuclear Regulatory Commission, 1975).

24. "Report to the American Physical Society by the Study Group on Light-Water Reactor Safety," *Reviews of Modern Physics* 47, supp. no. 1 (1975).

25. Luther Carter, "NRC Panel Renders Mixed Verdict on Rasmussen Reactor Safety Study," *Science* 201 (1978): 1196-97.

26. Eliot Marshall, "Meteorites and Nuclear Power," *Science* 203 (1979): 529.

27. Michael Flood, "Nuclear Sabotage," *Bull. Atomic Scientists* 32 (1976): 29-36.

28. William Lanuette, "No Longer Can the NRC Say. . . ," *Bull. Atomic Scientists* 35 (June 1979): 6-8.

29. Reports on Three Mile Island appear in *Science* 204 (1979): 152-55, 280-81, 594-96; and 206 (1979): 796-98.

30. Mason Willrich and Richard Lester, *Radioactive Waste: Management and Regulation* (New York: Free Press, 1977).

31. "Panel Throws Doubt on Vitrification," *Science* 201 (1978): 599; and Luther Carter, "Nuclear Wastes: The Science of Geologic Disposal Seen as Weak," *Science* 200 (1978): 1135-37.

32. "Interagency Group Cautious on Nuclear Waste Disposal," *Science* 203 (1979): 1,320-21.

33. Alvin Weinberg, "Social Institutions and Nuclear Energy," *Science* 177 (1972): 27-34.

34. Gene Rochlin, "Nuclear Waste Disposal: Two Social Criteria," *Science* 195 (1977): 23-31; and Todd LaPorte, "Nuclear Waste: Increasing Scale and Sociopolitical Impacts," *Science* 201 (1978): 22-28; Roger Kasperson, "The Dark Side of the Radioactive Waste Problem," in *Progress in Resource Management and Environmental Planning*, ed. T. O'Riordan and R. d'Arge (New York: Wiley, 1979); and R. and V. Routley, "Nuclear Energy and Obligations to the Future," *Inquiry* 21 (1978): 133-79.

35. M. Willrich and T. B. Taylor, *Nuclear Theft: Risks and Safeguards* (Cambridge, Mass.: Ballinger, 1974); and Michael Hamilton, ed., *To Avoid Catastrophe: A Study in Future Nuclear Weapons Policy* (Grand Rapids, Mich.: Eerdman's, 1977).

36. Nuclear Energy Policy Study Group, *Nuclear Power: Issues and Choices* (Cambridge, Mass.: Ballinger, 1977). A similar conclusion is reached in the report of the Committee on Nuclear and Alternative Systems (CONAES) (Washington, D.C.: National Academy of Sciences, 1979).

37. John Walsh, "Fuel Reprocessing Still the Focus of U.S. Nonproliferation Policy," *Science* 201 (1978): 692-97; and Harold Feiveson, "Proliferation Resistant Nuclear Fuel Cycles," *Annual Review of Energy* 3 (1978): 357-94.

38. Atomic Energy Commission, *Comparative Risk-Benefit-Cost Study of Alternative Sources of Electrical Energy* (Washington, D.C.: AEC, 1974); M. Granger Morgan et al., "The Social Costs of Producing Electric Power from Coal," *Proceedings of the IEEE* 61 (1973): 1431-42.

39. Barrager et al., op. cit.

40. David Rose et al., "Nuclear Power—Compared to What?" *American Scientist* 64 (May 1976): 291-300.

41. Nuclear Energy Policy Study Group, op. cit., p. 17.

42. Joel Yellin, "Judicial Review and Nuclear Power: Assessing the Risks of Environmental Catastrophe," *George Washington Law Rev.* 45 (1977): 969-93. Christopher Hohenemser et al., "The Distrust of Nuclear Power," *Science* 196 (1977): 25-34; and Dorothy Nelkin and Susan Fallows, "The Evolution of the Nuclear Debate: The Role of Public Participation," *Annual Review of Energy* 3 (1978): 275-312.

43. William Ramsey, *Unpaid Costs of Electrical Energy* (Baltimore: Johns Hopkins University, 1979), p. 39.

44. Ibid., chap. 12.

45. John Holdren, "Fusion Energy in Context," *Science* 200 (1978): 168-80.

46. Dorothy Nelkin and Michael Pollack, "The Politics of Participation and the Nuclear Debate: A Comparative Study," *Public Policy* 25 (1977): 333-57; three articles on the debate in Sweden appear in *Bull. Atomic Scientists* 35 (Nov., 1979).

47. Ian G. Barbour, "Justice, Participation and Sustainability at MIT," *Ecumenical Review* 31 (1979): 380-87; Roger Shinn, ed., *Faith and Science in an Unjust World*, 2 vols. (Philadelphia: Fortress Press, 1980).

48. For an overview of solar energy, see Daniel Behrman, *Solar Energy: The Awakening Science* (Boston: Little, Brown, 1977); Hayes, op. cit.; and William Metz and Allen Hammond, eds. *Solar Energy in America* (Washington, D.C.: Amer. Assoc. for the Advancement of Science, 1978). Stanford Research Institute, *Solar Energy in America's Future*, 2d ed. (Washington, D.C.: Energy Research and Development Administration, 1977).

49. Henry Kelly, "Photovoltaic Power Systems," *Science* 199 (1978): 634-43; and Martin Wolf, "Photovoltaic Solar Energy Conversion," *Bull. Atomic Scientists* 32, no. 4 (1976): 26-33.

50. Office of Technology Assessment, *Applications of Solar Technology to Today's Energy Needs* (Washington, D.C.: OTA, 1978).

51. Bent Sorensen, "Wind Energy," *Bull. Atomic Scientists* 32, no. 7 (1976): 38-45; and idem, "Solar Energy: Unsung Potential for Wind and Biomass," *Science* 200 (1978): 636.

52. Allen Hammond, "Photosynthetic Solar Energy: Rediscovering Biomass Fuels," *Science* 197 (1977): 745-46; and C. C. Burwell, "Solar Biomass Energy," *Science* 199 (1978): 1,041-48.

53. Demand and Conservation Panel, Committee on Nuclear and Alternative Systems, National Academy of Sciences, "U.S. Energy Demand: Some Low Energy Futures," *Science* 200 (1978): 142-52.

54. Stanford Research Institute, op. cit., p. 42.

55. Council on Environmental Quality, *Solar Energy: Progress and Promise* (Washington, D.C.: CEQ, 1978); and Luther Carter, "A Bright Solar Prospect Seen by CEQ and OTA," *Science* 300 (1978): 627-30.

56. Herbert Inhaber, "Risks with Energy from Conventional and Nonconventional Sources," *Science* 203 (1979): 718-23; replies by Rein Lemberg, Richard Caputo, and John Holdren in *Science* 204 (1979): 454 and 564-68; and John H. Herbert et al., "A Risky Business," *Environment* 21, no. 6 (1979): 28-33.

57. Barry Commoner, *The Poverty of Power* (New York: Knopf, 1976); and idem, *The Politics of Energy* (New York: Knopf, 1979).

58. Amory Lovins, "Energy Strategy: The Road Not Taken?" *Foreign Affairs*, (October 1976): 65-96; idem, *Soft Energy Paths* (Cambridge, Mass.: Ballinger/Friends of the Earth, 1977); and Sandy Eccli et al., *Alternative Sources of Energy* (New York: Seabury, 1975).

59. Norman Brown and James Howe, "Solar Energy for Village Development," *Science* 199 (1978): 651-56; Arjun Makhijani, "Solar Energy and Rural Development for the Third World," *Bull. Atomic Scientists* 32, no. 6 (1976): 14-24; James Howe, "Toward a Global Approach to the Energy Problem," in *The U.S. and World Development: Agenda for Action 1976*, ed. Roger Hansen, (New York: Praeger, 1976); Philip Palmedo et al. *Energy Needs, Uses and Resources in Developing Countries* (Upton, N.Y.: Brookhaven National Lab., 1978); and A. K. N. Reddy, "Energy Options for the Third World," *Bull. Atomic Scientists* 34, no. 5 (1978): 28-33.

60. William Metz, "Energy Storage and Solar Power: An Exaggerated Problem," *Science* 200 (1978): 1471-73.

61. Samuel Florman, "Small is Dubious," *Harper's*, August 1977, pp. 10-12; see replies to Lovins by George Pickering and others in *Alternative Long-Range Energy Strategies: Additional Appendices* (Select Committee on Small Business, and Committee on Interior

and Insular Affairs, U.S. Senate, 1977); and Allen Hammond, "'Soft Technology' Energy Debate," *Science* 196 (1977): 959-61.

62. William Pollard, "The Long-Range Prospects for Solar Energy," *American Scientist* 64 (1976): 424-29.

63. "Principle Conclusions of the American Physical Society Study Group on Solar Photovoltaic Energy Conversion," (New York: American Physical Society, 1979).

64. Council on Environmental Quality, *Solar Energy*, op. cit.

65. Michael Yodell, "The Role of Government in Subsidizing Solar Energy," *Amer. Economic Rev.* 69 (1979): 357-61.

66. Robert Stobaugh and Daniel Yergin, eds., *Energy Future* (New York: Random House, 1979), summarized in *Foreign Affairs* 57 (1979): 836-71.

67. See notes 50 and 55.

68. Bruce Hannon, "Energy Conservation and the Consumer," *Science* 189 (1975): 95-102; and Herman Daly, *Steady-State Economics* (San Francisco: W. H. Freeman, 1977).

69. For overviews of conservation, see Robert Williams, ed., *Energy Conservation Papers* (Cambridge, Mass.: Ballinger, 1975); and Marc Ross and Robert Williams, "The Potential for Fuel Conservation," *Technology Review* 79, no. 4 (1977): 49-57.

70. Eric Hirst and Janet Carney, "Effects of Residential Energy Conservation Programs," *Science* 199 (1978): 845-51; and Eugene Eccli, *Low-Cost Energy-Efficient Shelter for the Owner and Builder* (Emmaus, Pa.: Rodale, 1976).

71. Hayes, op. cit., chap. 8.

72. Commoner, *The Poverty of Power*, chap. 7; Robert Oelhaf, *Organic Agriculture: Economic and Ecological Comparisons with Conventional Methods* (New York: Wiley, 1979); and M. B. Green, *Eating Oil: Energy Use in the Food System* (Boulder, Colo.: Westview, 1978).

73. Hayes, op. cit., chap. 5; and Cooperative Extension Service, *Efficient Energy Management for Consumers* (Ithaca, N.Y.: Cornell University Press, 1977).

74. Lovins, op. cit., chap. 4.

75. Eric Hirst, "Transportation Energy Conservation Policies," *Science* 192 (1976): 15-20.

76. Gerald Leach, *A Low Energy Strategy for the United Kingdom* (London: International Institute for Environment and Development, 1979).

77. Allen Hammond, "Conservation of Energy: The Potential for More Efficient Use," *Science* 178 (1972): 1079-81; and Williams, op. cit., chaps. 2 and 3.

78. John Steinhart et al., "A Low Energy Scenario for the U.S.: 1975-2050," in Ruedisili and Firebaugh, op. cit.

79. CONAES report (NAS), note 53 above. For an earlier low-demand estimate, cf. Ford Foundation Energy Policy Project, *A Time to Choose* (Cambridge, Mass.: Ballinger, 1974).

80. Stobaugh and Yergin, op. cit.

81. Joel Darmstadter et al., *Energy in America's Future* (Baltimore: Johns Hopkins University Press, 1979).

82. Allan Mazur and Eugene Rosa, "Energy and Life Style," *Science* (1974): 607-9; and Joy Dunkerley, ed., *International Comparisons of Energy Consumption* (Baltimore: Johns Hopkins University Press, 1978).

83. Laura Nader and Stephen Beckerman, "Energy as it Relates to Quality and Style of Life," *Annual Review of Energy* 3 (1978): 1-28.

84. Lee Schipper and Allen Lichtenberger, "Efficient Energy Use and Well-being: The Swedish Example," *Science* 194 (1976): 1001-13; see also J. Darmstadter et al., *How Industrial Societies Use Energy: A Comparative Analysis* (Baltimore: Johns Hopkins University Press, 1978); and Hayes, op. cit., chap. 8.

85. Richard Grossman and Gail Baneker, "Jobs and Energy" (Washington, D.C.: Environmentalists for Full Employment, 1977); and *Creating Jobs Through Energy Policy* (hearings before the Joint Economic Committee, U.S. Congress, March 15-16, 1978).

86. Leslie Ellen Nulty, *Understanding the New Inflation: The Importance of the Basic Necessities* (Washington, D.C.: Exploratory Project for Economic Alternatives, 1977), appendix A; and Bob Swierczek and David Tyler "Energy and the Poor" *Christianity and Crisis* 38 (1978): 242-44.

87. Dorothy Newman and Dawn Day, *The American Energy Consumer* (Cambridge, Mass.: Ballinger, 1975); see also Eugene Grier, *Colder . . . Darker: The Energy Crisis and Low-Income Americans* (Washington, D.C.: Center for Metropolitan Studies, 1977).

88. John Palmer et al., "The Distributional Impact of Higher Energy Prices," *Public Policy* 24 (1976): 545-68.

89. "Energy Bill: The End of an Odyssey," *Congressional Quarterly*, October 21, 1978, pp. 3039-42.

90. Michael Corr and Dan MacLeod, "Home Energy Consumption as a Function of Lifestyle," in *Human Welfare: The End Use of Power*, ed. Barry Commoner (New York: Macmillan, 1975.)

91. Ivan Illich, *Energy and Equity* (New York: Harper & Row, 1974).

92. On the World Council of Churches debate, see Paul Abrecht, ed., *Faith, Science and the Future* (Philadelphia: Fortress Press, 1979); also Barbour, op. cit. and Shinn, op. cit. On the National Council of Churches, see the study document, *Energy and Ethics* (New York: National Council of Churches, 1979); and the statement, "The Ethical Implications of Energy Production and Use," adopted by the NCC Governing Board, May 11, 1979. See also Dieter Hessel, *Ethics and Energy* (New York: Friendship Press, 1979).

11
FOOD AND POPULATION

There is no more serious global problem than the provision of sufficient food for rapidly growing populations. At least 10,000 people die every day as a result of starvation and malnutrition.[1] What are the prospects for increased agricultural production and more equitable distribution of food? The benefits of improvements in agriculture will be short-lived unless population stabilization also is achieved. Can birthrates be reduced by family planning programs, specific incentives, or economic and social development? What development strategies and international assistance policies offer some hope in dealing with the crises of food and population? Within these broad and complex questions, we will look primarily at the ethical issues and the implications for U.S. policy.

I. FOOD AND AGRICULTURE

During the 1960s there were high hopes of significant progress in the global fight against hunger. The Green Revolution (new high-yield strains of wheat, corn, and rice, together with increased fertilizer and irrigation) brought dramatic increases in production in several countries, and great expectations in others. But continued population growth, several years of bad harvests, and a fourfold increase in the price of oil used for fertilizer and irrigation resulted in the food crises of the mid-1970s. Good growing weather in the late 1970s has disguised the precariousness of the balance between food production and population growth, and we have closed our eyes to the continuing prevalence of malnutrition. Overall, a 3% average annual increase in grain production in developing countries has barely kept ahead of the 2½% annual increase in population.[2]

The great diversity among nations today should be noted at the outset. The first world includes the Western free-market democracies of North America and Western Europe, plus Japan. The second world refers to the communist nations of the Soviet Union and Eastern Europe. Together they constitute the industrial nations, or "the North," with one-fourth of the world's population and 75% of the total GNP. The remaining three-fourths is designated Third World or less developed countries (LDCs), or "the South." It includes rapidly developing nations with extensive manufacturing and world trade, such as Brazil, Mexico, Taiwan, and South Korea. The OPEC countries also are developing rapidly, thanks to huge oil revenues. China, with a quarter of humanity, has a low GNP, but social and economic benefits are more evenly distributed than in any other nation. Finally there are the 40 poorest countries, with another quarter of humanity, in which hunger and malnutrition are most severe.

The Causes of Hunger

It is estimated that 10 million people died from starvation-related causes in 1975. But even in "good" years, hunger and malnutrition are widespread. Six hundred million people bear severe effects of malnutrition—half of them children in whom mental retardation and lowered resistance to disease are frequent results.[3] There are 70 million more people to feed each year—equivalent to adding to the globe every three years the entire population of the United States. Some of the causes of hunger are directly related to overpopulation while others are not.

1. Ecological Stress

Deforestation to obtain wood for cooking, heating, and building has led to severe soil erosion in many parts of Asia. Overplowing and the use of marginal land (such as slopes without terracing) have produced extensive erosion. In Africa and Latin America, more intensive agriculture has replaced traditional practices of shifting cultivation and alternate-year cropping; soil fertility declines if fallow periods are reduced. The United States has lost half its topsoil since the Civil War by farming and wind erosion. Overgrazing of African grasslands has hastened erosion in semi-arid regions. In the African Sahel, large cattle populations and drought conditions destroyed fragile grasslands in the early 1970s. Both the northern and southern edges of the Sahara desert have been advancing more rapidly because of deforestation and overgrazing, and "desertification" is occurring in many parts of the world.[4]

2. Low Agricultural Yields

There are many reasons why crop yields remain low in LDCs. Higher priority often has been given to industrial than to agricultural development. Agriculture in the tropics is complex, and most research to date has been on temperate zone conditions. Fertilizer is in short supply and expensive. The energy crisis has had a devastating impact on Third World farming. United States productivity per acre is three times that of India—but at a high cost in capital and energy. A ton of fertilizer used in India would yield three or four times as much addtitional food as a ton added to the high levels already applied in the U.S. There also have been insufficient efforts to improve traditional agricultural methods, using labor-intensive rather than capital- and energy-intensive techniques. Water is scarce and irrigation equipment is costly to install and operate. Water management plans, including dams, wells, and irrigation networks are crucial in agricultural development.[5]

3. Land Ownership

In many developing countries, a small percentage of the population owns most of the land. This is partly a legacy of colonialism under which one-crop plantations were established for export crops rather than for local food. Foreign aid policies often have strengthened ruling elites and provided little help for the peasant and small farmer. In Mexico, Brazil, and other Latin American countries, tax policies and access to credit have forced small farmers to sell out, further concentrating ownership in a few wealthy landowners. Even the Green Revolution has disproportionately benefited large owners who could afford the fertilizer and equipment it required.[6] If one has neither land for subsistence farming nor a dependable source of income, one's family is unlikely to get the minimum dietary requirements. Malnutrition is largely a reflection of patterns of land ownership and rural poverty.

4. Consumption by Affluent Nations

Since colonial days much of the land in LDCs has been devoted to such nonfood crops as tea, coffee, tobacco, cotton, jute, and rubber. In Central American countries where half the children are malnourished, more than half the agricultural land is used for export crops. In some parts of Colombia, carnations have replaced corn and wheat on prime agricultural land. Gambia grows peanuts, Ghana cocoa, and Senegal vegetables, mainly for Europe. Half the winter and spring tomatoes sold in the United States comes from Mexico. The strawberry industry in Mexico is controlled by U.S. companies who own or have contracts on some of the best land. Luxury foods for export to the well-fed do bring badly needed foreign exchange, but at a high cost in local health and nutrition. Prices paid for export crops are kept low because a few companies dominate the market. One company controls 80% of world trade in oil seed, another one 40% of world cocoa, still another 35% of all bananas. Countries that are dependent on a single cash crop are highly vulnerable to price fluctuations. When diversified agriculture has been replaced by plantations for export crops, much of the food for the local populace has to be imported—often at high prices, and with additional profits going to foreign companies.[7]

Meat consumption in affluent nations is particularly wasteful of world grain supplies. It takes 11 pounds of feed to make one pound of beef.[8] Other proteins in addition to grain are fed to cattle. Before the fish catch in South

America dropped from overfishing, 83% of it was exported (mainly as fish meal for cattle); only 17% was used locally among protein-deficient populations. The United States also is the world's leading beef importer (mainly for household pets and fast-food chains). Ninety percent of the annual world protein deficit could be met by the grain and fishmeal fed to U.S. cattle alone.[9]

A more equitable distribution of existing food supplies would eliminate hunger and malnutrition. Current food production is equal to more than twice the minimum calorie and protein requirements for every man, woman, and child alive today.[10] But redistributing existing food would involve horrendous transportation and balance-of-payments problems. If we look at causes and not symptoms, the basic problem is the distribution of agricultural production. Food production is determined by those who can pay; luxury foods for the affluent often displace subsistence needs of the poor. Famine relief to victims of starvation is important as a short-run act of charity, but political and economic changes are required as long-run acts of justice. The politics of hunger are inseparable from the politics of international and domestic power.

There are thus three kinds of response to the food crisis, all of which will be needed if humankind is to alleviate hunger and malnutrition. 1) *Increased production.* improvements in crop yields and acreage can make more food available. Agricultural technologists are likely to stress this approach. 2) *Better distribution.* beyond short-term measures (food aid, international grain reserves, and so on), the long-term distribution of production capacity can be improved by development assistance and changes in international trade, land ownership, and tax and credit policies. This approach (along with the first) is advocated by many development experts and Third World leaders. 3) *Reduced consumption.* the growth of population in LDCs and the growth of consumption in affluent nations both must be curbed. This approach is emphasized by many environmentalists. In the past we have given higher priority to food production than to distribution or consumption. I will argue that all three are essential today.

The Green Revolution

In the United States, high-yield wheat and hybrid corn, along with intensive use of fertilizer, pesticides, and herbicides, resulted in soaring production after World War II. There were large grain surpluses and land was taken out of production mainly to keep prices up. During the 1960s a strain of wheat with short, stiff stalks was developed in Mexico by Norman Borlaug and others, which increased yields and allowed two crops per year in some areas. These strains made more efficient use of fertilizer and moisture. India doubled its wheat crop in six years (1966–71), and despite population growth it was approaching self-sufficiency before bad weather and the oil crisis hit. Pakistan and the Philippines saw similar dramatic increases in production. Borlaug holds that these yields could be doubled again if adequate fertilizer is available and wise economic policies are followed. He thinks that production of maize, sorghum, and rice could be similarly doubled in ten years.[11]

Despite these impressive achievements, there are several *scientific and technical limitations* that should qualify our hopes for the Green Revolution. The new strains are more susceptible to pests and epidemics; they lack the diversity of native varieties that provides some protection against the rapid spread of such a disease as corn blight. The dose of a pesticide has to be increased as pests develop resistance to it. More research needs to be done on specifically tropical crops, diseases, and soil and weather conditions. New strains often have had high yields but lower protein value than the crops (such as pulses) they replaced; high yields of high-protein crops (soybeans, for example) have not been achieved. Above all, the new strains require high energy inputs to sustain their productivity; fertilizer production and fuel for irrigation pumps and tractors have been severely limited by the high price of oil.[12]

In addition, the Green Revolution has had some dubious *social impacts.* Benefits have accrued disproportionately to the more prosperous regions and the large landowners who could afford seed, fertilizer, irrigation, and machinery. This has increased income disparities and the polarization of owners and laborers. The small farmers could not obtain credit for seed and chemicals and were driven off the land, while the power of the large landowners increased. Many peasants were taken back as day laborers, without the security and direct access to food that land ownership provided.[13] In theory the new methods are equally useful on large or small farms, but in the prevailing social context the required inputs were not equally available. The Green Revolution could be adapted to labor-intensive conditions, but in practice it has accelerated capital- and energy-intensive mechanization and extensive unemployment has resulted. In the late 1960s, for example, Pakistan purchased 18,000 large tractors with a World Bank loan. By 1975, average farm size had doubled, labor use per acre had dropped 40%, yet yields per acre were unchanged.[14]

There have been attempts to adapt the Green Revolution to *local conditions* and low-energy, labor-intensive methods. At research institutes in the third world, experts in scientific agronomy have tried to start from indigenous practices and soil conditions. In India, regional institutes have experimented with the modernization of small-scale farming, using improved equipment powered by bullocks or small motors. The Institute of Tropical Agriculture in Nigeria has worked out modifications of the prevalent system of shifting cultivation. Mixed cropping, legumes, mulches, deep-rooted fallow vegetation, minimum tillage, and intermediate-scale equipment have been used.[15] The use of solar pumps for irrigation appears particularly promising as fuel prices escalate.

Rural development policies also can help to strengthen *small farms.* These include land distribution, communal land tenure, wider access to credit, tax subsidies to small rather than large farms, producer cooperatives, farmers' associations, and better agricultural extension services. There should be provisions for the participation of the farmers themselves in development planning. Small farms contribute to employment and to social equality, avoiding the polarization of large landowner and landless laborer. Smaller farms usually have higher productivity measured in yield per acre or yield per dollar (but not yield per worker). In Taiwan, output per acre on farms less than 1.25 acres is nearly twice that on farms of more than 5 acres. In India and Brazil, yields on smaller labor-intensive farms are from one-half to two-thirds higher than on larger mechanized ones.[16]

In addition to efforts to improve yields per acre, efforts to bring *new land* under cultivation will increase food production. Some oil-rich nations (including Nigeria, Ecuador, Indonesia, Iran, and Iraq) have potential land, energy, and financial resources for greatly increased agricultural production; other OPEC nations are in semi-arid regions where ecosystems are too fragile for expanded cultivation. There is little potential for new arable land in Asia, but large areas could be cleared in Latin America and Central Africa (though in some regions the soils are thin or would deteriorate if farming were too intensive). It sometimes is claimed that world cultivated acreage could be doubled, but this would involve marginal land and high development costs; moreover, it does not allow for loss of agricultural land through desertification, soil erosion, and urban growth. Fishing, including new methods of aquaculture, appears unlikely to make more than a small addition to land-based food sources, and the prospects for synthetic foods are too uncertain to count on.

All told, it appears that a *doubling of world food production* in a generation is by no means impossible.[17] But a tripling seems dubious because of environmental, energy, and land constraints. Since some estimates project a tripling of world population before it is stabilized, there is a strong case for giving attention to population, consumption, and distribution as well as production. We can be thankful for the Green Revolution, which has bought additional time and still has considerable potential for the future if it is integrated with wise social, economic, and land-use policies. But we cannot expect agricultural technology alone to provide a long-run solution to the hunger problem.

The Ethics of Food and Agricultural Aid

What ethical responsibilities do people in affluent nations have in relation to world hunger? Starting from the equal value of all persons (Chapter 4), I would argue that all persons have a right to life and therefore *a right to the basic necessities of life*, including food adequate for survival. Correspondingly, people who could prevent starvation, without sacrificing anything of comparable moral significance, have a duty to do so.[18] Our policies over the years have contributed to famine and malnutrition today, of which many of the victims are innocent children. Our present actions—or our failure to act—will result in the preventable deaths of other human beings.[19] Granted, we have more responsibility for the situations we can do more about (such as a starving person on our doorstep) than for the distant and indirect repercussions of our own actions. Granted, our social roles create particular duties (to our own children, for example) which may outweigh more universal duties. But if, as I have maintained, we can prevent starvation without even sacrificing our own health, we have a duty to take such action.

Principles of justice are difficult to apply between nations. The Rawls principle—that inequalities are acceptable only if they maximize benefits to the worst-off—can be applied only to members of a common social order in which one can predict the distributional consequences of one's actions. Yet we know enough about the inequalities of food and the mechanisms for redistribution to judge that present patterns strongly violate the Rawls principle. Perhaps at some point we could claim that further assistance to other countries would jeopardize our environment and our productivity, thereby harming hungry nations too, but that point is a long way off. The argument that inequalities provide incentives for hard work and for greater efficiency is simply not applicable to food; malnutrition, in fact, greatly reduces a person's agricultural productivity. We are reluctant to apply ethical principles across national boundaries, which would threaten our privileged position; yet if we accept the principle of justice we have to acknowledge its universal applicability.

In the name of *individual freedom* and *property rights*, some writers have maintained that people have the right to use fairly acquired wealth in any way they want (including the purchase of food), as long as there is no direct harm to others. Feeding the starving is a commendable act of charity, they say, but it is not a duty.[20] I would reply that the wealth of affluent nations has not been fairly acquired, but is in part a product of colonial exploitation (which provided a headstart in technology and capital accumulation) and of continuing economic neocolonialism that prolongs injustices.[21] Any claim of rights must be consistent with the rights of others; our consumption habits impinge on the freedom of hungry people, if not their right to life. Property rights are not absolute; the uses of property may be restricted when they inflict harm, even indirectly, on other parties.

The international economic order perpetuates *radical inequalities* between relative luxury and direst need. Food resources are adequate to raise the level of the hungry without significant deprivation of the affluent, but there are no effective redistributive mechanisms. Within our nation, redistributive taxation serves to correct the tendency of economic disparities to increase, and welfare measures are designed to prevent anyone from actually starving. There are no comparable political institutions to set limits on the international economic disparities that result in food inequalities. We must work for institutional changes, but in the meantime we can provide some corrective to the injustices of the international order through food aid and agricultural assistance.[22]

In Judaism and Christianity, there are specific teachings about *feeding the hungry*. The Deuteronomic laws upheld the right of the needy to glean in the fields.[23] Isaiah said that God seeks social justice before religious observances: "Is not this the fast that I choose: to let the oppressed go free . . . and to share your bread with the hungry?"[24] Matthew included responses to hunger in his portrayal of the last judgment: "I was hungry and you gave me food."[25] Equally relevant is the biblical conviction that God is on the side of the poor and works for the liberation of the oppressed. In both the Old and New Testaments, the coexistence of extremes of wealth and poverty is attacked; the rich are indicted for their unconcern about human need. In the biblical view there are no absolute rights to property; the earth's resources are intended for all.[26]

In the biblical tradition, *institutionalized injustice* is criticized as well as individual wrongdoing. The abuses of economic power by a privileged class represent a structural evil, an expression of group self-interest and greed. The prophets spoke of judgment on the nation and called for national repentance for complicity in injustice. In a similar vein, church leaders today insist that we must move beyond famine relief to deal with the causes of hunger. They call for national as well as individual action, long-term as well as short-term policies, justice as well as charity. Denominational agencies, local hunger task forces, and interdenominational organizations (such as Bread for the World) have promoted political action to increase agricultural assistance in order to enable people to feed themselves. Some also have urged institutional reforms aimed at a more just international economic order (see Section III below).[27]

Let us look at some specific food policy options in the light of these principles. *Food aid* can reduce starvation in the short run. U.S. food programs in the 1950s and 1960s were in part a humanitarian response to victims of famine and hunger. Under the Food for Peace program, extensive grain shipments were given or sold overseas. But much of the food went for political purposes to support friendly nations, rather than where the need was greatest. The program also helped U.S. farmers by reducing stored surpluses and raising farm prices. In the 1970s grain was sold mainly to rich nations (Europe, Russia, Japan) to improve U.S. balance-of-payments, especially to offset huge oil imports; in effect, we traded food for crude. Most of the food that did go to LDCs was financed by loans; current aid funds are pitifully small. "We spend more money on dog food than we do on the 600 million people in this world who are malnourished," says USAID director John Gilligan.[28]

Emergency food aid is an important short-run response to human suffering. It can be provided most effectively through international grain reserves, which the 1974 U.N. Food Conference proposed primarily as a buffer against weather fluctuations. But famine relief does nothing to correct the long-term conditions that produce hunger and starvation. It deals with symptoms rather than causes and leads to dependency rather than self-reliance. According to a Chinese proverb: "If you give someone a fish, he will eat for a day; if you teach him to fish, he will eat for the rest of his life."

U.S. development aid—especially agricultural and technical assistance—could make a significant long-run contribution, but it has been steadily declining and often has been misdirected. After World War II, we put 3% of our GNP into aid for one area (the Marshall Plan in Europe); today we put only 0.26% of our

GNP into development aid for all countries. Each U.S. citizen pays $450 in annual taxes for defense, and only $6 for development assistance.[29] Many LDCs actually pay back more on previous loans than they receive in new loans. Though the level of our aid remains low (compared to the 1% of GNP recommended by the Pearson Commission, for instance), greater efforts are being made to direct it to the areas of greatest need. Because previous aid funds often had been used for projects that benefited privileged minorities more than the impoverished majority, congressional action in 1975 established "new directions" in development assistance aimed at the rural poor and the small farmer.

Aid through multilateral channels, such as the U.N. or the World Bank, is preferable to bilateral aid from one country to another. The "soft loan window" of the World Bank makes long-term, low-interest development loans, and the U.N. administers the International Fund for Agricultural Development. The 1974 U.N. Food Conference recommended that this fund be increased from $1.5 billion to $5 billion per year, and that plans benefiting small farm holders and landless peasants should be emphasized. Multilateral aid avoids direct donor-recipient relations, and allows the international community to select the recipients and specify conditions for assistance.

Global justice also requires that affluent nations *reduce their own consumption*, especially of grain-fed meat and of nonfood and luxury food crops grown in LDCs. More grain is consumed by livestock (mainly in feedlots) in the United States and the Soviet Union than by the entire human population of the LDCs.[30] Meat production should be confined to pasture and rangeland unsuited for crops. Reduction of meat consumption is a matter for both individual decision and national policy (including measures to facilitate the production and the distribution of grains for human use, and to minimize the impact on farmers). Strenuous individual and national efforts at energy conservation also would benefit world agriculture; our ravenous appetite for oil drives up its price and escalates the cost of fertilizer everywhere.

The shift in LDCs from export crops to *diversified agriculture for local consumption* will not be easily accomplished. It appears, however, that if a country can control its own agricultural resources and encourage diversified crops to meet its own food needs first, it still can set aside enough land for export crops to earn substantial foreign exchange. Cuban agriculture, for example, was once dominated by foreign sugar companies (as is still the case in the Dominican Republic). But Fidel Castro's policy first stressed self-sufficiency in basic foods, and only then promoted trade. Cuba is the world's largest sugar exporter today, but not at the expense of its people's nutrition. Land redistribution and agrarian reform resulted in more food for local consumption in countries as varied as Japan, Taiwan, and China.[31]

There is no solution to the food crisis apart from wider *development strategies* and policies for *international trade*. Justice requires not only action within prevailing domestic and international systems, but also efforts to reform those systems. In Section III we will examine food and population policy in the broader context of policies for development and trade.

II. POPULATION STABILIZATION

The world population is 4.3 billion and growing at 1.7% annually. There are 200,000 more people to feed each day. At this rate, the population would double in 41 years. The global fertility rate is beginning to fall, but the population will continue to grow even after the replacement is achieved (roughly two children who will reach child-bearing age in coming decades. Even with major efforts to reduce birth rates, the world total is likely to reach at least 10 billion. Moreover, the growth is very unevenly distributed. For example, annual growth rates and doubling times range from Mexico, 3.4% (20 years) and India, 1.9% (36 years), to China, 1.2% (58 years), the United States, 0.6% (116 years) and Western Europe 0.1% (693 years).[32]

Birthrates and Development

At the 1974 U.N. Population Conference in Bucharest, delegates from industrial nations talked mainly about the importance of family planning programs. Delegates from LDCs, on the other hand, held that *social and economic development* should receive the highest priority. They defended the demographic transition theory: birthrates will fall when living standards rise, as happened in Europe after the Industrial Revolution. Parents in poor countries have many children because they expect some to die young, and they want several surviving children (especially sons) to support them in their old age. If infant mortality drops and economic security improves, there will be stronger motivation for small families. The final documents at Bucharest advocated both family planning and economic development, but put more emphasis on the latter.[33]

Is the *demographic transition* theory convincing? In Europe, death rates slowly fell with improvements in health and sanitation, and populations grew; but finally birthrates also fell as part of the process of industrialization and modernization—without any specific efforts at family planning. The situation in developing countries today is so different that the historical parallel is ambiguous. In LDCs, the decline in mortality has been much more rapid, the initial fertility levels higher, the typical age of marriage lower, and the population explosion more rapid than it ever was in Europe. Moreover, industrialization and population growth occurred in Europe when energy was cheap, overseas colonies provided inexpensive raw materials, and other continents such as North America served as safety valves for population pressures. On the other hand there are some hopeful factors today that were not present in Europe: technological knowledge could facilitate more rapid development, and more reliable methods of contraception are available.[34]

Preliminary results of the *World Fertility Survey* sponsored by the U.N. were published in 1979. Data from 15 developing countries showed birthrates falling much faster than had been expected. In Costa Rica, for example, the average number of children born to women today by the time they reach 49 years is 7.2; younger women are expected to average only 3.8 births by the same age. Birthrates declined dramatically in all of the countries surveyed except Bangladesh, Nepal, and Pakistan. The report cites evidence for three significant trends: later marriage, the desire for smaller families, and greater availability of contraceptives. These results are encouraging but they do not justify complacency; birthrates are still a long way from the replacement rate.[35]

Birthrates have dropped most sharply in those LDCs that have both *family planning programs* and a *relatively equitable distribution* of the benefits of development. When the majority of the population has benefited from improved literacy, nutrition, health services, and greater economic security, the motivation for small families has been present. These factors are a common denominator, despite the diverse political systems and levels of economic growth, among the countries showing rapid fertility decline, including Taiwan, China, South Korea, Sri Lanka (Ceylon), Uruguay, and Costa Rica. By contrast, birthrates have remained high in countries with considerably higher average per capita incomes but wide disparities in income and in access to social services. In Taiwan, the birthrate has fallen to 24 per thousand, while in Mexico it remains at 41, though Mexico has twice the per capita income. Many factors are involved here, but it is significant that in Mexico income is very unevenly distributed and there is a wide gap between landowners and landless peasants. In Taiwan, income is more evenly distributed and land ownership is limited to 7.5 acres.[36]

Broad access to *health care* and *basic education* is a feature of all of the countries in which major declines in birthrate have occurred recently. Sri Lanka, for instance, has made extensive use of paramedical and public health workers in village clinics. Turkey has a much higher GNP per capita and more modern hospitals, but the infant mortality per thousand is 117, compared to 50 in Sri Lanka. Education in Sri Lanka is aimed at wide diffusion of basic literacy rather than advanced degrees for the few. The education of women also has led to delayed marriage and new employment options as alternatives to childbearing. Sri Lanka may have expanded welfare services faster than its limited economy can afford, and inflation is creating serious problems (aggravated by long-standing social conflicts), but the reduction in birthrate is impressive.[37]

A final example is China, where *family planning* and *social change* together have achieved a dramatic drop in birthrates. Contraception and abortion are provided free. Recent visitors report that basic levels of food, health care, education, and economic and social security are available to all. Women participate extensively in education and the work force, and late marriage (in the mid-20s) is encouraged. There also are strong community pressures for small families; some commune brigades decide which couples should be eligible to have a child each year. It appears, then, that family planning plus widely distributed social development contribute to falling birthrates in free-market, socialist, and communist countries alike.

But can such social changes be introduced rapidly enough? Time is running out in many places where rapid population growth is hindering the very development process that would reduce birthrates. Economic and social development are slow processes, and there are strong forces resisting political change. Some population experts think that, with limited time and funds, governments should introduce *specific birth control programs* without waiting for broader social changes. Population growth in industrial countries also should concern us, even though it is slower than that in LDCs. If each U.S. citizen uses from 20 to 40 times the resources of a citizen of India, even a small increase in U.S. population, or in per capita consumption, adds significantly to world resource use. Slowing the growth of per capita consumption is the main challenge to affluent societies (see Chapter 12), but policies for achieving zero population growth also are important.

Justice, Freedom, and Birth Control

Let us consider some birth control policies from the standpoints of justice, freedom and effectiveness.

1. Voluntary Family Planning

The goal of the family-planning movement has been the freedom of parents to determine the number and spacing of their children by deliberate choice rather than biological accident. If information and birth control methods are available, each couple can make its own decision about family size. Women are protected from unwanted pregnancy, and family life is strengthened because all children are wanted. Justice is fulfilled if family-planning services are available to all through government subsidy. They can be linked to programs of maternal health and child care in rural and urban clinics. The pill is highly effective, but it has to be used regularly and it is not cheap; there is need for intensive research on cheaper, safer, and more long-lasting pills, vaccines, or time-capsules, including contraception for men. The key question is motivation, since family planning reduces unwanted births but not wanted births.

2. Education for Smaller Families.

Educational efforts can go beyond birth control information to portray the seriousness of the population crisis and to promote the small-family ideal. Pakistan recently has initiated such an educational campaign.[38] The public media and the schools can be enlisted in the legitimation of small-family patterns and new cultural symbols of family success; TV and radio are becoming major vehicles of social change around the world. There is no clear line that distinguishes education and persuasion from propaganda, manipulation, or social pressures that jeopardize human dignity. But there is some protection from abuse if the objective is informed and responsible choice, and not simply conformity to governmental directives or community pressures.

3. The Emancipation of Women

Most societies have been dominated by men; women have had few alternatives to motherhood and childrearing. Policies upholding new roles for women are desirable in the interests of freedom and justice, and they also lead to lower birthrates. The movement for women's rights around the world is demanding equal access to education, career options, and employment opportunities. Pregnancy often has been involuntary; women are now asserting the right to control their own reproduction. In many traditional societies, daughters have been married off at an early age; later marriage can be encouraged both by education and by raising the legal age of marriage. Child-care centers enable both parents to pursue careers outside the home. Birthrates fall when women are not confined to domestic roles and have more choices and greater control over their own lives.[39]

4. Social Welfare

Apart from general economic development and modernization (on which the demographic transition theory relies), policies can be targeted on more specific social programs that are inherently desirable and particularly conducive to lower birthrates. Old age security is important because children are seen as a source of economic security in later life, especially by the poor. Even minimal old age benefits would remove one of the major motives for having large families; people would perceive smaller families as in their own interests. Widespread literacy also contributes to a reduction in the desired number of children. Health care could be particularly targeted on infant mortality and maternal health. Such programs would be far more expensive than family planning, and their effects would be slower and more indirect. But they would eventually alter social conditions so that the family size that people want is more congruent with population stability.

5. Economic Incentives

More rapid results could be obtained with specific *positive* incentives, such as bonuses or tax credits for late marriage, for each childless year, or for a small family. In India, cash bonuses and transistor radios were offered to men as rewards for voluntary sterilization in vasectomy clinics. Even without direct coercion, such incentives exert great pressure on the poor; to a person with an annual income of $70, a $14 bonus is very persuasive. Individuals should be fully informed of the consequences and feel that they really do have some choice. Among *negative* incentives are taxes or fines on large families. Singapore uses income tax laws to favor small families. In the United States, where income taxes favor large families, the Packwood Bill (1970) proposed a cut-off of deductions after the second child. Such negative incentives would have little effect on the rich and would be a far heavier burden on the poor. They would indirectly penalize children in large low-income families, and further perpetuate the cycle of poverty. Would it be more just to have an equal tax on all children, a tax after a given number of children, or a tax that varies with income? In a society with large income disparities, any economic incentives would be unjust; but a strongly progressive child tax, which increases with both income and number of children, seems to offer a balance among social control, individual choice, and equitable distribution of sacrifices.[40]

6. Compulsory Sterilization or Abortion

In 1976 the state of Maharastra in India legislated compulsory male sterilization after three children, with arrest and jail sentences for noncompliance. Couples practicing contraception could receive exemption, subject to compulsory abortion if there was a fourth pregnancy. Several other Indian states were considering similar legislation. At the national level, sterilization programs still were supposedly voluntary, but there were many reports of overzealous promotion, harassment, and even forcible induction into vasectomy camps. Widespread resistance and public protests developed, and were a major factor in Indira Ghandi's defeat in 1977. Minority groups and the poor were particularly fearful that sterilization laws would be enforced more strenuously against them, and there were few safeguards against such abuses. There also were fears that the use of informants and excessive surveillance might lead to other repressive measures.[41]

How far along the spectrum from *voluntary* to *compulsory* measures should we move? Policy decisions here involve both facts and values. Among the factual judgments are: the *effectiveness* of different policies in reducing birthrates within various time-spans, as estimated from past evidence and plausible models of causal relationship; and the *consequences* of further population growth in relation to food production, resources, and environmental limits. The main *value judgment* is the relative importance of freedom, justice, and the benefits of lower birthrates. In general, the more coercive measures appear more effective in bringing rapid results, but at a greater cost in freedom and justice.[42]

Leaders in the family-planning movement have stressed *freedom* and are reluctant to go beyond the first one or two steps above. Bernard Berelson, for example, favors an expansion and energetic implementation of voluntary family planning programs, which he thinks might lower birthrates by 15 or 18 per thousand in twenty years.[43] Some ethicists also give priority to freedom. Thus Daniel Callahan advocates greater efforts at voluntary family planning and education before turning to social changes and positive economic incentives. Only if the latter clearly have failed, and survival is demonstrably at stake, should coercive measures be considered—and then only if the burden falls equally on all. Individual rights, he asserts, must be strenuously protected from intervention by the state.[44]

I would reply that this position overemphasizes freedom at the expense of *the common good.* Individual rights always are related to the rights of others and to the social context in which they arise. The right to procreate may conflict with the right of other people to adequate food, or the right of future generations to a habitable environment. Governments have the right to intervene for the sake of the general welfare and to adopt policies regulating individual behavior when it harms others. In times of emergency—during war or an epidemic, for instance—we accept restrictions on our activities in the common interest.[45] I have urged that freedom should be identified with participation in decision making, rather than with the absence of government regulation. Moreover, I believe that family planning alone is not rapid enough, and it neglects the social factors that influence the family size that people want.

At the other extreme, a number of biologists and ecologists have insisted that only *coercive measures* can bring results rapidly enough to avoid widespread catastrophe, and that freedom must be sacrificed for the sake of survival. The willingness of these authors to recommend strong measures is based partly on their pessimistic appraisals of the prospect for expanding agricultural production. Kingsley Davis advocates compulsory sterilization, and the Ehrlichs write: "Compulsory control of family size is an unpalatable idea to many, but the alternatives may be much more horrifying."[46] In the name of human survival, they defend compulsory abortion to terminate any pregnancy after a second child. Garrett Hardin believes that compulsory sterilization and abortion are justified to prevent the disaster of overpopulation that would endanger the survival of humankind. "Freedom to breed will bring ruin to all." Hardin advocates "mutual coercion, mutually agreed on by the majority of the people affected."[47]

I would maintain that the situation is not as hopeless as these authors hold, and that *family planning* plus *social and economic measures* offer a realistic prospect of balancing population against food production. The emancipation of women, old age security, and literacy are crucial, but I would add a child tax (small at low income levels, but rising rapidly with income) in countries where it would be practical. However, if major efforts have been made to avoid catastrophe by these intermediate measures, and they clearly have proved ineffective, then more coercive measures should be introduced, provided they are designed to give some consideration to freedom and justice. If survival is really at stake, it can be assigned higher priority than other values without totally excluding or preempting them.[48]

If *compulsory measures* do seem necessary in the future, they should, where possible, be democratically legislated; the burdens and benefits should fall equitably on all citizens, and there should be provision for as much individual choice as possible among the ways of achieving the socially desired goal. There should be minimal violation of human dignity and bodily integrity, and minimal extension of police powers that could be used for other purposes. There should be specific safeguards against the discriminatory use of coercive programs against particular groups or individuals.[49] Compulsory abortion is deeply offensive, morally and religiously, to many people around the world, and would be so bitterly opposed and resented that it could be enforced only with strong police measures. Sterilization is more widely acceptable, but the history of eugenic proposals points to the dangers of its misuse to protect class interests or to manipulate the poor, welfare mothers, and minority groups. If it proves necessary to legislate family size, there should be a choice of contraception or sterilization, with clear provisions for appeal and review to protect minorities from exploitation. Prison sentences or fines should provide sufficient sanctions, without resorting to compulsory abortion or the requirement that illegal children be given up for adoption.

These measures do not seem justified to me until much more strenuous efforts have been made with family planning and *social and economic measures.* I disagree with the demographic transition theorists who expect lower fertility as an automatic by-product of economic development. But the particular social measures I have advocated are consistent with equitably distributed development oriented toward human needs—which also is the most promising response to the food crisis. The family-planning movement assumes that individual choice will coincide with the social good, once the means of contraception are provided. The advocates of compulsion, at the other extreme, assume that individual choice will be contrary to the social good, so coercion will be necessary. The middle path tries to alter social conditions so that smaller families will be in the interest of individuals as well as society, and it offers incentives that still allow some choice. But the path is not an easy one, social and attitudinal changes are slow, and time is indeed running out.

Lifeboat Ethics

Garrett Hardin has proposed linking food and population policies; he advocates *withholding food aid* to countries that face famine but are unwilling to control population growth. Rich nations, he suggests, are like people in crowded lifeboats, with the poor of the world swimming outside. If we let them into our boat by sharing food, we will swamp the boat and all will go down. Affluent countries, he holds, have only a small safety factor now; in some respects they already have exceeded the carrying capacity of their own lands. Sending food aid would lead to further population growth in LDCs and only postpone the final catastrophe. The need for energy as well as food puts an ever-mounting burden on the environment. Hardin cites the example of Nepal, where food aid encouraged population growth, leading to extensive deforestation to obtain wood for fuel—followed by soil erosion and rapid runoff that contributed to the disastrous floods in Bangladesh in 1974. Every life we save only hastens environmental destruction. Hardin says that questions of justice are irrelevant; we cannot remake the past, we live in an unjust world, and we cannot achieve pure justice. "The criterion is survival. Injustice is preferable to total ruin."[50] Paul Ehrlich and the Paddocks have compared selective food aid to the triage system used in allocating scarce medical resources on the battlefront; the doctor does not try to help the hopeless cases, but only those that can be saved.[51]

I would reply as follows. First, the lifeboat metaphor is misleading since it suggests a *fixed carrying capacity* that cannot be changed. Hardin talks about food aid, which is indeed limited and temporary; he says nothing about assistance in agricultural development, which could significantly increase the carrying capacity of LDCs. I have suggested that there is time to build larger lifeboats, if we really work at it.

Second, people in *affluent nations* are not in a lifeboat situation in which helping someone else would jeopardize their own lives or even their health. Their situation is excessive consumption and great waste, not lifeboat rations. Moreover, affluent nations bear some past and present responsibility for the disparities; a more even distribution would fulfill the nutritional needs of everyone. Our consumption of oil makes it less available for fertilizer in LDCs, and thereby reduces their carrying capacity.

Third, starvation is not the only way to slow *population growth.* Withholding food aid actually might lead to higher birthrates as parents sought to offset higher infant mortality. Starvation would have to occur on a massive scale to be effective as a method of population reduction; to allow such a catastrophe when it could be prevented would require a callousness to human suffering that would undermine moral sensitivity in other relationships. I have suggested that there are more humane ways of bringing food and population into balance than by raising death rates.

Fourth, lifeboats are independent; when one sinks, other boats are not affected. But we live in an *interdependent* world. The United States is a net importer of 26 out of 36 raw materials critical for industry. Industrial nations are particularly vulnerable to terrorism and blackmail; amid social upheaval, desperate people will resort to violence from which no one is immune in an age of nuclear proliferation. There surely would be intense resentment if the rich and powerful were to claim the right to decide who shall survive—especially if they imply that Western countries are more deserving. Lifeboat ethics would encourage nationalism and isolationism rather than internationalism and cooperation. If we want to use a metaphor, it would be more appropriate to think of the world as an ocean liner in which, like the Titanic, we all go down together. Citizens of the United States are first-class passengers living in luxury on the top deck, with a crowded hold and an impoverished crew stoking the boilers below.[52]

I do not believe that *survival* for all is impossible; but even if it were, survival should not preempt all other values. In Chapter 5 I suggested that if it were clear that resources were not sufficient for everyone, the decision as to who should survive would still be subject to criteria of *justice.* If some people have to be thrown off an overcrowded lifeboat, those on whom everyone's life depends might be exempted (a doctor, for instance, or a strong rower), and the others might draw lots. One also would expect everyone to make any sacrifice, short of life itself, which would reduce the number of lives lost. As indicated earlier, Rawls holds that under conditions of radical scarcity it is justifiable to set aside the principle of liberty, but not the principle of justice.

Finally, there is an issue here concerning the balance between *freedom* and *coercion.* I have defended a positive concept of freedom as self-determination and participation in the decisions that affect one's life, rather than the right to do as one wishes, or the absence of external coercion. With Hardin, I believe that "mutual coercion, mutually agreed on" often is necessary in an interdependent world. But the coercion of starvation is not mutual, nor is it mutually agreed on. If other nations have to be encouraged to adopt stronger birth control measures, I would accept conditions on development assistance, given through international agencies in which all nations have a voice, before I would accept preventable starvation as a unilateral weapon used by one nation on another. As Reinhold Niebuhr says, a combination of coercion and persuasion may be needed to effect social change: "Sentimental moralism which under-estimates the necessity of coercion, and cynical realism which is oblivious to the possibility of moral suasion, are equally dangerous to the welfare of mankind."[53] Both within nations and between nations, then, population policies should use coercion only when there are provisions for justice and participation, and only if other methods have been vigorously tried and found to be ineffective.

III. TECHNOLOGY AND DEVELOPMENT

Behind the food and population crises lies the basic problem of Third World poverty. Growing populations, the demands of rich nations, and the escalating costs of energy, water, and productive land are pushing the cost of food for an adequate diet out of reach for a large portion of humanity. I have suggested that little progress on food and population problems can be expected apart from the wider issue of development. This is a vast topic, many aspects of which lie beyond the scope of this volume, but we can at least look at some basic strategies and the role of technology and international trade.

Alternative Development Strategies

Three strategies for development have been prominent in recent years, centering respectively on economic growth, basic human needs, and national self-reliance. Each is distinctive in its explanation of the causes of underdevelopment and its policy for change.

The *economic growth strategy* dominated development policies in the 1960s. Underdevelopment was equated with an inadequate industrial base, and the proposed cure was industrialization. All countries, it was held, would go through stages of development similar to those followed in Europe and the United States, and the poor countries would catch up with the rich ones.[54] Agriculture was considered less important than capital-intensive industry on the Western model, and huge showcase projects such as steel mills were built. Distributional measures should be avoided, it was said, because they might hinder capital accumulation and economic growth. The benefits of an expanding GNP would automatically "trickle down" to the poor. Impressive growth was indeed achieved by LDCs, averaging 5% a year during the 1960s—higher than the sustained rate of any country in earlier history—though half of this was soaked up by population growth.

However, economic growth has been accompanied by *widening gaps*, both between and within nations. Far from catching up, the poor nations find themselves further behind rich nations. The ratio of GNP per capita in the United States to that in India, for instance, has increased to 50:1. The disparities within LDCs also have widened. A privileged minority has benefited greatly from economic growth, but there has been little "trickle down" to those at the bottom. As capital-intensive techniques replaced labor-intensive ones, unemployment increased. Policies of land ownership, credit, taxation, education, political power, and technology transfer have combined to distribute the benefits of growth very unevenly.[55] In most LDCs there is an urban bias in the allocation of funds for education, health care, food subsidies, and capital investment.[56] These widening internal inequalities are overlooked if development is measured in aggregate measures such as GNP.

Economic growth evidently has not helped *the poorest groups* in LDCs. A study commissioned by the World Bank concludes that though per capita income increased 50% in a decade, there was little or no benefit to the bottom third of the population.[57] Wealth was concentrated in the hands of a small group. A study of 48 social, political, and economic indicators in 74 countries concluded that "the position of the poorest 60% typically worsens, both relatively and absolutely."[58] Other authors suggest that from 20 to 40% of the population in LDCs has suffered an actual decline in living standards. A striking example is Brazil, whose GNP grew at an 8.8% annual average in 1965-72. Yet most of the benefits went to large landowners, industrial owners, and skilled workers; the vast majority, especially in northeast Brazil, were no better off. In Ecuador, the top fifth of the population receives 73% of the income while the poorest fifth receives 2.5%.[59]

In response to these trends, a second strategy of development, focusing on *basic human needs*, emerged in the 1970s. In this view, poverty and unemployment should be directly attacked. The goal is not welfare payments to the poor, but programs that will increase their continuing productivity and income, in order to fulfill their needs for food, housing, and health. Since the majority of Third World populations is rural, policies that help the landless peasant and the small farmer will have the most far-reaching effects. Social indicators (such as life expectancy, nutritional level, and literacy) are better criteria of a nation's development than overall economic growth. Foreign investment and the domestic market have favored those with greatest purchasing power, putting luxury goods for the few ahead of necessities for the many. Instead, a more equitable distribution of the benefits of growth should be deliberately sought. Robert McNamara, president of the World Bank, since 1973 has advocated loans to support development plans targeted on the poorest 40% of the population.[60]

Such *equitable and need-oriented growth* has been advocated by many development experts and organizations. Mahbub ul Haq, a World Bank economist, has said that the distribution and content of the GNP are as important as its size. To him, the goal of development is not economic growth but the elimination of malnutrition, disease, and illiteracy. Production should be directed to essential commodities (food, housing, clothing) and public services (health, education, transportation). The productivity and income of the landless laborer can be improved only by greater access to land, credit, employment, and education.[61] The U.N.'s International Labor Office also has endorsed the basic needs approach. It holds that this requires a shift from urban to rural investment, from nonessential to essential goods, and from capital-intensive to labor-intensive production. It has urged land reform, institutional decentralization, mass participation in planning, and locally controlled appropriate technology.[62]

A third strategy, *national self-reliance*, has been articulated by a number of Third World representatives. Political independence was achieved by many LDCs after World War II, but the economic dependency of colonialism was perpetuated by governments, multinational corporations, and patterns of aid and trade dominated by rich nations. Tariffs and quotas in the North have kept the South in its role as supplier of raw materials. Movements for national self-determination and liberation from foreign exploitation have sought models of development more compatible with local needs and indigenous cultures. There have been increasing doubts about both the feasibility and the desirability of trying to imitate the resource-consumptive industrial nations. There have been calls for the redistribution of land and wealth, as well as the expansion of health and educational services.[63]

Adherents of this position share many of the goals of the basic needs approach, but think that they can be achieved only by more radical *internal political change* and greater *national independence*, not by integration into a world economy dominated by the North. Third World production should cater neither to the privileged domestic minority nor to export markets, but to the needs of the masses. According to Paulo Friere, Celso Furtado, and other South American social critics, domestic elites are supported by foreign governments and corporations; genuine human development requires the political empowerment of the people and liberation from dependency.[64] Dennis Goulet rejects dependency because it has led to economic and cultural vulnerability. He defends cultural identity and diversity against technological standardization and conformity to the Western model of modernization. Aid and trade within structures of unequal power, he says, have hindered the political and social reforms that would make justice and freedom possible.[65] The liberation theology movement in the Roman Catholic church has taken a similar stand.[66]

One example of a *self-reliant strategy* is Tanzania. Under Julius Nyerere's leadership, rural development plans are designed to fulfill basic needs without destroying African communal values and indigenous culture. Only those technologies are imported that contribute to these goals. Village water projects, small-scale farming, and rural cooperatives make use of labor-intensive, locally controlled technologies and local skills and resources. China provides another model of self-reliant development through which basic levels of food, health care, and education are available to all citizens. After Soviet aid ended in the late 1950s, Chinese development was carried out with no help from other nations and little interaction with them. Only since 1978 has this policy of self-reliance been modified by the initiation of foreign trade.

Government officials in many LDCs continue to give priority to economic growth and industrialization. Along with large landowners and industrial leaders, they often strongly resist efforts at land reforms or more revolutionary political change. But there is evidence of a new nationalism and recognition of the dangers of economic dependency. The basic needs approach, advocated by many international agencies, has been gaining influence among Third World leaders. Some hold that isolation and national self-sufficiency are impractical, but that "collective self-reliance" through regional agreements would lessen dependency on the North. Others hope that greater national self-determinism can be achieved within a new framework of international cooperation in which LDCs have a stronger voice in setting the terms of trade. I submit that such a combination of the *basic needs strategy* with *greater self-reliance* (but not isolationism) offers the best prospects for the type of development that would alleviate the food and population crises.

What effects do alternative development policies have on *the environment*? At the U.N. environmental conference in Stockholm (1972), many LDC leaders insisted that it is the rich nations that are responsible for most of the world's pollution and resource depletion. For poor nations, they said, urgent present needs have priority over future generations, and people come before wildlife. When life itself is threatened, the quality of life seems less important. The LDCs were reluctant to enact regulations or to accept international standards that might slow economic growth. They totally rejected the idea of "limits to growth," which they believed would freeze their living standards at present levels.[67] Yet at Stockholm there was the beginning of global environmental awareness, and the first steps were taken on international action. The U.N. Environmental Program was initiated to promote cooperation on the global environment (atmosphere, oceans, climate, and so on), including pollution monitoring, data exchanges, and research.[68]

By the late 1970s, the LDCs had recognized increasingly the *environmental impacts* of agriculture and development projects. They had seen the evidence of ecological stress on renewable resources (including soil erosion, deforestation, and overgrazing). Irrigation has resulted in waterlogging (from rising water tables) and salinization (from evaporation). The Aswan Dam harmed the fisheries of the eastern Mediterranean and led to increased schistosomiasis, a debilitating disease spread by aquatic snails.[69] Most governments are beginning to consider the environmental impacts of development programs, and such international agencies as the World Bank are requiring impact statements on assistance projects. LDCs are becoming more wary of multinational corporations that want to locate their factories in countries without pollution standards. Above all, they are beginning to look at the quality as well as the quantity of development, and to seek technologies that are less destructive—environmentally as well as socially—than those of the industrial nations.

Technology Transfer

Throughout this volume I have maintained that technology is an instrument of power and leads to the further concentration of economic and political power. Nowhere is this more evident than in the relationship between rich and poor nations. Technology transfer has reinforced economic neocolonialism, perpetuated international patterns of dominance and dependence, and strengthened ruling elites within LDCs.

Technology transfer has made many *positive contributions* to LDCs. It has been a major force in the economic growth and increased productivity of developing countries. LDCs can draw on an existing body of knowledge without having to repeat time-consuming research. Much of the transfer of technology occurs through multinational corporations (MNCs) that provide technical and managerial expertise as well as equipment. They have experience, organizational skills, geographical mobility, a worldwide communications network, and capital that can be invested in plants and machines. They provide jobs and training, pay local taxes, and in some cases promote local citizens to managerial and executive roles. Defenders of MNCs see them as agents of a new internationalism, an economic globalism transcending national boundaries.[70]

But *Western technologies* were developed under particular historical circumstances. Energy, raw materials, and capital were relatively abundant and labor relatively scarce (the opposite of the situation in LDCs). With a good transportation and distribution system and cheap fuel, economies of scale were substantial and large firms dominated production for mass markets. The organization of production in turn influenced the kinds of research that were pursued. These technologies frequently have been inappropriate in LDCs and have tended to perpetuate a dependent relationship. Western experts and suppliers are needed to install and maintain imported equipment. When the interests of elite groups in LDCs are allied with those of foreign investors, the control of technology links internal privilege and international dependency.

Many *multinational corporations* sell only complete packages of goods and services: machines, expertise, management, parts, trademarks, and market agreements. By holding patents on one part of a production process they are able to control the whole process or to get high prices for licenses and royalties, often under monopoly conditions that keep out local competitors. The corporations see this as a legitimate return on their original expenditures for reasearch. The LDCs claim that the R & D investment usually has been paid off already by sales within industrial countries, and overseas royalties should therefore be lower;

they say that portions of the production package in most cases could be more cheaply supplied locally. They view as excessive the MNC profits from investment in LDCs, which average twice the rate of return on investment in industrial countries. They also are critical of the political influence of MNCs in protecting their overseas holdings, often by supporting repressive regimes and opposing democratic reforms, especially in Latin America.[71]

Western-trained experts in government and industry are predisposed to favor capital-intensive technologies. Consider, for example, the choice made by the government of Bangladesh in the early 1970s concerning the construction of 20,000 *tubewells for irrigation.* One option would have used local labor and materials and small diesel pumps. Its advantages were lower cost, the creation of jobs, the training of indigenous experts, and the encouragement of small industries for producing the parts. A pilot project by an area cooperative using these labor-intensive techniques was highly successful, and local farmers took an active part in well construction. However, a team of experts from the World Bank—from which part of the loan for the project was sought—preferred power-drilling rigs run by foreign contractors and the installation of imported pumps with electric motors. With these methods, the cost was twice as much for the same output, and little local labor was used. But the experts were familiar with these techniques and considered them more reliable (though under the conditions prevailing in Bangladesh this judgment appears dubious). They also felt that a program with a few contracted drilling teams would be easier to administer than a more decentralized system. In the end, the preferences of the aid agencies were followed, even though they violated many of the objectives of development.[72]

In many cases there is a range of technological options, but the *selection mechanisms* of both the marketplace and government agencies have favored capital intensity. The market, as we have seen, responds to the demands of people with purchasing power, which heavily weights a small portion of the population. Government leaders in LDCs frequently identify "modern" with the most advanced technology, which in general is capital intensive. Tax policies, access to credit, and direct or indirect subsidies have favored large-scale technologies. Yet in many cases greater progress toward development goals could be achieved by improving the efficiency of traditional techniques, by adapting older Western technologies, or by creating new appropriate technologies.[73]

In some processes there are *economies of scale* that may outweigh other considerations, for example, mining, extraction, and the production of steel and chemical fertilizer. In other areas (including agriculture and textiles) economies of scale in LDCs are marginal or nonexistent, and the social advantages of labor-intensive decentralization are great. Taiwan and South Korea have built their export trade around small industries (textiles, electronics, wood products, and so on). Japan has selectively imported advanced technologies but has kept control of them, modifying and adapting them and avoiding some that compete with indigenous production. As we shall see in Chapter 13, regional institutes have been experimenting with intermediate-scale technologies suitable for rural development.

The *U.N. Conference on Technology for Development* (Vienna, 1979) made little progress on these issues. The Third World pressed for more favorable terms of technology transfer, while U.S. industry said that it did not intend to give away patents or proprietary rights. A $250 million U.N. Science and Technology Fund was established, but the amount is minute in relation to the need, and is less than 0.2% of the $150 billion that the world spends on science and technology each year. Scientific and governmental delegates from industrial countries were more sympathetic toward the attempts of Third World nations to build up their own science education and R & D capabilities, and plans for cooperation were initiated. Some LDC representatives advocated large-scale technologies for rapid industrialization. Others urged the selective purchase of foreign technologies and efforts to combine advanced and indigenous techniques appropriate to local conditions.[74] The latter approach seems to be most promising and most consistent with the strategy of development outlined above.

The International Economic Order

International trade could contribute significantly to Third World development. However, the terms of trade have favored the industrial nations, whose leaders talk about the virtues of the free market but in practice put up various protective walls. The United States, for example, has many tariffs and quotas that limit imports of manufactured goods. There is a 12% duty on copper wire, but unprocessed copper is duty free. The average duty on U. S. imports from LDCs is twice that from advanced countries. There are quotas on textiles, leather goods, and other products. Processing, shipping, and distribution are mainly controlled by the rich countries. LDC leaders are seeking a larger role in processing and manufacturing and not just in supplying raw materials. Some oil-rich countries are constructing refineries and fertilizer plants. Other countries are seeking to follow the example of Taiwan in exporting such labor-intensive manufactured goods as textiles, shoes, toys, wood products and electronic devices.

Third World leaders have been vocal in advocating a *New International Economic Order.* At its special sessions in 1974 and 1975, the U.N. General Assembly recommended: 1) the reduction of tariffs on international trade; 2) higher prices for raw materials exported from LDCs (including an index system that would tie these prices to the prices of manufactured goods that LDCs import from industrial nations); 3) renegotiation of the debt burden carried by the poorest LDCs; 4) a code of conduct for MNCs that would provide for greater disclosure, accountability, and limits on profits; 5) greater voting rights for LDCs on the International Monetary Fund and other international agencies; and 6) U.N. control of certain international resources, such as the seabed, which are "the common heritage of mankind," with the royalties to be used for international development. Third World leaders believe that these changes would create a more just and democratic international order in which they could have a greater voice in their own destiny.[75]

The initial *confrontation* over these proposals has been replaced by *negotiation* over more limited issues, but the gains to date have been meager. The International Monetary Fund has increased the voting rights of OPEC members, who along with other LDCs now can outvote the industrial nations. There has been some progress in negotiating a common fund to create buffer stocks that would help to stabilize the prices of 18 commodities exported by LDCs (including coffee, sugar, cotton, copper, and tin). The Law of the Sea negotiations have sought a compromise, a system of taxes or royalties on seabed mining (mainly of nodules rich in manganese, nickel, and copper) that would provide payments to the U.N. without discouraging private companies from investing in the equipment needed. The exposure of widespread corporate bribes and interference by MNCs in local political struggles has underscored the need for greater accountability. It appears that such modest, piecemeal changes are more likely than sweeping structural reforms.[76]

The *bargaining power* of the LDCs has varied considerably. In 1975 OPEC had achieved dramatic success, and the prices of many other commodities were high because they were in great demand. By 1978 economic recession was worldwide and the price of many raw materials had dropped. It is evident that OPEC is a special case, since the demand for oil remains high, substitutes take decades to develop, and the producing countries have maintained a united front. Yet Morocco was able to quadruple the price of phosphate, and Jamaica raised the levy on bauxite. The success of nations that hope to raise prices of copper, chrome, coffee, bananas, and other commodities depends on their solidarity in bargaining, the level of demand, and the availability of substitutes. The United States imports more than half its supplies of 20 critical minerals. Inflation, price fluctuations, huge balance-of-payments deficits, and the falling exchange rate of the dollar make clear our economic interdependence and vulnerability. U.S. dominance has been eroded by the Vietnam War and the increasing industrial competition of Europe and Japan. The continued frustration of the hopes of LDCs could lead to political instability and violence, and perhaps desperate disruptive actions. In the long run, both rich and poor nations have more to gain from negotiation than from confrontation.

It may seem that I have strayed from the crises of food and population, but I am convinced that they are part of the wider problem of poverty and development. Equitably distributed and need-oriented development is essential for bringing food and population into balance. In LDC policies and in international assistance programs, high priority must be given to agricultural and rural development, land reform, and appropriate technology. Affluent nations must reduce their use of energy and luxury goods imported from malnourished nations. Patterns of domination and dependency must be broken by a combination of greater self-reliance and steps toward a more just and participatory international order, which is explored further in the next chapter.

NOTES

1. William Aiken and Hugh LaFollette, eds., *World Hunger and Moral Obligation* (Englewood Cliffs, N. J.: Prentice-Hall, 1977), p.1.

2. Good introductions are Georg Borgstrom, *World Food Resources* (New York: Intertext Educational Publications, 1973); Arthur Simon, *Bread for the World* (Grand Rapids, Mich.: Eerdmans, 1975); the May 9, 1975 issue of *Science*; J. David Edwards, *U.S. Food Policies: A Primer* (Washington, D.C.: Amer. Assoc. of University Women, 1977); and Sterling Wortman and Ralph Cummings, *To Feed the World* (Baltimore: Johns Hopkins University Press, 1978).

3. C. Dean Freudenberger and Paul Minus, *Christian Responsibility in a Hungry World* (Nashville: Abingdon, 1976).

4. Erik Eckholm, *Losing Ground* (New York: W. W. Norton, 1976).

5. See Lester Brown, *By Bread Alone* (New York: Praeger, 1974); idem, "The World Food Prospect," *Science* 190 (1975), 1053-59, and idem, *The Twenty-Ninth Day* (New York: W. W. Norton, 1978).

6. Keith Griffin, *The Political Economy of Agrarian Change: An Essay on the Green Revolution* (Cambridge, Mass.: Harvard University Press, 1974).

7. Frances Moore Lappé and Joseph Collins, *Food First* (Boston: Houghton Mifflin, 1977); and Susan George, *How the Other Half Dies* (Montclair, N.J.: Allenhold Osmun, 1977).

8. U.S. Dept. of Agriculture, *Agricultural Statistics 1975* (Washington, D.C.: USDA, 1975), p. 356. It takes 17 lbs. of feed *protein* to make 1 lb. of beef protein; see David Pimental et al., "Energy Constraints in Food Protein Production," *Science* 190 (1975): 754-61. For total *plant* input (including forage), the ratio is somewhat lower.

9. Frances Moore Lappé, *Diet for a Small Planet* (New York: Ballantine, 1975).

10. Ervin Lazlo, ed., *Goals for Mankind* (New York: E. P. Dutton, 1977), p. 275. Figures for food consumption are lower than for production, partly because of the waste in converting grain to meat: the U.N. Food and Agriculture Organization reported that in 1970 world average consumption of calories was 101%of minimum requirements, and of proteins, 173%; see Paul Abrecht, ed., *Faith Science and the Future* (Philadelphia: Fortress Press, 1979), p. 126.

11. Norman Borlaug, "Civilization's Future: A Call for International Granaries," *Bull. Atomic Scientists* (October 1973): 7-15; also idem, "The Fight Against Hunger," in *Finite Resources and the Human Future*, ed. Ian G. Barbour (Minneapolis: Augsburg, 1976). See also Sudhir Sen, *Reaping the Green Revolution: Food and Jobs for All* (Maryknoll, N.Y.: Orbis, 1975).

12. Nicholas Wade, "Green Revolution," *Science* 186 (1974): 1093-96 and 1186-92; Thomas Poleman and Donald Freebairn, eds., *Food, Population and Employment: The Impact of the Green Revolution* (New York: Praeger, 1973); and B. H. Farmer, *The Green Revolution* (Boulder, Col.: Westview, 1977).

13. Griffin, op. cit.; also idem, *Land Concentration and Rural Poverty* (New York: Holmes and Meier, 1976); and Bruce Johnston and Peter Kilby, *Agriculture and Structural Transformation* (New York: Oxford University Press, 1975).

14. Mahmood Hasan Khan, *The Economics of the Green Revolution in Pakistan* (New York: Praeger, 1975); L. Nulty, *The Green Revolution in West Pakistan* (New York: Praeger, 1972); and John P. McInerney et al., "The Consequences of Farm Tractors in Pakistan" (Washington, D.C.: World Bank, 1975).

15. D. J. Greenland, "Bringing the Green Revolution to the Shifting Cultivator," *Science* 190 (1975): 841-44; and Uma Lele, *The Design of Rural Development: Lessons from Africa* (Baltimore: Johns Hopkins University Press, 1975).

16. William Rich, *Smaller Families through Social and Economic Progress* (Washing-

ton, D.C.: Overseas Development Council, 1973), p. 32. In India the average yield per acre for farms of less than 50 acres is 40% higher than for farms of more than 50 acres (Brown, *By Bread Alone*, op. cit., p. 214).

17. See Wortman and Cummings, op. cit.; and Pimental, op. cit. But Lester Brown is doubtful that world food production can be doubled (Brown, *The Twenty-Ninth Day*, op. cit., p. 156).

18. Cf. Richard A. Watson, "Reason and Morality in a World of Limited Food," and Peter Singer, "Famine, Affluence and Morality," in Aiken and LaFollette, op. cit.

19. Onora O'Neill, "Lifeboat Earth," in ibid.

20. John Arthur, "Rights and the Duty to Bring Aid," in ibid.; see also Peter Brown, "Food as National Policy," in *Food Policy*, ed. Peter Brown and Henry Shue (New York: Free Press, 1977).

21. Michael Slote, "The Morality of Wealth," in Aiken and LaFollette, op. cit.

22. Thomas Nagel, "Poverty and Food: Why Charity is not Enough," in Brown and Shue, op. cit.

23. Deut. 14:28.

24. Isa. 58:6.

25. Matt. 25:35.

26. Ronald Sider, *Rich Christians in an Age of Hunger* (New York: Paulist Press, 1977); and Bruce Birch and Larry Rasmussen, *The Predicament of the Prosperous* (Philadelphia: Westminster Press, 1978).

27. Sider, op. cit.; Simon, op. cit.; and Abrecht, op. cit.

28. John Gilligan, quoted in *Time*, March 26, 1979, p. 48.

29. 1976 budget, cited in Simon, op. cit., p. 124.

30. Shue and Brown, op. cit., p. 2.

31. Lappé and Collins, op. cit.

32. "1979 World Population Data Sheet" (Washington, D.C.: Population Reference Bureau, 1979). For surveys of the population problem, see the September 1974 issue of *Scientific American*; and P. Reining and I. Tinker, eds., *Population: Dynamics, Ethics and Policy* (Washington, D.C.: Amer. Assoc. for the Advancement of Science, 1975).

33. John Walsh, "U.N. Conferences," *Science* 185 (1974): 1143-44; and W. Parker Maudlin et al., "A Report on Bucharest," *Studies in Family Planning* 5 (1974): 357-93.

34. Michael Teitelbaum, "Relevance of Demographic Transition Theory for Developing Countries," *Science* 188 (1975): 420-25.

35. Maurice Kendall, "The World Fertility Survey: Current Status and Findings," *Population Reports* 7 (1979): 73-101.

36. Rich, op. cit.; see also James Kocher, *Rural Development, Income Distribution and Fertility Decline* (New York: Population Council, 1973); and Ronald Freedman and Bernard Berelson, "The Record of Family Planning Programs," *Studies in Family Planning* 7 (1976): 1-40.

37. Robert Repetto, *Economic Equality and Fertility in Developing Countries* (Baltimore: Johns Hopkins University Press, 1979).

38. Milton Viorst, "Population Control: Pakistan Tries a Major Experiment," *Science* 191 (1976): 52-53.

39. Linda Gordon, *Woman's Body, Woman's Right: A Social History of Birth Control in America* (New York: Grossman, 1976).

40. Robert Veatch, "Population Policy: Governmental Incentives" in *Encyclopedia of Bioethics* (New York: Macmillan, 1978); and Ronald Green, *Population Growth and Justice* (Missoula, Mont.: Scholar's Press, 1976), p. 228ff. A proposal for baby licenses that could be bought and sold is set forth in Kenneth Boulding, *The Meaning of the 20th Century* (New York: Harper & Row, 1964), chap. 6; and Herman Daly, *Steady-State Economics* (San Francisco: W. H. Freeman, 1977).

41. Kaval Gulhati, "Compulsory Sterilization: The Change in India's Population

Policy," *Science* 195 (1977): 1300-05; and Michael Henry, "Compulsory Sterilization in India," *Hastings Center Report* 6 (June 1976): 14-15.

42. Robert Veatch, "An Ethical Analysis of Population Policy Proposals," in *Population Policy and Ethics: The American Experience*, ed. R. Veatch (New York: Irvington, 1976).

43. Bernard Berelson, "Beyond Family Planning," *Science* 163 (1969): 533-43.

44. Daniel Callahan, "Ethics and Population Limitation," *Science* 175 (1972): 477-86. See also J. Philip Wogaman, ed., *The Population Crisis and Moral Responsibility* (Washington, D.C.: Public Affairs Press, 1973).

45. Yale Task Force on Population Ethics, "Moral Claims, Human Rights and Population Policies," *Theological Studies* 35 (1974): 83-113; and World Council of Churches, Working Committee on Church and Society, "Population Policy, Social Justice, and the Quality of Life," *Study Encounter* 9, no. 4 (1973): 1-12.

46. Paul and Anne Ehrlich, *Population, Resources, Environment*, 2d ed., (San Francisco: W. H. Freeman, 1972), p. 340. See also Paul Ehrlich, *The Population Bomb* (New York: Ballantine, 1968); and Kingsley Davis, "Zero Population Growth," in *The No-Growth Society*, ed. M. Olson and H. L. Landsberg (New York: W. W. Norton, 1973).

47. Garrett Hardin, "The Tragedy of the Commons," *Science* 162 (1968): 1243-48; also idem, *Exploring New Ethics of Survival* (Baltimore: Penguin, 1973).

48. Roger Shinn, "Ethics and the Family of Man" in *This Little Planet*, ed. Michael Hamilton (New York: Scribner's, 1970); also idem, "The Population Crisis: Exploring the Issues," *Christianity and Crisis*, August 5, 1974, pp. 170-75.

49. Drew Christiansen, "Ethics and Compulsory Population Control," *Hastings Center Report* 7 (1977): 30-33.

50. Garrett Hardin, "Lifeboat Ethics: The Case Against Helping the Poor," *Psychology Today* September 1974, pp. 38-43 and 123-24; idem, "Living on a Lifeboat," *BioScience* 24 (1974): 561-68; and idem, *The Limits of Altruism* (Bloomington: Indiana University Press, 1977).

51. Ehrlich, op. cit.; and William Paddock, *Famine—1975!* (Boston: Little, Brown, 1967).

52. See articles by James Sellers and Donald Shriver in *Soundings* 59 (1976): 100-19 and 234-43, reprinted in *World Famine and Lifeboat Ethics* ed. George Lucas (New York: Harper & Row, 1977).

53. Reinhold Niebuhr, *Essays in Applied Christianity*, ed. D. B. Robertson (New York: Meridian, 1959), pp. 80-81.

54. W. W. Rostow, *The Stages of Economic Growth*, 2d. ed. (Cambridge: Cambridge University Press, 1971). For an overall view of development policies, see Charles Wilber, ed., *The Political Economy of Development and Underdevelopment* (New York: Random House, 1973); and Barbara Ward, *Rich Nations and Poor Nations* (New York: W. W. Norton, 1974).

55. Charles Elliott, *Patterns of Poverty in the Third World* (New York: Praeger, 1975). See also Edgar Owens and Robert Shaw, *Development Reconsidered* (Lexington, Mass.: Lexington, 1972).

56. Michael Lipton, *Why Poor People Stay Poor* (Cambridge, Mass.: Harvard University Press, 1977).

57. Hollis Chenery et al., *Redistribution with Growth* (London: Oxford University Press, 1974).

58. Irma Adelman and Cynthia Taft Morris, *Economic Growth and Social Equity in Developing Countries* (Stanford, Calif.: Stanford University Press, 1973).

59. Dennis Goulet, *The Uncertain Promise: Value Conflicts in Technology Transfer* (Washington, D.C.: Overseas Development Council, 1977), chap. 6.

60. Robert McNamara, "The Third World: Millions Face Risk of Death," *Vital Speeches* 41 (October 15, 1974): 13-20.

61. Mahbub ul Haq, *The Poverty Curtain: Choices for the Third World* (New York: Columbia University Press, 1976). See also John Sewell, ed., *The United States and World Development: Agenda 1977* (New York: Praeger, 1977); and Mary Jegen and Charles Wilber, eds., *Growth with Equity* (New York: Paulist Press, 1979).

62. International Labor Organization, *Employment, Growth and Basic Needs: A One-World Problem* (New York: Praeger, 1977); and Franklin Lisk, "Conventional Development Strategies and Basic-Needs Fulfillment," *International Labor Review*, March 1977, pp. 175-91.

63. Guy Erb and Valeriana Kallab, eds., *Beyond Dependency: The Developing World Speaks Out* (New York: Praeger, 1975).

64. See essays by Celso Furtado and Theotonio Dos in Wilber, op. cit.; Frank Bonilla and Robert Girling, eds., *Structures of Dependency* (Stanford, Calif.: Stanford University Press, 1973); and J. D. Cockcroft et al., eds., *Dependence and Underdevelopment* (Garden City, N.Y.: Doubleday, 1972).

65. Dennis Goulet, *The Cruel Choice: A New Concept in the Theory of Development* (New York: Atheneum, 1971); and idem, "On the Ethics of Development Planning," *Studies in Comparative International Development* 11, no. 1 (1976): 25-43.

66. Gustavo Gutierrez, *A Theology of Liberation* (Maryknoll, N.Y.: Orbis, 1973); and Rosino Gibellini, ed., *Frontiers of Theology in Latin America* (Maryknoll, N.Y.: Orbis, 1979).

67. See the September 1972 issue of *Bull. Atomic Scientists.*

68. Clayton Jensen et al., "Earthwatch," *Science* 190 (1975): 432-38.

69. M. Taghi Farvar and John P. Milton, eds., *The Careless Technology: Ecology and International Development* (Garden City, N.Y.: Doubleday, 1972); and Eckholm, op. cit.

70. Orville Freeman, "Multinational Companies and Developing Countries," *International Development Review* 16, no. 4 (1974): 17-19; and Richard Farmer, *Benevolent Aggression* (New York: McKay, 1972). See also Abdul Said and Luiz Simmons, eds., *The Sovereigns: Multinational Corporations as World Powers* (Englewood Cliffs, N.J.: Prentice-Hall, 1975); and George W. Ball, ed., *Global Companies* (Englewood Cliffs, N.J.: Prentice-Hall, 1975).

71. Richard Barnet and Ronald Müller, *Global Reach: The Power of Multinational Corporations* (New York: Simon & Schuster, 1974); Goulet, *The Uncertain Promise*, op. cit.

72. John Thomas, "The Choice of Technology for Irrigation Tubewells in East Pakistan" in *The Choice of Technology in Developing Countries*, ed. C. Peter Timmer et al. (Cambridge, Mass.: Harvard Center for International Affairs, 1975).

73. Frances Stewart, *Technology and Underdevelopment* (Boulder, Col.: Westview, 1977); see also *Choice and Adaptation of Technology in Developing Countries* (Paris: OECD, 1974); and Jairam Ramesh and Charles Weiss, eds., *Mobilizing Technology for World Development* (New York: Praeger, 1979).

74. Anne Roark, "The Vienna Conference," *Environment* 21, no. 8 (1979): 2-4; Richard Lyons, "Technology Separates the Rich from the Poor," New York *Times*, August 26, 1979, p. 24E; see the November 1979 issue of *Bull. Atomic Scientists.*

75. Resolutions of the sixth and seventh special sessions of the U.N. General Assembly are reprinted in Erb and Kallab, op. cit. See also Jagdish Bhagwati, ed., *The New International Economic Order* (Cambridge, Mass.: MIT Press, 1977); and Karl Sauvant and Hajo Hasenpflug, eds., *The New International Economic Order* (Boulder, Col.: Westview, 1977).

76. Robert Rothstein, *The Weak in the World of the Strong* (New York: Columbia University Press, 1977); Albert Fishlow et al., *Rich and Poor Nations in the World Economy* (New York: McGraw-Hill, 1978); Roger Hansen, *Beyond Stalemate: North-South Relations in the 1980s* (New York: McGraw-Hill, 1978); and Martin McLaughlin, ed., *The United States and World Development: Agenda 1979* (New York: Praeger, 1979).

12
RESOURCES AND GROWTH

Will resource depletion and environmental degradation limit future industrial growth? In this chapter the debate over limits to growth is first summarized. What are the ethical and political implications of slower growth? The relation of growth to distributive justice and the problem of freedom and coercion under conditions of scarcity are explored. A position between "no-growth" and "pro-growth" advocates is developed, namely "selective growth." The goal is the achievement of global growth that is equitably distributed and environmentally sustainable.

I. THE DEBATE OVER GROWTH

Some studies have concluded that continued industrial growth at rates typical of recent decades will lead to resource exhaustion and ecological catastrophe within the lifetime of our children. Other studies maintain that market mechanisms can adjust to scarce resources, and that technological ingenuity can vastly extend carrying capacity. How can the experts disagree? It appears that in this debate, as in others we have analyzed, the divergence arises from differences in assumptions, models, and values.

The Limits to Growth

Industrial and economic growth in Europe and North America was aided by cheap fuel and abundant resources obtained either domestically or from overseas colonies. New technologies greatly increased the productivity per man-hour, while the economic institutions of capitalism facilitated capital accumulation and reinvestment. Water, air, and land were adequate to absorb industrial wastes, despite local environmental deterioration. This industrial and economic expansion was supported by a strong ideology equating growth with progress, as we saw in Chapter 2. A high GNP per capita and high standards of living for a substantial portion of the population were achieved. The American dream has imagined ever-rising levels of production and consumption. Since World War II, many Third World nations have sought rapid industrial development along similar lines.

By the late 1960s environmentalists were warning that the world's population would exceed the global carrying capacity—not only because of food shortages but also because of resource depletion and the generation of pollutants by industrial production. In *The Limits to Growth* (1972), a team at MIT, under the sponsorship of the Club of Rome, concluded that if population and industrial production continued to grow exponentially, global limits would be exceeded within a few decades, The team, headed by Dennis Meadows, described three kinds of limits: agricultural production (land area and yield per acre); nonrenewable resources (mainly mineral ores); and the capacity of the environment to absorb pollutants. A distinctive feature of the study was the use of computer simulations to plot the future values of a large number of variables under various sets of assumptions about the interactions between the variables.[1]

As long as either global population or consumption per capita *grows exponentially* (that is, at a constant percentage rate), any given extension of the limits provides only a temporary solution, according to these simulations. If it is assumed that agricultural production can be doubled, for example, the food crisis is only postponed by one population doubling time. Moreover, there are considerable time delays—in the accumulation of pollutants, for instance, or in the latent effects of carcinogens, or in the stabilization of population size after birthrates have fallen. Once the limits are exceeded, the environment deteriorates very rapidly; the population curves typically show an "overshoot and collapse" pattern. When changes are introduced to deal with one kind of crisis, there are economic and environmental repercussions that create other crises. Even when population is stabilized, the authors conclude, industrial production cannot continue to grow.

The basic constraints arise not from literally running out or encountering absolute limits, but from *diminishing returns* in the use of scarce resources. After the richer and more accessible ores are used, the mining and extraction costs of other deposits will be higher, diverting capital needed elsewhere. The cost per pound to remove an air or water pollutant goes up rapidly as one approaches 100% removal. Even if it is assumed that mineral reserves are five times current estimates, they will be scarce and more costly within 25 to 100 years if consumption grows exponentially at current rates. Similarly, a fourfold improvement in pollution abatement and recycling technologies postpones but does not avert an environmental and health catastrophe.

The only way in which disaster can be avoided in these computer projections is to halt industrial growth as well as population growth. An *equilibrium state* can be achieved with constant levels of capital and population (each at not more than twice 1970 levels), and with low levels of resource use and pollutant emission (roughly one-fourth 1970 levels per unit of production). Such a steady-state society would have to shift its emphasis from resource-intensive material goods to services such as health and education, which use few natural resources. There could be growth in cultural activities, the arts and sciences, and the technologies of conservation, but no growth in heavy industry.

The Limits to Growth evoked some sharp rebuttals from economists, who said that *market mechanisms* will provide an automatic adjustment to resource scarcities. When a resource becomes scarce its price will rise, thereby reducing the quantity demanded and increasing the quantity supplied. A rise in the price of a resource will discourage its use, encourage the search for new reserves and more efficient extraction technologies, and provide an incentive to use substitute materials. It is claimed that the market is thus an efficient feedback system for allocating and regulating scarce resources.[2]

I find this criticism unconvincing. The Meadows's computer runs do take *rising resource prices* into account; a larger fraction of capital is allocated to obtaining resources as they become scarcer. However, there is a loss in the capital available for other purposes; the substitute materials are themselves often scarce, and the technological improvements entail mounting environmental costs to achieve diminishing returns. Moreover, price responses are short term and delayed. The market heavily discounts the future, as we have frequently observed, and it is not an effective instrument for conservation. The world market is far from free; there are monopolies, cartels, tariffs, and quotas in addition to differences in political power among nations. As food becomes scarce, the affluent bid up the prices until millions cannot afford enough to avoid starvation.

The disagreement between the Meadows team and its economic critics arises in part from the use of *different conceptual models*. Every model is selective, portraying the features of the world and the kinds of causal relationships that its advocates think are most important. Every model leaves out a lot and therefore has its limitations. The *economic model* is useful in predicting short-term adjustments in a relatively free market when supply and demand are strongly price dependent. It can incorporate constraints introduced through political processes to internalize environmental costs. Yet the model has severe limitations when environmental and social costs cannot be adequately internalized into economic costs, or when demand or supply are not strongly price dependent. The market model optimizes short-term advantages and virtually ignores costs to future generations.

The *environmental model*, on the other hand, starts from a finite carrying capacity and resource base. Market mechanisms can stimulate new technologies to extend the carrying capacity somewhat, but rapidly diminishing returns and increasing environmental costs are expected. This model, which governs the assumptions of *Limits to Growth*, seems most relevant globally in considering long time-horizons, or in dealing with fragile ecosystems in the short run. But it tends to neglect the potentialities for creative technological responses in extending resource reserves and reducing environmental impacts. Both models are aggregative and do not deal with the distribution of resource use among groups or nations, which is a crucial issue today. Both models neglect the urgent needs of developing nations.

Technological Solutions

A number of critics claim that the assumptions of *Limits to Growth* about resource reserves and technological advances are unduly pessimistic. Even the computer runs based on a fivefold increase in *reserve estimates* appear conservative in the light of past history. Iron ore estimates increased fivefold from 1954 to 1966 alone, for instance. Moreover, the price of most minerals has stayed fairly constant because improvements in exploring, mining, and extraction techniques have compensated for the need to use less accessible and lower-grade ores. New technologies can turn previously useless substances into useful resources. Virtually inexhaustible minerals (iron, aluminum, and magnesium) can be substituted for scarcer ones such as copper, though at the cost of additional energy.[3]

In *pollution control*, too, the critics claim that we can do a lot better than a fourfold improvement. After all, some technologies have already achieved emission reductions by factors of 20 or 50, and further reductions are likely, though again at some cost in energy. Currently, pollution abatement costs the United States 1½% of our GNP, and this might have to go up to 2 or 3%, but with a growing economy we could afford it. A direct attack on emissions would be far more effective than a ceiling on industrial production. A group in England reran the Meadows computer program with the assumption of 2% annual improvements in resource extraction and pollution control technologies. The world population curves leveled off at 20 billion without overshoot and collapse. The group concludes that

> continued growth is possible provided that the total costs of resource extraction and pollution do not exceed about one-quarter of total industrial investment. . . . Even with a very high population figure, continued growth, at least for the next two centuries, will not inevitably stop at an early date because of physical limits.[4]

An even more optimistic prognosis is offered in *The Next 200 Years* by Herman Kahn and his colleagues at the Hudson Institute. He sees no significant limits to growth as each of the alleged constraints is pushed back by technological advances. He anticipates "a huge surplus of land, energy and resources." Within two centuries the world population will have leveled off at 15 billion (almost four times the current level) with 15 times the current per capita income—a 60-fold increase in total world income. Economic growth eventually will slow down, not because of resource limits, but because people will be satisfied with their affluence. Kahn asserts that "the resources of the earth will be more than sufficient—with a wide margin of safety—to sustain, for an indefinite period of time and at high living standards, the levels of population and economic growth we project."[5]

Kahn holds that there are *abundant raw materials* for future generations. Far from harming other nations or generations, the rich nations that develop new technologies will help them by making additional raw materials useful. High-grade aluminum ores may run out, but low-grade aluminum compounds are virtually inexhaustible. Deeper mines and improved extraction methods will open up vast new sources. Eventually the extraction of metals from seawater will provide "an essentially infinite resource base." Further, "every environmental hazard can be corrected by technology with sufficient time and a reasonable amount of money." In particular, Kahn is confident that new technologies will make abundant energy available at today's prices; this would be essential since most of the other technologies on which he is counting are energy intensive. (He would rely mainly on coal until new solar techniques, and perhaps fusion, have been developed.) Kahn maintains that without any basic changes in current national or international policies, the whole world can achieve a high level of material prosperity.

For Kahn, *the gap between rich and poor countries* is not a problem. He expects the rich to get superrich, but the lot of poor nations also will improve. He holds that higher consumption rates in industrial nations will stimulate the economies of developing nations and everyone will benefit. It should be evident from the previous chapter that I strongly disagree with Kahn's assumptions here. I suggested that with current patterns of economic dominance and dependence, the industrial growth of the North often is detrimental to the interests of the South. Kahn has rationalized the perpetuation of exploitation and injustice by assuming that benefits to the rich will indirectly help the poor also. Both Kahn and Meadows have ignored the existing structures of economic power between nations.

I find myself closer to Meadows's view of technology than to Kahn's, though I believe that ecologically sound technologies can considerably extend resource limits. Kahn has greatly overestimated the prospects for technical fixes by playing down their environmental and human impacts. Our ecological knowledge is too sketchy to be sure that massive environmental assaults will not have serious delayed and unintended consequences. The 60-fold increase in world income and expansion of industry on the scale he projects would require a huge increase in energy use, which I believe is a highly unrealistic expectation (see Chapter 10). Moreover, as the scale and complexity of technical systems increase, they tend to be more vulnerable to human errors and institutional failures. The constraints that we now face are social and political as well as environmental and technological. There are sufficient resources for the foreseeable future, but only if they are wisely used and equitably distributed. I will try to show that selective growth is both environmentally sustainable and ethically desirable.

II. POLITICAL AND ETHICAL ISSUES

What are the political and social implications of rapid growth, slow growth, and no growth? Rapid industrial growth has in the past improved the lot of the poor, but also has produced great inequalities between rich and poor. It appears that a no-growth policy would make distributional justice even more difficult to achieve. A further question is whether democratic institutions can respond effectively to resource scarcities, or whether scarcities will lead to increasing authoritarianism and coercion.

Growth and Distributive Justice

Advocates of growth point out that in the past the poor usually have gained more from national economic growth than from any efforts at redistribution. Even though the rich benefit most from growth, many benefits eventually *trickle down* to those at the bottom. In another metaphor, the economy is an escalator that carries everyone upward; the absolute level of those at the bottom improves, even if their relative position remains the same. Again, the economy is compared to a pie that is growing in size; even those with the smallest slices are better off, without any change in their relative shares. But if the whole pie is not growing, the only way in which the least advantaged can gain is by changes in their relative shares.

In industrial countries, the poor have indeed benefited more from *overall economic growth* than from measures for greater distributional equity. The only periods in which relative income differences decreased significantly in the United States were during the Depression and World War II, when temporary sacrifices were accepted in response to national emergencies. In general, growth creates new jobs, better wages, higher productivity, and better living standards for all. Conversely, slower growth leads to unemployment from which the disadvantaged suffer most. During the 1974–75 recession, minority groups were the first laid off. The prospect of a permanent no-growth society probably would make redistribution even more difficult, and would intensify social conflicts as each group defended its own interests. In the past, economic growth has covered over many potentially divisive issues and class conflicts. Inequalities are bearable if there is hope that things will improve. But if the poor feel locked into poverty and the rich try to solidify their privileges, social unrest, violence, and authoritarian responses can be anticipated.[6]

In earlier chapters I noted that *fear of unemployment* is a major factor in opposition to environmental and conservation measures, and suggested that this linkage might be reduced by full employment legislation, a guaranteed annual wage, or a negative income tax scheme. Similarly, fear of unemployment is a major factor in pressures for high growth rates. It would be very difficult to secure funds for a guaranteed annual wage during times of severe recession, unless there were a deep national commitment to economic equality. During crises people are willing to accept sacrifices if they are shared equally by other people, but this does not include the loss of one's job. With modest economic growth, full employment legislation could create a framework in which efforts to curb resource consumption would not be seen as a threat to job security.

Developing nations believe that their only hope lies in economic and industrial growth. They think that a no-growth world would freeze present inequalities and leave them in permanent poverty. They are understandably skeptical of the idea that affluent nations might reduce their standards of living voluntarily. On the whole, they think that the Third World could benefit indirectly from modest growth in industrial nations, though the evidence here is more ambiguous. The Third World does receive many benefits from trade and aid, despite the often unfavorable terms and the perpetuation of dependency. Slower economic growth and inflation in the United States during the 1970s contributed to a revival of isolationism and a reduction in foreign aid. Some countries exporting raw materials were hard hit by falling prices during this recession. In a no-growth world, the rich nations perhaps would hang onto their advantages even more tenaciously.[7]

According to the Rawls criterion (Chapter 4), inequalities are just only if they maximize *benefits to the disadvantaged.* Wage differentials, for example, create work incentives and increased productivity from which all benefit. But inequalities of the magnitude that exist between nations could not conceivably be justified on such grounds. Between nations, the "trickle down" has been relatively small, and has mainly helped the ruling classes in poorer countries. Many aspects of growth in affluent countries have been at the expense of impoverished nations—especially when tariffs and quotas in the North have hindered the industrialization of the South. I will suggest that increased consumption within an affluent society provides diminishing returns in human welfare; the same resources would make a far greater contribution to human welfare in meeting basic needs in developing countries. On ethical grounds, it is clear that future growth should be concentrated mainly in the Third World.

It was noted in Chapter 5 that resource depletion and environmental destruction constitute a transfer of benefits from *future generations* to the present. The Rawls principle might justify such temporal transfers within poor countries, in which the present generation may be the least well off of the generations; the transfer must not go so far as to make a future generation the worst off. But in industrial countries accumulated environmental degradation soon may reach the point where their own future generations will be worse off than the present one—quite apart from the depletion of world resources of which future generations in developing countries would be deprived. Thus intergenerational as well as international justice dictates a halt to the growth of resource consumption in affluent countries.[8]

The poor of the world, in short, will benefit most from *environmentally sustainable growth* coupled with *more equitable distribution.* There is considerable "trickle down" to the poor within rich nations, but relatively little between rich and poor nations; the redistribution of growth prospects between nations is crucial in a world where gaps are increasing. The present dominant values of rich nations will lead them to resist such redistribution more strongly if their own growth is halted. This suggests that some economic growth in industrial nations is important, both for domestic "trickle down" and to enhance the chances for domestic and international redistribution. It also suggests the need for value changes if a more equitable international order is to be created.

A policy of *selective growth* seems to offer some prospect of moving toward greater justice in global resource allocation. The growth of resource-intensive industry would be concentrated mainly in developing countries. There would be some economic growth in affluent nations, which would prevent a buildup of social tensions. Growth would be slower than in the past and would be oriented toward human services that are not resource intensive. Such a pattern would be environmentally sustainable for several generations, assuming population levels are stabilized. The shift to renewable resources would occur gradually as basic human needs were met around the world. With a shift in growth rates, the gap between rich and poor nations would be reduced. Such selective growth will be very difficult to achieve; some institutional changes that might make it possible will be discussed in Section III.

Scarcity, Freedom, and Coercion

It has been asserted by some writers that there are *social and political limits to growth* that are as important as physical limits. Rufus Miles, for one, maintains that large-scale social and technical systems are increasingly difficult to predict and to manage. Mistakes in planning or human errors can have unforeseen and far-reaching consequences. Complex centralized systems, such as nuclear power plants, are highly vulnerable to sabotage by alienated and disaffected groups. Risks of large-scale disasters lead to more extensive regulation and bureaucracy, and the public feels incompetent to participate in crucial policy decisions. The political limitations in the human capacity to manage large systems are, for Miles, the main grounds for opposing industrial expansion.[9]

William Ophuls argues that in a world of resource scarcities *democracy is doomed.* The relative abundance of the last 200 years produced individualism, freedom, democracy, and laissez-faire economics in Western nations. But the future prospects for food, energy, and other resources are dismal, in Ophuls's view. Only governments with strong powers to regulate individual behavior can deal with the destruction of the environmental "commons". In crises, people accept authoritarian governments with powers to compel obedience. When there are dangers that they do not understand, they assent to decisions by elites based on competence and expertise. The most competent experts are put in charge of a hospital or a ship in a storm. In the name of survival, says Ophuls, we will have to accept drastic restrictions of personal liberty. Faced with catastrophe, the sacrifice of freedom is the lesser evil.[10]

Ophuls holds that democratic political systems are *incapable of responding to scarcity.* Powerful special interests promote growth, and short elected terms make political leaders reluctant to advocate strong actions or long-term planning. Steady-state politics, Ophuls maintains, will require authoritarian measures and a technical elite. However, he thinks that such a system need not be completely totalitarian; there can be limits to power under a rule of law. There can be some social diversity if decentralized communities will practice self-restraint. But if we do not voluntarity give greater authority to central governments, we will end with total tyranny, whether we like it or not.

An even gloomier prognosis is offered by Robert Heilbroner. He starts with a grim picture of population growth, widespread starvation, and environmental deterioration—and the threat of nuclear blackmail and terrorism by desperate nations. He attacks the obsession with productivity, efficiency, technology, and economic growth that he sees in capitalist industrialism and Soviet socialism alike. Heilbroner anticipates intensified competition for dwindling resources and increasing regimentation by all governments. Only *authoritarian regimes* could organize society to face such catastrophes. Democratic institutions will be unable to cope with internal strife, international conflict, and threats to survival. Dissent and freedom of expression will be looked on as the obsolete luxury of self-indulgent intellectuals.[11]

Heilbroner holds that we cannot avoid *the collapse of industrial civilization.* He thinks it is unrealistic to hope for a restraint in growth, or a shift in values toward simpler life-styles, or a more equitable allocation of resources. He foresees massive crop failures, environmental disasters, social breakdown, and wars of redistribution. Only after catastrophe, convulsive change, and the disintegration of industrialism might a very different civilization emerge from the ruins, perhaps a society oriented toward inner states of consciousness rather than toward science and technology. Heilbroner pictures a future culture that is traditional and ascetic, with the self-renewing vitality of primitive societies, rather than the technological utopia of which modern society has dreamed.

I do not believe that *the prospects for peaceful change* are as hopeless as Heilbroner thinks. The earth's physical limits will permit substantial global growth if technology is wisely used. Scarcities are as much the product of maldistribution as of resource exhaustion or ecological limits, and new directions in national and international policy are possible. Heilbroner's outlook is almost fatalistic; disaster unfolds like a Greek tragedy that has to play itself out. Such despair can be a self-fulfilling prophecy if it prevents creative initiatives. Despite the rigidity of institutional structures, the presence of widespread crises and social turmoil actually may facilitate institutional change. In the next chapter I will suggest that there is ground for hope in the coalescence of resource crises, political pressures, and a new vision of a just and sustainable society. The future is open and there are encouraging signs as well as discouraging ones.

Nor do I believe that *freedom* and *democracy* are doomed. Granted, the individualistic view that equates freedom with the absence of government interference is no longer tenable. Yet freedom as participation in the decisions that affect one's life is still the only defense against the abuse of power. Authoritarian leaders initially might act in the common good, but there is ample historical precedent for expecting that they soon would become a new ruling elite protecting their own interests. Granted, a technological society is dependent on the competence of experts. However, technical expertise and citizen participation are not mutually exclusive, as was emphasized in Part Two. Granted, the "commons" can be protected only by stringent regulations. Yet the regulations can be democratically legislated if enough people are concerned. Granted, sanctions are needed. But the harsher forms of coercion can be avoided. Economic incentives and strong regulations with legal sanctions can alter behavior patterns without undermining civil liberties or creating a totalitarian society.[12]

Even where centralized regulation is necessary, there is room for considerable *diversity and local initiative.* The decentralization of decisions about policy implementation is consistent with the ideals of local participation and regional self-reliance discussed earlier. Instead of repudiating technology, as Heilbroner does in his vision of the future, we should seek the right kind of technology—especially the appropriate technologies of intermediate scale that are decentralized and ecologically sound. Solar energy, for example, avoids most of the dangers of centralization, vulnerability, and resource depletion that Heilbroner fears. Resource scarcities and environmental dangers will lead to more extensive governmental regulation, but they do not require the abandonment of political democracy or participation in decisions.

In the history of nations, periods of growth have often stimulated the search for freedom, and periods of decline have led to a search for order and security. Periods of uncertainty and anxiety have encouraged the search for meaning and purpose. Looking to the future, can we still hope for new forms of growth, even if resource-consuming forms must be limited? Can we develop internal restraints for the sake of common goals and avoid the external imposition of totalitarian forms of order? Can we find meaning and purpose in less resource-intensive activities?

III. SELECTIVE GROWTH

I have maintained that the *no-growth* position is unduly pessimistic about environmental constraints and technological improvements, and that it would entail high social and political costs, especially for the poor around the world. I have said that *pro-growth* advocates are unrealistically optimistic about resources and technology, and are mistaken in thinking that growth makes redistribution unnecessary. In this section a policy of *selective, sustainable growth* is developed. The important questions are: Whose growth? What kinds of growth? It is proposed that industrial growth should take place mainly in the Third World, while growth in the affluent nations should occur primarily in goods and services that are not resource intensive. Economic and political policies to achieve these objectives are outlined.

Sustainable Growth in Industrial Nations

Three kinds of questions can be asked about further growth in rich nations: 1) Is it *possible* without violating environmental constraints? 2) Is it *ethical* in the context of huge and growing global inequalities? 3) Is it *desirable* for affluent nations themselves, by criteria of genuine human fulfillment? I have suggested that further industrial growth is environmentally and technically possible in rich countries, but only by their use of an increasingly disproportionate share of world resources. The higher costs of energy, resources, and pollution control will enforce slower growth rates than were typical of the 1960s. There are practical as well as ethical grounds for redistributing growth prospects; affluent societies are vulnerable to the effects of social unrest, political instability, and retaliatory action by nations that feel trapped in poverty.

A policy of selective growth would encourage expansion mainly in *service areas* such as health and education, rather than in resource-intensive consumer goods. This would permit a growing economy, but the composition of the GNP would shift, and resource depletion and environmental pollution would be reduced. Although there would be temporary unemployment during the transition, the long-run impact on employment could be favorable, even with a lower growth rate in the GNP, because many services are labor intensive. Such a policy would reverse the trend to substitute capital and energy for labor, which has contributed to persistent unemployment. The possibility of more satisfying work would justify at least some loss of productivity per man-hour, but the mass production processes that offer significant economies of scale would be retained.

New technologies suitable for a world of resource constraints would be promoted. Resource use would be a major criterion in technology assessment and in government R & D policy. Engineering design criteria would give greater emphasis to product durability, initial cost in energy and materials, and operating energy efficiency. The technologies of recycling and waste utilization would be fostered more vigorously. We noted earlier the potential for energy conservation, especially in transportation; smaller cars and public transit conserve materials as well as fuel. There also is a wide range of postindustrial information technologies that are not resource intensive, including electronic communication systems and the miniaturization of computers. In the past, industrialization has involved a shift of emphasis from food to material goods and then to human services. Technological policy in the future deliberately would accelerate this latter trend.

Economic policies also would be designed to direct growth away from resource-intensive production. Herman Daly and others have proposed depletion quotas or severance taxes on nonrenewable raw materials. For example, licenses to mine a given quantity of ore could be auctioned off, or a tax levied on each ton mined. Such quotas or taxes would limit both resource depletion and pollution; they would encourage resource-saving and recycling technologies, and a shift to renewable resources. These measures also would raise the price of resource-intensive goods and discourage consumption of them. The quotas or taxes could be introduced gradually to allow time for readjustments to occur, and the levels would reflect a social decision about acceptable depletion rates. The scheme relies on market mechanisms once the policies have been established. Social controls are introduced with a minimum sacrifice of personal freedom. There are economic incentives for conservation, but there is scope for individual decision.[13]

Increased prices of resource-intensive goods would be a heavy burden on the poor, so they should be accompanied by *redistributive measures.* Daly proposes lower and upper limits on personal income, with a sharply progressive tax on income between these limits. There is considerable political support today for the idea of a minimum income (that is, a negative income tax below this minimum). Above the maximum, the tax would be 100%. The ratio of maximum to minimum would have to be sufficient to provide some incentives and rewards. The very idea of a maximum recognizes the importance of distributive justice in a slow-growth economy. Daly contends that a maximum also should be placed on accumulated wealth in order to limit the concentration of economic power and to spread more widely the ownership of private property.

A task force under a contract with 14 agencies of the Canadian government has prepared a report on *The Conserver Society.*[14] It presents three scenarios representing varying degrees of reduction in material throughput in order to reduce resource use, pollution and waste disposal. The first scenario is *growth with conservation*, or industrial expansion with increased efficiency. Its motto is "Do more with less," following the "Scotch model" of thrift, and requiring minimum changes in social values. Products would be designed for greater energy efficiency, durability and recyclability. Peak loads in transportation and electricity would be reduced by better load distribution in time (such as staggered work hours). Solid wastes would be reduced by cutting down packaging and by recycling materials. Rental or cooperative ownership of autos and farm equipment would reduce the number of machines needed. By means of taxes the prices of products would internalize the full social costs of depletion, pollution and disposal. Tax breaks would encourage resource conserving, nonpolluting and recycling industries and processes.

The second scenario, *an affluent steady state*, allows some selective growth up to fixed ceilings. Its motto is "Do the same with less." It follows a "Greek model," Plato's ideal of moderation and the curbing of unlimited desires beyond basic needs. All the measures of the first scenario would be more rigorously applied in order to achieve the goals of zero growth in population, urbanization, energy use and material throughput. There would also be a minimum and a maximum on personal income. There would be a bar against any advertising which stimulates artificial wants; ads would be limited to information rather than persuasion, and anticommercials would ensure a more even-handed use of the media. This path would entail considerable behavioral and institutional change, but no major changes in fundamental values.

The third scenario, *a decentralized steady-state economy*, would yield a greater reduction in material throughput and require a more extensive change in values. Its motto is "Do something else." The authors refer to it as a "Buddhist model" of personal growth and harmony with nature. It stresses self-realization, communally oriented relationships, local self reliant and meaningful work. It calls for appropriate technology, small decentralized industries, and rural development. It would release resources to meet basic human needs and would greatly reduce the inequalities among nations in resource use. The authors favor the first scenario because it requires the minimum value change. In Chapter 13 I will defend a mixture of the second and third alternatives because I believe value changes are necessary in the interest of long-term sustainability, global justice, and human fulfillment.

Slower economic growth, more equitably distributed within the nation and oriented toward services, might actually enhance *genuine human fulfillment*. Imagine a graph of "quality of life" plotted against "individual income." At first the curve would rise rather rapidly; in very poor countries, a person's welfare increases dramatically with income level. As the curve continues, it rises less rapidly; there seem to be diminishing returns from further income after basic needs have been met. A given increment to the income of people in a rich country adds less to their well-being than it would to the well-being of people below the subsistence level. Does the curve level off, or perhaps even turn down, at high levels of affluence? Does the quality of life actually decline as the environmental and human costs of sustaining additional affluence mount? Does competitiveness and anxiety in the pursuit of increased consumption lead to decreased satisfaction? European countries with half of the U.S. GNP per capita outrank us on most of the indicators of quality of life (health care, housing, literacy, and so on).[15]

There have been a number of studies of the correlation between *income* and *happiness*. Richard Easterlin has summarized 30 surveys in a variety of rich and poor countries and finds that, within every country studied, perceived happiness (subjective well-being) is correlated with income level. For example, in the United States in 1970 only one-fourth of the lowest income group (under $3,000) said they were "very happy," as compared to more than half in the highest income group (over $15,000). But in comparing nations, or the same nation at different times, there is no relation between perceived happiness and average national income. Easterlin concludes that happiness depends not on the absolute level of income but on the level relative to other people and to social norms and personal aspirations. As a nation's economy grows, expectations and pressures to consume also escalate; people do not feel any better off, even though their standards of living have risen. These findings suggest that industrial populations would be no less happy if growth were slower, as long as the society was more egalitarian.[16]

Another study indicates that a significant shift from *materialist* to *post-materialist values* is already under way in industrial nations. In a series of surveys from 1970 to 1976 in the United States and Western Europe, respondents were asked to rate the importance of nine issues. Many respondents gave high ratings to "materialist" issues (rising prices, national defense, economic growth, controlling crime); others were more dedicated to what the author calls "post-materialist" values (more voice in government, freedom of speech, a less impersonal society). "Materialists" were predominant among older people, but "post-materialists" were almost as common as materialists among younger people. The author maintains that a generation that grew up in unprecedented prosperity, with wide access to education, is less concerned about material consumption than about intellectual activity, political participation, and the quality of life. He agrees with Maslow that when lower levels of the hierarchy of needs are satisfied, self-realization plays a larger role (see Chapter 4 above). The study concludes that such value shifts will facilitate the transition to a steady-state society.[17]

Other surveys in the early 1970s documented *value shifts* from "materialistic and status goals" to "humanistic and spiritual values" and "concern for the quality of life and community interaction." These shifts were found not just among college students but also among younger corporate executives and blue-collar workers.[18] Of course, answers to survey questions do not necessarily correspond to actual behavior in real life, and the idealism of youth may be lost at an older age. Moreover, the trend may have been reversed by the late 1970s when neoconservative movements flourished and the tax revolt, expressed in California's Proposition 13, was strong in many states. Dissatisfaction with the materialism of industrial society would be less common in times of scarcity or unemployment. Nevertheless, some of these value shifts seem to be continuing and could be deliberately encouraged in the future (see Chapter 13).

Sustainable Growth in Developing Nations

In the previous chapter a number of policies for developing countries were advocated. It was argued that population stabilization can be achieved by a combination of family planning, social and economic incentives, and wide access to health care and education. It was suggested that development strategies should be oriented toward basic human needs in food, shelter, and health. It was urged that the goals of economic development and distributional equity should be pursued together if growth is to benefit the poorest sectors of the population rather than ruling elites. What additional policies would be needed for sustainable growth in the Third World?

The second Club of Rome report, *Mankind at the Turning Point*, concludes that worldwide catastrophe can be averted only by major assistance from rich to poor countries. Whereas *Limits to Growth* uses global aggregates, the second computer study uses ten regional submodels that permit diverse scenarios to be explored for diverse regions of the world. In the second study, the most serious constraints (assuming significant progress in population control) arise from capital and energy for agricultural and industrial development—rather than from land, mineral resources, or pollution. The authors advocate "balanced organic growth," including rapid industrial growth in the Third World.[19]

According to this study, the gap between rich and poor nations could be permanently reduced by *substantial development assistance.* After an annual transfer of $100 billion for 25 years, the poorer nations would be self-sufficient, and the per capita income ratio between the richest and poorest regions, currently 15:1 and growing, would be reduced to 5:1. This report offers to the poor nations some hope for an improvement of their living standards without an actual reduction of standards in rich countries. But even the second report would require large-scale transfers that would slow the growth of rich nations for a generation in order that poor nations could grow more rapidly. The proposed $100 billion transfer seems large, but it is only a quarter of current world military expenditures. The authors recognize that a new level of global awareness would be a prerequisite for the adoption of such policies. Similar conclusions are reached in an economic study for the U.N.[20]

A more recent report to the Club of Rome, *Reshaping the International Order (RIO)*, goes beyond development assistance and recommends changes in the institutions of international power. Its authors believe that fears of resource exhaustion have been exaggerated and that balanced growth should be sought. Economic growth rates of 3.3% in the North and 5% in the South would slowly narrow the gap between them. (These are close to the actual figures for 1977: 3.5% for industrial nations and 4.9% for LDCs excluding OPEC).[21] The ratio between the richest tenth of the world's population and the poorest tenth would fall from 13:1 to 7:1 in 40 to 50 years. This would require self-restraint in the overconsumptive and wasteful patterns of affluent nations. It also would require internal reforms in developing countries, such as greater emphasis on basic human needs, rural development, land reforms, and appropriate technology.[22]

The *RIO* report is explicit in its value commitments, including *distributional justice in the international order.* It is sympathetic to Third World critiques of the economic structures that perpetuate dependence. The report recommends the abolition of tariff barriers in the North and the growth of raw material processing and manufacturing in the South. It urges that ocean beds and other essential resources be viewed as the common heritage of mankind. A world tax on all mining would slow depletion and provide funds for a World Development Authority. The *RIO* authors advocate a strengthening of U.N. agencies, with centralization at the policy level but decentralization at the operational level. They present a functional rather than a territorial view of sovereignty; jurisdictional authority would apply to specific activities rather than geographical areas.

Such a *strengthened U.N.* and the provision for *international taxation* would be major instruments for global justice. Within our own nation we recognize the need for distributive justice. We know that economic power is self-perpetuating, so that from market forces alone the gaps between rich and poor people increase. We counteract this tendency by legislating progressive income taxes, unemployment compensation, social security, and minimum levels of education, food, and health care. We use political power to ameliorate imbalances of economic power. There are no comparable mechanisms at the international level to prevent the gaps between rich and poor from increasing. In the long run, only a more just world will be a stable world.

But would not a stronger U.N. involve a *centralization of authority* that would diminish freedom? Some forms of freedom would be restricted, but other forms would be enhanced. The overall structure would remain participatory. The developing nations would achieve greater self-determination and a more democratic voice in international decision making. The areas of global authority would be limited and functionally defined, in contrast to the unlimited powers of an all-embracing world government. Policy would be centrally coordinated, but administration would be as decentralized as possible. Many of the desired results would come from the design and automatic operation of policies, such as taxation, that do not require continual central intervention. There should be deliberate provision for cultural diversity and national identity, and for diverse regional emphases consistent with global sustainability and justice.

A Global Compact

There have been several proposals that would tie together changes in industrial nations, changes in developing nations, and changes in the international order. Harlan Cleveland of the Aspen Institute has urged a *planetary bargain*, a compact between rich and poor countries to achieve a more stable world. Such a compact would be based on common interests, on mutual interdependence, and on negotiated trade-offs. Interactions between countries, he holds, should not be viewed as a "zero-sum game" (in which one side's gain is the other side's loss), but as a "positive-sum game" in which both sides can benefit, though perhaps in differing ways.[23]

Among the *common interests* that Cleveland lists are avoidance of nuclear war, reduction of the burden of arms expenditures, prevention of environmental breakdown, and a reliable international monetary system. At other points, rich and poor countries have divergent interests that might be paired in *negotiated agreements.* The developing nations' interest in higher prices for exported raw materials is obviously linked to the industrial nations' interest in dependable supplies of raw materials. Beyond that, the poor nations would agree to formulate development strategies and make internal reforms aimed at basic human needs. The rich nations, for their part, would agree to the stabilization of resource consumption. For the sake of global stability, self-restraint in the North would be correlated with internal reforms in the South. Eventually, limits to poverty and limits to affluence would be correlated through the establishment of minimum and maximum levels of consumption of food, energy, and other essential resources. The minimum levels would be based on studies of nutritional, medical, and social data, such as those assembled by the McHales.[24]

Such a planetary bargain would require *a new international order*, according to Cleveland, but the changes could occur gradually through expansion of existing U.N. agencies and the addition of new functional agencies. In place of bilateral agreements and unpredictable aid from donors to recipients, there would be an international development program with independent sources of long-term funding. One source would be the international taxation of common resources such as the seabed, ocean fisheries, and international communication and shipping. The funds would be used to support national development plans that had been reviewed by international agencies in terms of their consistency with the planetary bargain. In this proposal, more balanced global growth would be sought through international negotiation in the framework of an expanded U.N.

Several of the volumes in the *World Order Models Project* have described further steps in the reform of the existing pattern of rivalry between nation-states.[25] Most of the authors call for functional global agencies with greater authority over specific policy areas. Nations would have to give up sovereignty over some decisions, but they would have more real autonomy at other points. Conflict-resolution procedures and peacekeeping forces of the U. N. would be expanded into an effective world security system. A World Court would have power to adjudicate major disputes between nations. Richard Falk advocates a system in which military forces, the protection of the planetary environment, and the allocation of critical resources would be internationally rather than nationally controlled. National sovereignty would not be abolished but would be modified within a framework of world law and strengthened functional agencies.[26]

Falk recognizes that *a new international outlook* is a prerequisite of such a world order. He proposes a program of education for action, starting with consciousness raising through existing citizens groups and public interest networks. Coalitions would be formed among grass-roots constituencies as people realized that many issues (women's rights, civil liberties, environmental preservation, peace and antiwar activities, and so on) are tied to the international order. Domestic reform movements thus would be mobilized for political action in support of the creation of new international institutions. As citizens increasingly recognize that national welfare is dependent on the health of the international system, they would support government leaders willing to take steps toward a more stable world order.

Arms reduction and disarmament would be the most significant step toward world order, but the most difficult to achieve as long as nationalistic assumptions are dominant. Worldwide military expenditures are currently $400 billion, 30 times the total transfer of funds from rich to poor countries. In contrast, the budget for the U.N. and all its agencies and programs adds up to less than $2 billion annually, the equivalent of two days of military expenditures.[27] 40 to 50% of the world's scientific manpower and 40% of its research funds are devoted to arms.[28] Nuclear weapons already have a huge overkill capacity, equivalent to 30,000 lbs. of TNT for every person alive today. Control of the escalating arms race would be the most important step toward world peace and the release of funds for a major global development effort.

In summary, I have argued that *global resources* are adequate to meet the basic needs of humankind if population is stabilized at the levels projected in the previous chapter. For a couple of generations, at least, scarcities will be more the product of maldistribution and institutional deficiencies than of resource exhaustion, but the shift to greater dependence on renewable resources must be started now. World industrial production could increase considerably, without violating environmental constraints, if ecologically sound technologies are deployed. But the costs of pollution abatement, resource extraction, and energy are likely to increase, even with strenuous conservation efforts.

The *curtailment of freedom* in many areas can be expected. Both environmental protection and the allocation of critical resources require extensive powers of public regulation. I have maintained, however, that this does not imply that democracy is doomed or that authoritarianism is inevitable. The use of taxes and economic incentives leaves scope for individual choice. Even with stringent measures for resource conservation, there can be provisions for cultural diversity, local initiative, decentralized administration, participation in decision making, and the preservation of civil liberties.

I see reasonable prospects of *selective growth in the United States.* This would require effective measures for conserving energy and materials and for promoting recycling technologies, human services (such as health and education, which are not resource intensive), and postindustrial technologies (which process information rather than materials). But we should not underestimate the hold of materialistic values on the U.S. public, nor the power of economic institutions with a vested interest in perpetuating past patterns. In the next chapter I will suggest that major crises, shifts in political power, and a new vision of the future all will be necessary if the transition to a sustainable society is to occur without widespread violence or ecological catastrophe.

There also is great resistance in the United States to the *changes in the international order* that would improve the development prospects of poor countries. There is widespread skepticism about the U.N., and reluctance to alter the patterns of trade and overseas investment from which we benefit. Arms control and disarmament are crucial for peace—and for the direction of funds and technology toward human needs—but we are trapped by our self-perpetuating fears. Yet there are some signs of a new global awareness and a greater recognition of our interdependence. The gap between rich and poor nations is a mounting threat to world order and stability. Industrial nations are more vulnerable than in the past, and nations with natural resources are gaining greater bargaining power. Some of the hopeful signs, as well as the almost overwhelming obstacles, on the road to a more just, participatory, and sustainable society are explored in the final chapter.

NOTES

1. Donella H. Meadows et al., *The Limits to Growth* (New York: Universe, 1972); Dennis and Donella H. Meadows, eds., *Toward Global Equilibrium* (Cambridge, Mass.: Wright-Allen Press, 1973); and Dennis Meadows, ed., *Alternatives to Growth–I* (Cambridge, Mass.: Ballinger, 1977).

2. Wilfred Beckerman, *Two Cheers for the Affluent Society: A Spirited Defense of Economic Growth* (New York: St. Martin's Press, 1975); Peter Passell and Leonard Ross, *The Retreat from Riches: Affluence and Its Enemies* (New York: Viking, 1972); Andrew Weintraub, ed., *The Economic Growth Controversy* (White Plains, N.Y.: International Arts and Sciences Press); and Mancur Olson and Hans Landsberg, eds., *The No-Growth Society* (New York: W. W. Norton, 1973).

3. Ronald Ridker, "To Grow or Not to Grow: That's Not the Relevant Question," *Science* 182 (1973): 1315-18; and H. E. Goeller and Alvin Weinberg, "The Age of Substitutability," *Science* 191 (1976): 683-89.

4. H. S. D. Cole et al., eds., *Models of Doom* (New York: Universe, 1973), p. 131.

5. Herman Kahn, William Brown, and Leon Martel, *The Next 200 Years* (New York: William Morrow, 1976), p. 27.

6. Robert Stivers, *The Sustainable Society* (Philadelphia: Westminster, 1976); Denton Morrison, "Growth, Environment, Equity and Scarcity," *Social Science Quarterly* 57 (1976): 292-306; Lee Rainwater, "Equity, Income, Inequality and the Steady State" in *The Sustainable Society*, ed. Dennis Pirages (New York: Praeger, 1977); and Willard Johnson, "Should the Poor Buy No Growth?" in Olson and Landsberg, op. cit.

7. Jack Barkenbus, "Slowed Growth and Third World Welfare," in Pirages, op. cit.; Mahbub ul Haq, *The Poverty Curtain* (New York: Columbia University Press, 1976), chap. 5; and Denton Morrison, "The Growth-Equity Debate" (paper given at February 1978 meeting of the American Assoc. for the Advancement of Science).

8. Victor Lippit and Koichi Hamada, "Efficiency and Equity in Intergenerational Distribution," in Pirages, op. cit.

9. Rufus Miles, *Awakening from the American Dream: The Social and Political Limits to Growth* (New York: Universe, 1976). See also Fred Hirsch, *Social Limits to Growth* (Cambridge, Mass.: Harvard University Press, 1976).

10. William Ophuls, *Ecology and the Politics of Scarcity* (San Francisco: W. H. Freeman, 1977).

11. Robert Heilbroner, *An Inquiry into the Human Prospect* (New York: W. W. Norton, 1974). A symposium on Heilbroner appears in *Zygon* 10 (1975): 215-375.

12. See Stivers, op. cit., pp. 134ff.

13. Herman Daly, *Steady-State Economics* (San Francisco: W. H. Freeman, 1977); idem, ed., *Toward a Steady-State Economy* (San Francisco: W. H. Freeman, 1973); and Bruce Hannon, "Economic Growth, Energy Use, and Altruism," in Meadows, *Alternatives to Growth–I*, op. cit.

14. Kimon Valaskakis et al., *The Conserver Society* (New York: Harper & Row, 1979).

15. E. J. Mishan, "Growth and Antigrowth: What are the Issues" in Weintraub, op. cit.; and E. J. Mishan, *The Economic Growth Debate* (London: George Allen and Unwin, 1977).

16. Richard Easterlin, "Does Money Buy Happiness?," *Public Interest* (Winter 1973): 3-10.

17. Ronald Inglehart, *The Silent Revolution* (Princeton, N.J.: Princeton University Press, 1977).

18. Daniel Yankelovich, *The Changing Values on Campus* (New York: Washington Square Press, 1972); and idem, *The New Morality: Profile of American Youth in the 70's* (New York: McGraw-Hill, 1974).

19. Mihajlo Mesarovic and Eduard Pestel, *Mankind at the Turning Point* (New York: E. P. Dutton, 1974).

20. Wassily Leontief et al., *The Future of the World Economy: A United Nations Study* (New York: Oxford University Press, 1977).

21. "Vigorous LDCs: Cheer from the World Bank," *Time*, September 25, 1978, p. 69.

22. Jan Tinbergen, ed., *Reshaping the International Order* (New York: E. P. Dutton, 1976).

23. Harlan Cleveland, *The Third Try at World Order* (New York: Aspen Institute, 1977); and Aspen Institute, *The Planetary Bargain* (Palo Alto, Calif.: Aspen Institute, 1975).

24. John and Magda McHale, *Basic Human Needs: A Framework for Action* (Houston, Tex.: Center for Integrative Studies, 1977).

25. Saul Mendlovitz, ed., *On the Creation of a Just World Order* (New York: Free Press, 1975); and Rajini Kothari, *Footsteps into the Future* (New York: Free Press, 1974).

26. Richard A. Falk, *A Study of Future Worlds* (New York: Free Press, 1975); and Gerald and Patricia Mische, *Toward a Human World Order* (New York: Paulist Press, 1977).

27. Ruth Sivard, *World Military and Social Expenditures, 1978* (New York: Rockefeller Foundation, 1978).

28. Tinbergen, op. cit., chap. 3.

13
NEW DIRECTIONS

How can environmental and human values be realized in the world of resource constraints and global inequalities portrayed in the previous three chapters? In this chapter some new directions are delineated that represent significant departures from patterns prevailing in industrial nations. They are responses not only to resource scarcities, but also to the environmental and human impacts of large-scale technology today.

The first section takes up ways in which technology might be redirected toward ecological sustainability and human fulfillment. Some characteristics and examples of appropriate technology are given, and the pros and cons of centralization and decentralization are discussed. Section II asks "What can the individual do?" It depicts individual life-styles and alternative institutions that challenge the assumptions of a consumer society. Finally, some visions of the future are presented. What are the sources of change that might offer some hope for humanity? What are the prospects of achieving a more just, participatory, and sustainable society?

I. APPROPRIATE TECHNOLOGY

The Reorientation of Technology

In Part Two it was argued that criteria for decisions about technology should include environmental and human values as well as technical feasibility and economic efficiency. These criteria can be spelled out further in the light of the resource scarcities and global disparities that we have been considering. What are the characteristics of technologies that are compatible with environmental stewardship and human liberation today?

First, Technology for Basic Human Needs

A greater emphasis is needed on technologies related to survival, health, and minimal material welfare. These values correspond to the lowest levels of Maslow's hierarchy: survival and security needs (see Chapter 4). Adequate food, shelter, and health are the most universal needs and, when not fulfilled, the most urgent. They can be met in ways that do not ignore higher-level needs (belonging, esteem, and self-actualization). The Rawls principle (discussed in Chapter 4) tells us to maximize the welfare of the least privileged, while the biblical commitment to social justice demands action concerning the causes and not merely the symptoms of hunger, disease, and poverty (see Chapter 4). Technology as an instrument of justice must reduce rather than increase the gaps between rich and poor. Technologies of agriculture, public health, and low-cost housing—along with schools, hospitals, and public transportation—are relevant to developing nations as well as to low-income groups in industrial nations.

Second, Reordered Research Priorities

We have seen that both public and private research and development (R & D) have been dictated more by bureaucratic and industrial interests than by human needs or ecological wisdom. 40 to 50% of the world's scientists are in defense-related research, and many of the remainder are working on projects that will provide luxuries for the privileged. According to one estimate, "less than 2% of the world's R & D effort is devoted to the urgent agricultural, environmental, and industrial problems of developing countries."[1] If energy, food, and population are the most urgent global problems, higher priority should be assigned to such research areas as low-cost solar energy, high-protein crops, and family planning in all its dimensions.

Third, Ecological Soundness

Low levels of pollution and of use of energy and nonrenewable resources are important criteria in technological design. Recyclability cuts down on pollution and resource depletion, and in most (though not all) cases also conserves energy. Product durability—the opposite of planned obsolescence—also contributes to all these conservation objectives. Flexibility is achieved when one machine can be adapted to several tasks. Waste is reduced further when several processes can be integrated, as in the cogeneration of heat and electricity. I have said that in industrial nations selective growth should be concentrated in the technologies related to services, such as education, health care, and communications, rather than in the more resource intensive and more heavily polluting consumer-goods industries.

Fourth, Job Satisfaction

Work provides society with needed goods and services, and it provides the worker with income. But work also can be an important source of personal fulfillment. In Chapter 3 the alienating character of many industrial jobs was described. Most workers derive little personal reward from production; they work to obtain money to spend, seeking satisfaction primarily in consumption. Among the factors in job satisfaction are humane labor conditions, meaningful production objectives, participation in job-related decisions, and opportunity for some creativity, self-expression, or pride in one's work. Such factors are not irrelevant in technological policy decisions.

Fifth, Democratic Control

Technological decisions in the past have been made primarily on economic grounds. Often the structures of power (such as the alliance among an industry, a congressional committee, and a federal agency) have limited the effectiveness of controls through the political process. Several political mechanisms for wider participation in policy decisions were outlined in Part Two: citizen action through the courts, technology assessment that includes long-range environmental and social impacts, and wider information and awareness that enables an informed public to influence elected representatives. State and local governments also present many opportunities for the democratic redirection of technology toward broader human and environmental goals.

While some redirection can be achieved through these political processes, there are features of the *large-scale, capital-intensive technologies* typical of industrial countries that make the achievement of each of the above objectives particularly difficult. Being highly capital intensive, such technologies are controlled by large corporations and/or bureaucracies, which seldom give priority to human needs. Being large scale and usually resource intensive, they tend to have a massive impact on natural ecosystems. Their scale and mode of organization similarly militate against creativity and meaningful participation by workers. These technologies often are very complex and poorly understood by the public, so that crucial decisions may be made by a small group of technical experts. In addition, the centralized character of a technology and the magnitude of the risks involved encourage hierarchical organization and the centralization of control. Each of the objectives above can be achieved more readily with smaller-scale technologies.

Characteristics of Appropriate Technology

A technology may be called "intermediate" with reference to physical size, technical complexity, or economic cost. The designation "appropriate" is a favorable judgment of the effects of a technology in a social, cultural, and environmental setting. A "soft" technology is commended for the gentleness of its environmental, and perhaps also human, impact. An "alternative" technology usually is associated with alternative institutions that differ from prevailing patterns of work, consumption, and daily life. Each of these terms emphasizes particular aspects of smaller-scale technology in which there has been considerable interest in recent years in both industrial and developing countries.[2]

1. Intermediate Scale

Between the sickle and the combine is a range of machines of intermediate size and cost. They are more efficient than traditional methods, but cheap enough to be widely accessible. If the equipment for one worker costs $1 with traditional tools, and $1,000 with high technology, an intermediate technology might cost $10 or even $100 per worker. It could be obtained by scaling down a large industrial process. Traditional techniques also can be scaled up or adapted, and scientific knowledge can be applied to a middle range of community-based production. A variety of examples are given in E. F. Schumacher's *Small Is Beautiful*, such as a brick-making plant with one-fiftieth the capacity of standard plants, but a much higher productivity than can be achieved with handmade bricks.[3]

2. Labor Intensive

Most technologies today are laborsaving and capital intensive; the extreme case would be an automated factory or a highly mechanized farm. This is the wrong prescription for the Third World, in which labor is plentiful and capital scarce, and it is a mixed blessing in industrial countries plagued by unemployment. Centralized industry has accelerated migration to the cities and intensified the problems of urbanization around the world. Appropriate technology (AT) creates rural, village, and small-town jobs. It also is more likely to encourage the creativity and diverse skills that contribute to work satisfaction. It is, as Schumacher puts it, "technology with a human face."

3. Relatively Simple

The person using highly complex and sophisticated technology is dependent on experts for the production and repair of equipment. By contrast, AT is more understandable, less vulnerable to breakdown, and has less serious consequences when breakdowns do occur. Equipment can be locally adapted and repaired, with opportunities for diverse skills. This is a "self-help" or "people's" technology. Simplicity is a relative term, and in some cases reliable design depends on very sophisticated principles—as, for example, in the solar pump developed by French engineers at the University of Dakar in the Sahel and then widely introduced in Mexican villages.[4]

4. Environmentally Compatible

Whereas large-scale technology has a massive and often violent impact on the environment, soft technologies are gentler in their impact. They use local materials when possible, reducing the costs of transportation. To be sure, dispersed actions can severely damage fragile ecosystems, as overgrazing did in the Sahel, but they do not create the concentrated environmental assaults typical of large-scale technology. Again, AT design criteria include low energy use, minimal pollution, renewable resource use, and the integration of functions. All of these criteria are exemplified in the village-level digestors that convert animal, human, and agricultural wastes into methane gas for cooking and fertilizer for agriculture; by 1977 there were 4 million biogas digestors in use in China.[5]

5. Local Control

A major social goal of the AT movement is self-reliance and self-determination. With small-scale units there are opportunities for participation in decisions, rather than dependence on outside experts or huge organizations. Production for local markets is more likely to be directed to basic human needs. The use of local materials also encourages local autonomy. AT proponents have been concerned about social equality, rural development, and participatory democracy. Institutions facilitating such participation include producers' cooperatives, credit unions, village-level or municipal utilities, neighborhood associations, and small businesses.

Let us look more specifically at the opportunities for appropriate technology in *developing nations.* In an earlier chapter we noted some of the problems of technology transfer from rich to poor countries. One large factory can put hundreds of village shops out of business, accelerating unemployment and urban migration. Imported technology is expensive, adding to an already unfavorable balance of payments. It tends to strengthen the power of ruling elites within developing countries and perpetuates both economic and technical dependence on foreign experts, governments, and multinational corporations. Along with hardware, Western technology brings cultural values, attitudes, and patterns of life that have far-reaching and often disruptive social impacts. AT, by contrast, is less destructive of local cultures and more compatible with self-reliance in development. It receives much of its support from movements for national self-determination and liberation from foreign domination.[6]

Appropriate technology in *the Third World* has been promoted by a number of organizations. Some are international, like the London-based International Technology Development Group. There are regional technical institutes, government extension services, and information centers in India, Ghana, the Philippines, and other countries. There have been attempts to facilitate exchanges of information among local innovators and small businesses. From these various sources have come a variety of agricultural tools, small workshops, and miniplants. A community sugar-processing plant has been successfully established in the Philippines. Catchment tanks and plastic roof coverings have been used to collect rainwater in several countries, and small machines have been designed to make construction blocks for houses from local materials.[7] It has been estimated that, for the same cost, one large fertilizer plant (employing 1,000 people) or 26,000 village digestor units (employing 130,000 people) could be built in India. The fertilizer output in the two cases would be comparable, but the village units, instead of consuming energy, would be energy sources.[8]

There are several *distinctive problems* encountered by AT in developing countries. In the past, government policies usually have favored large industries and granted them subsidies or favorable tax or credit status. Political leaders have looked on high-visibility centralized plants as symbols of modernization. University scientists, often trained in the West and identifying themselves with urban elites, have been oriented toward large-scale technology. However, there has been considerable interest in AT among the faculty at a number of technical colleges, and among activist students concerned about human needs. They have insisted that AT does not involve second-rate science, nor a return to primitive methods, but rather the application of science and technology toward goals that differ from those of Western industrialism. Greater government support of regional research centers would help the development of low-cost technologies suitable for local conditions. There also has been new interest in AT among U.N. agencies.[9]

In *the United States* interest in AT is a product of environmental concern, disillusionment with industrial technology, the search for alternative life-styles, and commitment to participatory democracy and community development. There is a proliferation of informal AT activities, such as those displayed in the *Whole Earth Catalogue* and its successors. There are rural groups dedicated to AT, particularly in conjunction with organic farming and renewable energy sources. An example is the New Alchemy Institute, founded by John Todd on Cape Cod, which built integrated solar-powered greenhouse units combining fish farming with the growing of hydroponic vegetables.[10] An urban version is Community Technology, Inc., founded by Karl Hess in Washington, D.C., which promotes neighborhood self-help projects and small business enterprises. Its goals include community cooperation and self-reliance, as well as the development of local skills and sources of income.[11]

A sign of the increasing respectability of AT is its inclusion in the budgets of a number of *government agencies.* The National Science Foundation has provided funds for AT research and for small grants to local groups, and commissioned a survey of AT activities in the United States.[12] The Agency for International Development has allocated $20 million over a three-year period to AT in foreign aid. The Department of Energy budget assigns $10 million to AT. The Community Services Agency provided $3 million in 1977 to establish the National Center for Alternative Technology in Montana. The center is conducting research relevant to the food, housing, and energy needs of low-income families, and it makes grants to neighborhood groups for projects that will further community self-reliance and local control.[13]

Does AT sacrifice *efficiency* for the sake of other goals? In the U.S. context, does it represent a romantic and impractical longing for the simpler life of a bygone era? Some critics claim that only the high productivity of large-scale technology could support even the present world population at an acceptable standard of living. There are indeed many industrial processes, such as the production of steel or autos, in which economies of scale are so great that they should outweigh other considerations. But the alleged economies of scale often turn out to be illusory. We have seen that there are highly efficient small-scale energy sources. Small farms compare favorably with large ones in yields per acre and in output/capital ratios.[14] As costs of transportation and distribution rise, centralized factories for large market areas are relatively less advantageous. Often the choice of large-scale systems is determined by direct or indirect subsidies, or by the particular interests of corporations or bureaucracies.

Moreover, calculations of short-term efficiency neglect *long-term environmental and human costs* that eventually take their toll. Some of the broader social objectives in the choice of technology—such as the creation of rural jobs and the encouragement of local and national self-reliance—are receiving increasing recognition. Efforts have been made to strengthen small businesses and industries and to raise the technical level of the traditional sector. Some advanced countries have tried to improve productivity without losing cultural diversity. In Japan, for example, textile production often is subcontracted to small workshops that use indigenous designs. It appears that there are many situations in which there are diseconomies of scale, or in which the social costs of large scale outweigh any gains in efficiency.[15]

Decentralization Versus Centralization

Let us look more carefully at the relationship between technological systems and political systems. There is considerable evidence that centralized technologies contribute to *political centralization.* The Industrial Revolution brought not only capital-intensive machines but also hierarchical forms of organization and the demand for docile workers. The technostructure of the modern industrial corporation achieves a further concentration of economic power, managerial authority, and political influence. As the scale and complexity of a technology increase, so does the need for centralized planning and management. The vulnerability of systems also increases and the consequences of error or accident become more serious (as in power blackouts, supertanker oil spills, nuclear plant accidents—or the ultimate disaster, nuclear war). The complexity and risks of such systems become increasingly difficult to manage despite huge government bureaucracies created to plan and regulate them.[16]

Some centralized technologies share with nuclear power the requirement of *strict security.* Such security is difficult to guarantee in an age of social unrest. High-technology systems are highly vulnerable to disruption by sabotage or terrorism from alienated individuals and revolutionary political groups. Dissident students and disgruntled farmers prevented the opening of the $2.6 billion Tokyo airport for six years after it was completed, and in 1978 a small group destroyed the control tower while 15,000 police were guarding the airport.

Correspondingly, decentralized technology seems to encourage *decentralized politics.* Decisions can be made in a plurality of centers. Users as well as producers can have a direct voice. Appropriate technology, rather than imposing uniformity, is adaptable to local conditions and cultural differences. Such diversity is important ecologically—since it diversifies resource demands and environmental burdens—and also politically, since it facilitates political freedom and pluralism.[17] The stakes are lower in case of accidents, errors, or unintended consequences. On-site or community-level solar installations were cited earlier as an example of an adaptable, low-risk technology that would foster local self-reliance. Big systems can be managed only by hierarchical organization, but small ones can be more egalitarian.

But does decentralized technology in itself produce *economic and political decentralization*? Schumacher holds that appropriate technology will change the social system. Other authors maintain that technological and social patterns reinforce each other, so both must be changed at once. David Dickson gives considerable attention to the design of a decentralized social order.[18] The British ecological manifesto, *Blueprint for Survival,* advocates a pattern of self-supporting villages and small towns.[19] L. S. Stavrianos holds that the most hopeful sign today is the emergence of grass-roots participatory groups around the globe—from neighborhood community-development organizations in the United States to village producers' cooperatives in Tanzania.[20]

Advocates of *radical technology* have as their goal the total restructuring of society. In the tradition of the early utopian movements and philosophical anarchism, they have proposed autonomous rural communes and worker-run factories. Exponents of radical technology are as critical of the Soviet Union as of capitalist nations, claiming that both are obsessed with size, efficiency, and bureaucratic centralization. Paul Goodman, Peter Harper, and others see decentralized technology as a by-product of political decentralization.[21] Ivan Illich attacks almost everything big—industry, education, medicine—and rejects reliance on experts in any field. He wants to give people the tools to meet their own needs in housing, learning, health, and other areas through self-help communal associations and readily accessible paraprofessionals.[22]

Mainland China is an interesting combination of centralization and decentralization in both technology and politics. Mao initially followed the Soviet pattern of promoting large-scale, capital-intensive industry (in the first five-year plan of 1953, for instance). During the "great leap forward" of the late 1950s, some poorly conceived programs of decentralization were attempted, such as back-yard furnaces for iron production. Since the Cultural Revolution (1966-69), there has been a dual policy of "walking on two legs": centralized, high-quality, capital-intensive plants to produce industrial equipment and machine tools, and decentralized secondary industries to produce agricultural equipment and consumer goods (such as textiles, bicycles, and sewing machines). For the decentralized component, economic efficiency was considered less important than local participation and rural development. Industry was promoted primarily as a means of improving agriculture.[23]

China has been outstandingly successful in directing technology toward *basic human needs* in food, housing, and health. It feeds a quarter of the human race with only 7% of the world's cultivated land. Western observers report that hunger and starvation have been eliminated, and that every citizen is guaranteed basic levels of nutrition, shelter, medical care, education, and employment.[24] It is not surprising that the Chinese model of development has had great appeal in the Third World. However, industrialization has been very slow. Starting in 1978, Deng Xiaoping intiated a program of more rapid modernization, including the purchase of large-scale technologies from industrialized nations with which there previously had been virtually no contact. But the strong emphasis on decentralized technoligies is likely to continue, both for ideological reasons and from pragmatic necessity.

The political implications of the Cultural Revolution in China are more difficult to evaluate. There are many evidences of *political decentralization.* The early five-year plans had seen the emergence of a large bureaucracy and a party elite increasingly remote from the masses. In the mid-1960s Mao sharply attacked elitism and reliance on experts; managers, professionals, and teachers were told to work for varying periods in field and factories. Each commune or factory has considerable autonomy under the direction of an elected revolutionary committee. Each unit is supposed to be self-managing and to encourage grass-roots initiative and participation.

On the other hand, there is *centralized political control* of both policy and administration. Any fundamental criticism or deviation from the party line has been treated harshly. Leaders whose viewpoints were out of favor have been deposed and imprisoned or "reeducated," and dissident intellectuals have been silenced. Granted, the interaction between the party cadres and the masses is not simply in one direction. Ideas that are successfully tried out at the local level can have an influence on policy decisions at higher levels. Nevertheless, administrative directives from the top down set strict limits on the autonomy of local units. There is massive political indoctrination throughout the educational and production system. At the local level there is social regimentation and strong community pressure on individuals to conform. Insofar as this represents an ideal of "serving the people" by putting community welfare above personal gain, this may be ethically preferable to the prevailing Western individualism and competitiveness. But insofar as coercive measures and the repression of dissent are common, I believe the exercise of freedom by participation in decision making has been jeopardized. The decentralization of technology does indeed encourage local participation, but in a communist society the ultimate control is centralized in the party.

Let me summarize my own conclusions on centralization and decentralization in the United States. In Chapter 3 I tried to show that technologies are not socially neutral. In choosing types of technology we are not just choosing hardware; we are choosing *forms of social interaction* that influence the kind of society in which we will live.[25] In free-market countries, centralized technologies tend to increase the concentration of economic and political power; they are not readily accessible to public control and participation, and have given rise to a widespread sense of alienation and powerlessness. Decentralized technologies offer greater opportunities for participation, community control, local self-reliance, regional diversity, and a less hierarchical organization, all of which I consider desirable.

On the other hand, *centralization* sometimes offers advantages that may outweigh its disadvantages. In steel mills and oil refineries, for example, the economies of scale are very great; efficient, small units seem to be technically impossible.[26] The use of private autos is more decentralized than public transportation, but the cost in energy and pollution is too high. Environmental regulation has to be centralized because pollution crosses local jurisdictional lines. Similarly, resource allocation requires long-range comprehensive planning. However, even when policies are legislated nationally, the details of implementation can be left to state and local governments, with some latitude for diversity in the ways in which legislated goals and standards are met.

I myself would advocate *a mix of small and large technologies.* Diversity, including a range of sizes and alternatives, is valuable in itself. For some products, mass production is substantially more efficient, even after the main environmental and social costs are internalized. Centralized production may bring down the price of components that can be used locally (such as solar collectors). A device that is simple to operate may be complex to manufacture, such as photovoltaic cells. With many technologies, including those related to agriculture, community-level industries can make effective use of standardized, mass-produced parts. The relation of scale to efficiency and participation has to be examined for each portion of each technological process separately. There may be inescapable trade-offs among efficiency, environmental preservation, safety, and local control. Labor intensity may be incompatible with work satisfaction in particular tasks.

There also are some systems that by their nature cannot be decentralized. One of these is long-distance *transportation*, which must be planned and administered centrally. Because natural resources are not distributed uniformly, local and even regional self-sufficiency is not possible, and trade and exchange are essential. Again, *communication* can contribute to dynamic interaction and prevent isolation and stagnation. Sophisticated techniques of information exchange are neither energy nor resource intensive, and they must be centrally coordinated. They also can contribute to global awareness. Yet even electronic communications can be adapted to local control, such as citizens' band radios or community cable TV programs.

Such a mix of large and small technologies requires an *emphasis on the small* in our society. I have mentioned some of the forces that have favored the large: research funding, indirect subsidies, corporate interests, and the neglect of external costs in the marketplace. Even creative new possibilities, such as solar energy, tend to get absorbed into the prevailing structures of economic power. In developing countries, multinational corporations and domestic and international credit policies have exerted pressures toward the large. These imbalances can be corrected by deliberate national policies to foster appropriate technologies and improve their efficiency. Centralized political authority is necessary in many areas, including environmental preservation, resource allocation, and the regulation of corporate power. Such national policies can be instituted through democratic processes, and they can leave ample scope for local decisions. In sum, we should seek to reorient all technologies, large or small, toward human needs, ecological soundness, and democratic control. These goals are more readily achieved with small- or intermediate-scale technologies than with large-scale ones, and often with little, if any, sacrifice in economic efficiency.

II. INDIVIDUAL LIFE-STYLES

How can an individual in an affluent society live in accordance with the environmental and human values we have been discussing? A life-style is a pattern of daily life and work, a set of activities and relationships and ways of spending time and money. Every life-style expresses distinctive values. The dominant pattern in the contemporary United States may be characterized as a high-consumption life-style. In this section, some of the motives and actions in simpler and more personally fulfilling patterns of life are explored.

The Consumer Society

In previous chapters *the inequitable share* of the world's resources consumed by the United States has been repeatedly documented. Overall, a U.S. citizen is responsible for as much resource consumption and environmental pollution as at least 20 persons in India. 63% of the natural gas and 44% of the coal used worldwide each year is burned in the U.S. In a Harris poll 65% of the respondents agreed that it is morally wrong for this nation to be consuming a disproportionate share of the world's resources.[27]

There are powerful institutional pressures toward the *escalation of consumption.* Corporations promote increased consumption for the sake of their own growth and profits (return on investment). Mass production requires mass markets. Planned obsolescence, frequent style changes, and easy credit terms accelerate the turnover. In the United States each year $25 billion is spent on advertising, which comes to $115 per person—more than the total annual income of a quarter of the world's people.[28] Advertising promotes new products and stimulates demand for commodities that people did not know they wanted until they were told that they needed them. A barrage of TV commercials persuades us that deodorants, cigarettes, soft drinks, and breakfast cereals are vital to our well-being. The mass media hold before us the images of success of a consumer society. Personal identity and self-worth are defined by possessions, and happiness is identified with rising levels of consumption.

In theory, *needs* can be distinguished from *wants*. Needs are objective, universal requirements arising from the biological and social nature of human beings. Wants are subjective, individual desires influenced by social conditioning and personal values. Maslow holds that lower-level needs (survival and security) are finite; once these are satisfied, there are limitless possibilities for personal growth at higher levels.[29] In a consumer society, however, even self-actualization and the esteem of others are sought through commodities that compete for the limited resources that also satisfy biological needs. The insatiable wants of a consumer society create an ever-expanding demand for resources, which the richest countries have the greatest power to secure.

William Leiss has argued that our society encourages people to seek through *consumption* the satisfaction of *all their diverse needs*. We know, of course, that overeating often is an attempt to use food as a substitute for self-acceptance and acceptance by others. Similarly, we use other commodities to satisfy needs that could be met in less resource-consumptive ways. We misinterpret our real needs when we think they can be fulfilled by increased consumption. Leiss suggests that we must reorient our behavior toward less resource-consumptive sources of satisfaction, such as productive activity, human relationships, shared decision making, and the enjoyment of nature.[30] U.S. consumption patterns are not only excessive relative to the basic needs of other people, but they also reflect a misunderstanding of the character of genuine human fulfillment. In the long run we damage ourselves as well as other people and the environment by the pursuit of material affluence.

Thus one of the motives for a simpler life-style is *environmental preservation*, since resource extraction usually involves environmental damage, and resource use usually adds to pollution. But the main values at stake are the three that we have traced throughout this volume:

Social justice is violated by American overconsumption. A simpler life-style can be a personal act of solidarity with the world's poor, an acknowledgment of past exploitation and of the injustice of present power relationships. It can be an expression of individual commitment to a more equitable distribution of wealth and allocation of resources. We have traced some of the connections between overconsumption in the United States and underdevelopment in other countries. But along with the reduction of one's own consumption there must be political action if the resources saved are to benefit others, and a more just international order is to be created.

Individual freedom is asserted in taking greater control of one's own life, and in the refusal to be manipulated by the "hidden persuaders" and subtle compulsions of a consumer society. What can one person do? Aren't we helpless facing the massive institutions of industrial society? Each individual can at least begin with his or her own life. We can seek greater consistency between our beliefs and our actions. Alternative life-styles are themselves a means of consciousness raising. As examples of what is possible, they are a contribution to social change in the future, as well as to personal intergrity in the present. Both the exercise of purchasing power within existing institutions and the creation of alternative institutions can help redirect production toward the satisfaction of genuine human needs for all—rather than artifically generated wants and luxuries for the few.[31]

Personal fulfillment is perhaps the central consideration. New life-styles arise not only from a rejection of prevailing patterns but also from a positive vision of human fulfillment and a redefinition of the good life. A person's life can be built around the intrinsic satisfaction of activities and relationships rather than the extrinsic rewards of possessions. As Erich Fromm puts it, happiness comes from "being" rather than from "having."[32] One can seek significance in work itself, apart from earning income for consumption. Priority can be given to persons rather than things. Such a life-style is less competitive and more cooperative, less hectic and more serene, and provides opportunity for personal growth and exploration of the inner life.[33] It acknowledges the spiritual emptiness of a materialistic culture. Here is a response not simply to an environmental and resource crisis but also to the modern crisis of meaning. A simpler life frees time and resources for the human relationships that are most significant. Both poverty and affluence, it seems, may be inimical to personal fulfillment.

Toward Simpler Life-styles

What can a person do—as an individual or as a member of a family, group, or community—to live in harmony with global needs, personal fulfillment, and the earth's capacity to sustain us? Here are seven aspects of a simpler life-style today.

1. Resource Conservation

Many of the methods of energy conservation presented in Chapter 10 are open to individuals, quite apart from any national policy changes. Each of us can use smaller cars, car pools, bicycles, or public transportation. Home energy use can be reduced by insulation, solar water heating, and the elimination of air conditioning. Appliances can be judged by their energy consumption and by life-cycle costs (which allow for durability and operating costs as well as initial cost). Similarly, the individual can cut down on resource waste (including throwaway containers and excessive packaging) and recycle paper, glass, and aluminum. These are modest measures that entail no great sacrifices. A more radical reduction in consumption would involve getting along with a smaller array of household appliances (or sharing them with others), using fewer consumer products, and making more things for oneself.[34]

2. Food for Health—and for Justice

Over 70 million U.S. citizens are overweight. We have the diseases of overeating (such as heart disease), while other countries have the diseases of malnutrition. For the sake of health we should reduce our food consumption, particularly of fats and sugars. We should eat more natural foods in place of the overprocessed, overpackaged, and artifically flavored commercial products that fill the supermarkets. For the sake of justice we should reduce our use of such nonfood crops as coffee, tea, tobacco, and grains for alcohol, which tie up land in developing countries that could be used for food crops. In an earlier chapter I indicated that 11½ lbs. of grain are used to produce 1 lb. of beef; I urged a reduction in meat consumption or a vegetarian diet. Eating "lower on the food chain" can provide adequate nutrition with less waste and with less suffering to animals.[35] Here, clearly, individual action and national policy can support each other.

3. Significant Work

Work can indeed be a personally meaningful activity and a source of self-respect. Participation in work-related decisions and opportunity for creativity can enhance job satisfaction. So can the conviction that one is contributing to a socially useful product. Even boring work acquires significance if it is part of a larger purpose to which one is committed. Industrial societies have sharply separated production from consumption, isolating work from other activities. The dichotomy of work versus leisure is minimized when there is production for use rather than for wages (by growing a garden or building or repairing a house, for instance). Community-based industries and cooperatives offer opportunities for work that are related to other aspects of the community's life.

4. Community Self-Help

A variety of participatory self-help projects have been initiated to meet basic needs, especially among low-income families. In "urban homesteading," families can renovate old houses and apply their own labor toward purchase of the houses. Community gardening can provide better nutrition and lower food bills, as well as exercise and recreation. Community health-care programs stress preventive medicine, nutrition, family planning services, and individual responsibility for health. Through physician's assistants, paramedical workers, and public health nurses, access to health care can be more widespread. Decentralized energy sources amenable to community control have been mentioned already. Some of these community efforts have failed from poor organization, narrowly defined interests, or opposition from established economic or political institutions. But many have succeeded, and have brought social as well as economic benefits to local groups. Government support in helping to finance such community projects seems to be more effective in encouraging individual responsibility than the funding of large government projects planned by experts.[36]

5. Alternative Institutions

A variety of decentralized institutions in which people have greater control of their own lives has sprung up in recent years. There are producer cooperatives, worker-owned businesses, printing and craft collectives, and communal farms. On the consumption side, grocery and food co-ops are common. Neighborhood clinics and alternative schools usually are autonomous and self-managed. These local groups are in touch with each other through information networks and informal publications without any overall organization. They typically combine personal participation, shared decision making, and wider political action. Such small-scale institutions provide viable patterns of economic activity under local control, with some degree of independence from the larger structures of economic power.[37]

6. Intentional Community

A simpler life-style can be facilitated by sharing living arrangements, whether in extended families, ad hoc households, cooperative houses, or larger communes or communities. Some are short-lived and experimental, while others, such as the kibbutzim in Israel, have continued for decades. Some have a common religious commitment, for example, the Taizé Brothers, the Bruderhof communities, and Koinonia Farms.[38] Others, such as Findhorn and Lindesfarne, are united by a common set of ideals.[39] In these diverse communal ventures, resource use is low because life-styles are simple, and work and meals are shared. We noted the greatly reduced energy use per capita in urban communes where appliances, vehicles, and living space are shared. Rural communes often have tried to integrate work and leisure, and to live in harmony with nature. But above all, intentional communities permit an openness in personal interaction that is seldom possible in large impersonal organizations. Here wholeness and cooperation can replace fragmentation and competition. There may be a loss in privacy, but there is a gain in human relationships and a supportive community of caring and sharing.[40]

7. Political Involvement

Individual life-styles and political action can be mutually reinforcing. Reducing meat consumption conserves grain, but apart from national policies this does not benefit the malnourished. The gap between rich and poor nations is a product not simply of individual overconsumption, but also of national policies and the distribution of economic power. By both word and deed, each person can influence public attitudes and help to create new constituencies that can act through the political process. Political representatives do respond if enough citizens are concerned. In response to new attitudes today, there could be initiatives for national resource conservation and for changes in the international order toward a more equitable distribution of the planet's resources.

III. VISIONS OF THE FUTURE

In earlier chapters I suggested that we have made considerable progress in national policies to preserve the environment, but have barely begun to take resource sustainability and global justice seriously. There is much greater resistance to conservation measures than to pollution control, for example. We have not yet awakened from the American dream of limitless national progress and ever-growing consumption. The enormous disparities between rich and poor countries continue to grow, reinforcing existing patterns of domination and dependence. What are the prospects for reordering national priorities in a world of limited resources? How might a long-term global outlook be achieved? What are the sources of change that might offer some hope for the future of humanity on Spaceship Earth?

Justice, Participation, and Sustainability

Three values that have been traced through this volume are crucial for a global perspective on scarce resources. *Justice* is the most urgent demand in a world of radical inequalities. Global resources are indeed sufficient to meet the basic needs of all people—if they are wisely used and more equitably distributed. We have seen that in energy, food, and raw materials, issues of distribution are crucial. Selective growth in industrial countries entails a shift toward human services and resource-conserving technologies. Sufficiency for all must come before luxuries for the few. Justice also requires that development strategies be aimed at basic human needs. I noted that a more equitable distribution of the benefits of development also seems to contribute significantly to lowered birth rates.

Participation in decision making, I have suggested, is the most important form of freedom today. Such freedom is compatible with the extensive powers of public regulation that are needed for the protection of the environmental "commons" and for more just allocation of limited resources. Earlier I defended citizen participation in decisions about technology, as against dependence on experts. Policy in many critical areas has to be centralized, but the decisions can be made through democratic processes. I also have urged a greater emphasis on technological decentralization and on appropriate technologies that are adaptable to local control. Community-level organizations provide a path between individualism and collectivism. The task here is to increase awareness of local-global connections, and to motivate participatory local action in a global context.

Sustainability entails the conservation of energy and materials and slower depletion of nonrenewable resources. It also requires a shift to renewable resources and preservation of the environments on which such resources depend. Soil erosion, deforestation, overgrazing, and overfishing are worldwide problems. The goal of sustainability includes population stabilization in developing countries and the leveling off of resource-intensive production and consumption in industrial nations. Sustainability implies a long time-frame and the obligations to future generations that were discussed in Chapter 5.

While a greater emphasis on these three values could occur within the framework of prevailing cultural assumptions, they are unlikely to be given high priority apart from a broader change in assumptions and perceptions, which we may call a *paradigm shift.* In *The Structure of Scientific Revolutions*, Thomas Kuhn uses the term "paradigm" to refer to "standard examples of scientific work which embody a set of conceptual, methodological and metaphysical assumptions." Paradigms are conceptual models that define the kinds of entities there are in the world and the kinds of questions that can be legitimately asked. When a paradigm shift occurs—from Newtonian to relativistic physics, for example—the world is seen in a different way. There are perceptual changes and categories of interpretation based on new assumptions. Kuhn calls these changes "revolutions" because the choice of paradigms cannot be made by the rules of normal scientific research.[41]

Social scientists have used the term in an even broader sense. A *dominant social paradigm* is "the collection of norms, beliefs, values and habits that form the world-view most commonly held within a culture and transmitted from generation to generation by social institutions." It is "a mental image of social reality that guides behavior and expectations."[42] Willis Harman identifies the following components of *the industrial era paradigm*: expectation of unlimited material progress and ever-growing consumption; faith in science and technology to solve all problems; goals of efficiency, growth, and productivity; mastery of nature; and competition and individualism. Harman holds that this paradigm has led to environmental degradation, resource depletion, loss of meaningful work roles, inequitable distribution, and ineffective control of technology.[43]

Harman outlines a *postindustrial paradigm* that he thinks is beginning to emerge as a basis for a sustainable society: material sufficiency in the satisfaction of basic needs; frugality in resource use and transition to renewable resources; ecological ethics and stewardship of nature; goals of human development, self-realization, and growth in awareness and creativity; and cooperation and community solidarity, in place of competition and individualism. Other writers have described additional components in the shift in values and assumptions that they see occurring: from individual property rights to shared benefits (public health, quality of life, survival), from short-term to long-term and from national to global issues.[44]

The prospects for the emergence of *long-term global goals* is the subject of a report to the Club of Rome, *Goals for Mankind.* The report surveys the goals of nations around the world and finds them to be predominantly short term and narrowly national, based on mistaken expectations that rapid economic growth, industrialization, and defense expenditures will be in their own interests. Long-term global goals, by contrast, would include resource conservation, selective growth, and arms reduction and control. The report concludes that a new world order will require a new global ethics, a recognition of interdependence, and a sense of global solidarity. The authors hold that religious and intellectual leaders already have a more global viewpoint than either government or business leaders, and they could develop grass-roots support for a "solidarity revolution." The world religions could be a major force for globalism if each emphasized the universalistic and future-oriented elements in its own tradition.[45]

The Biblical Perspective

What contribution can the Western religious traditions make to a post-industrial paradigm for a sustainable global society? As indicated in Chapter 4, the U.S. churches usually have supported the status quo, though they also have contributed to social change (such as the civil rights movement and protests against the Vietnam War). The churches themselves will have to change drastically if they are to facilitate the transition to a sustainable world. A new religious awakening could be a strong force in social transformation. Biblical images still have a latent power to evoke response.[46] Consider, in particular, the message of the prophets of ancient Israel, who, like us, lived in times of national crisis and international conflict.

The prophets' commitment to *justice* was rooted in a belief in the fundamental equality of all persons before God. Speaking in the name of a God of justice, Amos denounced the inequalities of his day:

> For three transgressions of Israel, and for four, I will not revoke the punishment; because they sell the righteous for silver, and the needy for a pair of shoes–they that trample the head of the poor into the dust of the earth, and turn aside the way of the afflicted . . . But let justice roll down like waters, and righteousness like an everflowing stream.[47]

This same commitment to social justice is evident in recent statements from the Roman Catholic Church and the World Council of Churches (WCC) concerning a more equitable distribution of global resources. Here is one sentence from an address by the biologist, Charles Birch, at the last WCC assembly: "The rich must live more simply that the poor may simply live."[48] Today, as in ancient Israel, the sharing of resources is a demand of justice, not an act of charity.

The prophetic view of *a created order* that is inclusive in space and time also is relevant today. The whole creation is part of God's purpose. Because all forms of life are within His plan, we are accountable for the way we treat them. I suggested in Chapter 2 that the idea of humanity's stewardship of nature is more typical of the biblical viewpoint than the idea of dominion over nature that later was emphasized. Moreover, the prophets used an extended time-scale because they believed that God's purposes extend into the future. We have obligations to posterity and to a God who spans the generations. There is a solidarity in time, a covenant "from generation to generation." The idea of creation is a great unifying framework, encompassing all forms of life and all time from past to future. The theme of sustainability is prominent in recent WCC documents.[49]

Next, *a broad view of human fulfillment* is expressed in the prophetic literature. The good life is identified not with material possessions but with personal existence in community. The prophets asserted the dignity of the individual and upheld the importance of interpersonal relationships. They portrayed harmony with God and neighbor as the goal of life. They recognized the dangers of both poverty and affluence. They saw the harmful consequences of affluence—for the rich as well as for the poor. Jesus, in turn, stressed the importance of feeding the hungry, but he also said that "man does not live by bread alone," and he vividly pictured the dangers of wealth. The earliest Christian community, as described in Acts, "had all things in common." Distribution was made to each "as any had need."[50] Over the ensuing centuries the monastic orders preserved the ideals of simplicity and community. The Reformation and then the Puritan movement upheld frugality and simplicity, and were critical of "the luxuries of the rich."

Today, in an overconsumptive society, we need both this negative attack on materialism and the positive witness to the priority of the personal and the quality of the life of the community. New life-styles arise not only from a concern for global justice but also from a new vision of the good life, a focus on sources of satisfaction that are not resource consumptive. Here religious faith can speak to the crisis of meaning that underlies the pursuit of affluence. One example is the book *Enough is Enough* by the British theologian John Taylor, who urges restraint in consumption, a level of "material sufficiency" that is neither affluence nor poverty.[51] A U.S. group has expressed in the Shakertown Pledge a religious commitment to such goals as creative simplicity, ecological soundness, occupational accountability, and global justice.[52] I believe that the biblical vision of human fulfillment can strengthen the search for simpler life-styles today.

Finally, the Hebrew prophets brought a double message of *judgment and hope* that has particular contemporary relevance. On the one hand, they spoke of God's *judgment* on the structures of human greed. They even saw military defeat and national catastrophe as forms of divine judgment on the materialism, idolatry, and injustice of national life. The prophets were realistic about human sinfulness and aware of the dangers when economic and political power are concentrated in the hands of any group or nation. Their first word was a call to repentance and humility. Today such humility might be an antidote to the Promethean pride to which industrial nations are prone. A greater awareness of limits might help us to recover a sense of the sacred. Obviously the prophets said nothing about technological centralization, but I suspect that if we accepted their view of human nature we would be hesitant to rely too heavily on large-scale systems that are vulnerable to human frailty and the abuse of institutional power. Perhaps for us, too, catastrophe will be a form of judgment on an unrepentant nation.

The other side of the prophetic message is *hope.* Beyond judgment and repentance there is reconciliation and redemption. Reconciliation is restoration of wholeness, the overcoming of alienation from God, from nature, and from other persons. Redemption is creative renewal in response to God's redemptive activity. The ultimate symbol of hope is the vision of a future kingdom of peace and brotherhood. In Micah's words:

> They shall beat their swords into plowshares and their spears into pruning hooks; nation shall not lift up sword against nation, neither shall they learn war any more; but they shall sit every man under his vine and under his fig tree, and none shall make them afraid; for the mouth of the Lord of hosts has spoken.[53]

The prophetic imagination pictured a future harmony that would include all humankind and all nature. All people would be at peace with their neighbors and with the created order. There is a tension between particularism and universalism in biblical religion, as in all religions, but in the final vision universalism is dominant. The image of the kingdom gathers up the themes of justice, creation, and human fulfillment.

The idea of *the kingdom* took many forms in subsequent history. Some members of the early church expected it to come very soon on earth. Others visualized the kingdom as another world, a heavenly realm that is unrelated to this world. More commonly it has been understood both as the goal of history and as beyond history. The kingdom is indeed an imaginative vision, but it is not just an idle dream. As do all visions of the future, it influences the way we interpret the present. It leads us to see the world in a new way. We see both judgment and renewal in history, and we act in response to them. The recovery of these biblical themes, I suggest, could contribute significantly to a more just and sustainable society.

Sources of Change

There are some *hopeful signs* for the future. There is widespread recognition that industrial technology exacts a heavy toll in environmental and human damage. There are the stirrings of an equality revolution around the world, evident in the growing demands for equal rights for women and minorities. There are strong movements for national self-determination and liberation from dependency. There are shifts in the balance of power as resources become scarcer and industrial nations more vulnerable. There is some recognition of interdependence and of a common interest in a peaceful and stable world. There are people in every land who know that inequalities are unjust and explosive, that the arms race leads to national insecurity rather than national security, and that cooperation among nations promises greater benefits than confrontation.

But there are almost overwhelming *obstacles to change.* We should not underestimate the strength of individual or institutional greed, or the economic power of corporations and governments. The late 1970s have seen a resurgence of neoconservatism and isolationism in the United States, with groups campaigning to get us out of the U.N. and to spend more on weapons. Rich nations have become "resource addicts," hooked on consumerism. Poor nations harbor great resentment of past exploitation and increasing global disparities. It is a world of political instability and potential for violence. We already have gone so far down unjust and unsustainable paths that we cannot expect to avoid conflict and suffering. Changes in power structures will not be achieved without struggle, upheaval, and in some cases revolution.

My hope is that the inadequacy of old patterns plus the vision of positive alternatives will produce major changes without widespread violence. If we do not accept these changes voluntarily, they will be forced on us by environmental constraints or revolutionary movements—and at a very high cost in social disruption and human suffering. In their day, the Hebrew prophets were convinced that *catastrophe* would occur if their nation persisted in its ways. Perhaps we, too, can only prepare foundations for a new order beyond the catastrophe, or plant seeds that will grow later. I am hopeful but not optimistic. Hope refers to what is possible; optimism refers to expectation or probability. Despair is a self-fulfilling prophecy when catastrophe is accepted as inevitable, whereas hope provides motivation for an uphill struggle.

Hope arises from visions of the future and convictions about the possibility of change, but it also rests on *faith.* Throughout this volume I have expressed a political faith in renewed democratic participation, despite the obstacles of institutional power and the fragmentation of the electorate. I also have expressed a philosophical faith (influenced, no doubt, by my background as a scientist) that there are reasonable solutions that people will accept, despite human ignorance and greed. Ultimately, my hope derives from a religious faith in the power of reconciliation to overcome alienation—specifically, a biblical conviction that love and forgiveness can transform self-centeredness and isolation. In each case, faith can be realistic about the present and yet remain open to creative new potentialities.

What is needed, then, is not moral exhortation or appeals to altruism, but *individual and institutional change.* A new image of human fulfillment, based on alternative sources of satisfaction, must be embodied in specific examples of personally fulfilling life-styles. There are opportunities through existing institutions to redirect technology to basic human needs, but changes in economic and political institutions also will be required. Only in a new international economic order (Chapter 11) can historic patterns of domination and dependence be broken. I see four main sources of such individual and institutional change today.

First, *education* is a far-reaching instrument of cultural change. The schools have begun to spread environmental and global awareness, but they could do much more. New curriculum materials and class projects can be developed around environmental and resource issues. At both school and college levels, interdisciplinary courses can bring the insights of several fields to bear on a common problem or policy question.[54] However, the education of a new generation is a slow process, and teachers only gradually depart from what they themselves were taught. Adult education can be more rapid. Citizens' movements and voluntary community groups can provide information and help to raise the consciousness of the public. It was only seven years from Rachel Carson's *Silent Spring* to the National Environmental Policy Act. The media have reported on environmental and technological controversies, but could present more extensive analysis. The churches also have a considerable potential for fostering responsible stewardship. Education for an age of resource scarcities must deal with facts and values, information and attitudes, cognitive processes and affective responses.

Second, *crisis and disaster* can speed the process by challenging prevailing assumptions. Perhaps oil spills, gasoline shortages, electricity blackouts, and widespread famines will have to occur on a more massive scale before people will wake up. Taken alone, however, crises may lead to undesirable changes. Governments may rely on technical fixes whose indirect costs turn out to be high. When these fail the reaction may be pessimism and despair. But in combination with the vision of positive alternatives, crises can be catalysts for constructive action. In emergencies such as war or natural disaster, people will make sacrifices for the common good—provided the sacrifices are shared, the common good is clear, and a more hopeful future is envisaged. There is hope if along with resource crises there is new respect for the earth, new dedication to justice, and new visions of the good life.

Third, *political power* can greatly accelerate processes of social change. Sometimes a coalition formed around common or overlapping interests can become a new force for political change within a nation. I mentioned earlier the potential strength of joint action by environmental, consumer, and labor groups. In the international arena, I have noted the new bargaining power that is being acquired by raw material producers and by nations gaining access to nuclear weapons. Both domestic and international unrest in a world of scarcities and inflation will be potent sources of change. But again, the changes may be destructive. Domestic pressures for reform could lead to reaction and repression. Pressures from abroad could lead the United States to isolationism and a fortress mentality, an extension of "lifeboat ethics." The limited gains from the civil rights movement of the 1960s would not have occurred without political pressures from the black community; but without at least some national dedication to justice, black power probably would have simply evoked further repression and violence. So, too, with global resources; only a more inclusive vision can direct political power so that conflict becomes a step toward a just and peaceful world rather than a step away from it.

Fourth, *new visions* can provide the motivation and direction for creative social change. Moral exhortation seldom inspires action among those who are reluctant to change. Visions, on the other hand, present positive alternatives in an imaginative way. They summarize a set of values, using concrete images rather than abstract principles. Visions of alternative futures offer hope instead of despair, a sense of the possible rather than resignation to the inevitable. Most movements of social reform started from utopian imagination, new images of the good life, vivid portrayals of what might be.[55] It is the combination of education, crisis, power, and vision that might bring about a paradigm shift and institutional change.

I have mentioned some contributions of the biblical tradition to a *global vision.* Other religious traditions also will have distinctive contributions to make. Communications today enable people to be aware of what is happening in the "global village" in a way that was not possible in previous history. There is not only a revolution of rising expectations; there is a new demand for more just distribution. Events dramatize our interdependence. Worldwide economic crises or rises in oil prices leave no nation untouched. Global awareness will have to come from many sources.

As a symbol of such awareness, think of the pictures of the earth that were taken by astronauts on the moon. For the first time, the earth actually could be seen as a single unit. There it is, a spinning globe of incredible richness and beauty, a blue and white gem among the barren planets. It has been proposed that we should think of it as *Spaceship Earth.*[56] The earth is a fragile life-support system. Like a spaceship, it has limited resources that must be conserved and recycled. Its inhabitants are interdependent, sharing a mutual responsibility and a common destiny. This is a striking image that forcefully represents the importance of life support and cooperation for human survival.

But we must extend the spaceship image if it is not to mislead us. A spaceship is a mechanical, man-made environment, devoid of life except for human beings. *Planet earth*, however, is enveloped in a marvelous web of life, a natural environment of which humanity is a part and on which it is dependent. We must think not just of life support but of ecological sustainability on a long time-scale. So, too, the social order on a spaceship is relatively simple, with only a few people interacting in highly structured ways. On planet earth there are complex relationships between groups and nations, and there are crucial issues of distributive justice and participatory freedom in the allocation of scarce resources. If freedom means participation in decision making, then globalism must be combined with localism and decentralization. As I see it, we must move toward more globalism, and more localism, but less nationalism.

Let us keep before us that image of the spinning globe, but let us imagine its natural environments and its social order. It is still possible to achieve a more just, participatory, and sustainable society on planet earth.

NOTES

1. H. S. D. Cole et al., *Models of Doom* (New York: Universe Books, 1973), p. 11. See also Nicholas Wade, *Science* 189 (1975): 770. Twenty-eight industrial nations account for 97% of world R & D expenditures, New York *Times*, September 2, 1979, p. 11.

2. R. J. Congdon, ed., *Introduction to Appropriate Technology* (Emmaus, Pa.: Rodale Press, 1977); Robin Clarke, *Building for Self-Sufficiency* (New York: Universe, 1977); and Norman Colin, *Soft Technologies, Hard Choices* (Washington, D.C.: Worldwatch Institute, 1978). The October-December 1973 issue of *Impact of Science on Society* (UNESCO) is devoted to appropriate technology.

3. E. F. Schumacher, *Small is Beautiful* (New York: Harper & Row, 1973); idem, *Good Work* (New York: Harper & Row, 1979); and Richard Dorf and Yvonne Hunter, eds., *Appropriate Visions* (San Francisco: Boyd & Fraser, 1978).

4. Nicolas Jequier, *Appropriate Technology: Problems and Promises* (Paris: OECD, 1976), chap. 2; and editors of RAIN, *Rainbook: Resources for Appropriate Technology* (New York: Schocken, 1977).

5. New China News Agency, May 1977, cited in Bruce Stokes, *Local Responses to Global Problems* (Washington, D.C.: Worldwatch Institute, 1978).

6. Denis Goulet, *The Uncertain Promise: Value Conflicts in Technology Transfer* (Washington, D.C.: Overseas Development Council, 1977), chaps. 6 and 7.

7. Examples from Jequier, op. cit.; see also Richard Eckaus, *Appropriate Technology for Developing Countries* (Washington, D.C.: National Academy of Sciences, 1977).

8. A. K. N. Reddy, "The Trojan Horse," *Ceres* (March-April 1976): 40-43; and idem, "Alternative Technology: A Viewpoint from India," *Social Studies of Science* 5 (1975): 331-42.

9. Jequier, op. cit., chap. 4; Frances Stewart, *Technology and Underdevelopment* (Boulder, Col.: Westview, 1977), chap. 4; and International Labor Office, *Employment, Growth and Basic Needs: A One World Problem* (New York: Praeger, 1977), chap. 9.

10. Nancy and Jack Todd, eds., *The Book of the New Alchemists* (New York: E. P. Dutton, 1977); N. Wade, "New Alchemy Institute," *Science* 187 (1975): 727-29; John and Nancy Todd, *Tomorrow is our Permanent Address* (New York: Harper & Row, 1979); and Richard Merrill, ed., *Radical Agriculture* (New York: Harper & Row, 1976).

11. N. Wade, "Karl Hess: Technology with a Human Face," *Science* 187 (1975): 332-34; and Karl Hess, *Community Technology* (New York: Harper & Row, 1979).

12. Integrative Design Associates, *Appropriate Technology in the U.S.: An Exploratory Study* (Washington, D.C.: National Science Foundation, 1977).

13. C. Holden, "NCAT: Appropriate Technology with a Mission," *Science* 195 (1977): 857-59.

14. Stokes, op. cit., pp. 23-28.

15. Leopold Kohr, *The Overdeveloped Nations: The Diseconomies of Scale* (New York: Schocken, 1978).

16. Rufus Miles, *Awakening from the American Dream* (New York: Universe, 1976).

17. David Livingston, "Little Science Policy: A Research Agenda for the Study of Appropriate Technology and Decentralization," in *Science and Technology Policy*, ed. Joseph Haberer (Boston: D. C. Heath, 1977); and idem, "Global Equilibrium and the Decentralized Community," *Ekistics* (September 1976): 173-76.

18. David Dickson, *The Politics of Alternative Technology* (New York: Universe, 1977).

19. Edward Goldsmith et al., *Blueprint for Survival* (Boston: Houghton Mifflin, 1972).

20. L. S. Stavrianos, *The Promise of the Coming Dark Age* (San Francisco: W. H. Freeman, 1976).

21. Paul Goodman, "Can Technology Be Humane?," in idem, *New Reformation* (New York: Random House, 1969); Godfrey Boyle and Peter Harper, eds., *Radical Technology* (New York: Random House, 1976); and Murray Bookchin, "Towards a Liberatory Technology," in idem, *Post-Scarcity Anarchism* (San Francisco: Ramparts Press, 1971).

22. Ivan Illich, *Tools for Conviviality* (New York: Harper & Row, 1973).

23. Stavrianos, op. cit., chaps. 5 and 6; Jon Sigurdson, "The Suitability of Technology in Contemporary China," *Impact of Science on Society* 23 (1973): 341-52; C. H. G. Oldham, "Science and Technology Policies," in *China's Development Experience*, ed. Michael Oksenberg (New York: Academy of Political Science, 1973); and Jon Sigurdson, *Rural Industrialization in China* (Cambridge, Mass.: Harvard University Press, 1977).

24. Paul T. K. Lin, "Development Guided by Values: Comments on China's Road and Its Implications," in *On the Creation of a Just World Order*, ed. Saul Mendlovitz (New York: Free Press, 1975); and John Gurley, "Rural Development in China, 1949-72, and the Lessons to Be Learned from It," *World Development* (July-August 1975).

25. Langdon Winner, "The Political Philosophy of Alternative Technology," *Technology in Society* 1 (1979): 75-86.

26. An AAAS symposium on large and small technologies is summarized in *Bull. Amer. Acad. of Arts and Sciences* (February 1977): 4-14.

27. Louis Harris poll, September 16, 1975. See Ervin Laszlo, ed., *Goals for Mankind* (New York: E. P. Dutton, 1977), p. 39.

28. Laszlo, op. cit. On the effects of the media, see Herbert Marcuse, *One Dimensional Man* (Boston: Beacon Press, 1964), chap. 1.

29. On Maslow, see references in Chapter 4.

30. William Leiss, *The Limits of Satisfaction* (Toronto: University of Toronto Press, 1976).

31. Simple Living Collective, American Friends Service Committee, *Taking Charge: Personal and Political Change through Simple Living* (New York: Bantam, 1977).

32. Erich Fromm, *To Have or to Be?* (New York: Harper & Row, 1976).

33. Duane Elgin and Arnold Mitchell, "Voluntary Simplicity: Life-style of the Future?," *Futurist* (August 1977): 200-07.

34. Center for Science in the Public Interest, *99 Ways to a Simple Lifestyle* (Garden City, N.Y.: Doubleday, 1977); Albert J. Fritsch, *The Contrasumers: A Citizen's Guide to Resource Conservation* (New York: Praeger, 1974); Jim Leckie et al., *Other Homes and Garbage: Designs for Self-Sufficient Living* (San Francisco: Sierra Club, 1975); and Kimon Valaskakis et al., *The Conserver Society* (New York: Harper & Row, 1979).

35. Francis Moore Lappé, *Diet for a Small Planet*, rev. ed. (New York: Ballantine, 1975).

36. Stokes, op. cit.; and David Morris and Karl Hess, *Neighborhood Power: The New Localism* (Boston: Beacon Press, 1975).

37. See Simple Living Collective, op. cit., chap. 1; also Susan Gowan et al., *Moving Toward a New Society* (Philadelphia: New Society Press, 1976); and Hugh Nash, ed., *Progress As If Survival Mattered* (San Francisco: Friends of the Earth, 1977).

38. John Taylor, *Enough is Enough* (Minneapolis: Augsburg, 1977), chap. 5.

39. William Irwin Thompson, *Passages About Earth* (New York: Harper & Row, 1974), chap. 7.

40. Rosabeth Kantor, *Communes: Social Organization of the Collective Life* (New York: Harper & Row, 1973).

41. Thomas S. Kuhn, *The Structure of Scientific Revolutions*, 2d ed., (Chicago: University of Chicago Press, 1970). See Ian G. Barbour, *Myths, Models and Paradigms* (New York: Harper & Row, 1974), chap. 6.

42. Dennis Pirages and Paul Ehrlich, *Ark II: Social Responses to Environmental Imperatives* (San Francisco: W. H. Freeman, 1974).

43. Willis Harman, *An Incomplete Guide to the Future* (Stanford, Calif.: Stanford University Press, 1976); and idem, "Changing Society to Cope with Scarcity," *Technology Review* (June 1975): 29-35.

44. Kan Chen et al., *Growth Policy* (Ann Arbor: University of Michigan Press, 1974); George Cabot Lodge, *The New American Ideology* (New York: Knopf, 1975); and Ronald Inglehart, *The Silent Revolution* (Princeton, N.J.: Princeton University Press, 1977).

45. Laszlo, op. cit.; and idem, *Inner Limits of Mankind* (New York: Pergamon, 1979).

46. Frederick Ferré, *Shaping the Future* (New York: Harper & Row, 1976), chap. 7.

47. Amos 2:6 and 5:24.

48. Charles Birch, "Creation, Technology and Human Survival," *Ecumenical Review* (January 1976). See also Paul Abrecht, ed., *Faith, Science and the Future* (Philadelphia: Fortress Press, 1979).

49. "Science and Technology for Human Development," *Anticipation* no. 19 (November 1974); Report on Nairobi WCC Assembly, *Anticipation* no. 22 (May 1976); Ian G. Barbour, "Justice, Participation, and Sustainability at MIT," *Ecumenical Review* 31, no. 4 (1979): 380-87; and Roger Shinn, ed., *Faith and Science in an Unjust World*, 2 vols. (Philadelphia: Fortress Press, 1980).

50. Acts 2:44 and 4:35.

51. Taylor, op. cit., chaps. 3 and 4. See also Larry Rasmussen and Bruce Birch, *The Predicament of the Prosperous* (Philadelphia: Westminster Press, 1978); William Gibson, "The Lifestyle of Christian Faithfulness," in *Beyond Survival*, ed. Dieter Hessel (New York: Friendship Press, 1977); and Adam Finnerty, *No More Plastic Jesus: Global Justice and Christian Lifestyle* (New York: E. P. Dutton, 1978).

52. Reprinted in *99 Ways to a Simple Lifestyle*, op. cit. See also Arthur Gish, *Beyond the Rat Race* (Scottsdale, Pa.: Herald Press, 1973).

53. Mic. 4:3-5.

54. Lewis Perlman, *The Global Mind: Beyond Limits to Growth* (New York: Mason/Charter, 1976); Rodger Bybee, "Science Education for an Ecological Society," *American Biology Teacher* 41 (March 1979): 154-63; and Ferré, op. cit., chap. 8.

55. Frederick Polak, *The Image of the Future* (New York: Elsevier, 1973); Robert Bundy, ed., *Images of the Future* (Buffalo: Prometheus Books, 1976); Charles Erasmus, *In Search of the Common Good: Utopian Experiments, Past and Future* (New York: Free Press, 1977); and Dolores Hayden, *Seven American Utopias* (Cambridge, Mass.: MIT Press, 1976).

56. Kenneth Boulding, *The Meaning of the Twentieth Century* (New York: Harper & Row, 1965), p. 143; and Barbara Ward, *Spaceship Earth* (New York: Columbia University Press, 1966).

INDEX OF SELECTED TOPICS

Only topics discussed in more than one chapter are listed here. For other topics see the Table of Contents. The symbol "f" means "and the following page."

INDEX OF NAMES

ABOUT THE AUTHOR

IAN G. BARBOUR is professor of physics and professor of religion at Carleton College, Northfield, Minn. As director of the program in Science, Ethics, and Public Policy he has developed a series of interdisciplinary seminars on policy issues involving the environment and technology. He received his Ph.D. in physics from the University of Chicago and his B.D. from Yale. A recent Guggenheim Fellow, Professor Barbour is author of *Issues in Science and Religion* and *Myths, Models and Paradigms.* He is also editor of *Earth Might Be Fair, Western Man and Environmental Ethics,* and *Finite Resources and the Human Future.* He has served on the advisory board of the National Science Foundation's program on Ethics and Values in Science and Technology (EVIST).